Evgeny Larionov

Activity and selectivity of DMAP derivatives in acylation reactions

Evgeny Larionov

Activity and selectivity of DMAP derivatives in acylation reactions

Experimental and Theoretical Studies

Südwestdeutscher Verlag für Hochschulschriften

Impressum/Imprint (nur für Deutschland/only for Germany)
Bibliografische Information der Deutschen Nationalbibliothek: Die Deutsche Nationalbibliothek verzeichnet diese Publikation in der Deutschen Nationalbibliografie; detaillierte bibliografische Daten sind im Internet über http://dnb.d-nb.de abrufbar.

Alle in diesem Buch genannten Marken und Produktnamen unterliegen warenzeichen-, marken- oder patentrechtlichem Schutz bzw. sind Warenzeichen oder eingetragene Warenzeichen der jeweiligen Inhaber. Die Wiedergabe von Marken, Produktnamen, Gebrauchsnamen, Handelsnamen, Warenbezeichnungen u.s.w. in diesem Werk berechtigt auch ohne besondere Kennzeichnung nicht zu der Annahme, dass solche Namen im Sinne der Warenzeichen- und Markenschutzgesetzgebung als frei zu betrachten wären und daher von jedermann benutzt werden dürften.

Coverbild: www.ingimage.com

Verlag: Südwestdeutscher Verlag für Hochschulschriften GmbH & Co. KG
Dudweiler Landstr. 99, 66123 Saarbrücken, Deutschland
Telefon +49 681 37 20 271-1, Telefax +49 681 37 20 271-0
Email: info@svh-verlag.de

Zugl.: München, LMU, Diss., 2011

Herstellung in Deutschland:
Schaltungsdienst Lange o.H.G., Berlin
Books on Demand GmbH, Norderstedt
Reha GmbH, Saarbrücken
Amazon Distribution GmbH, Leipzig
ISBN: 978-3-8381-2767-5

Imprint (only for USA, GB)
Bibliographic information published by the Deutsche Nationalbibliothek: The Deutsche Nationalbibliothek lists this publication in the Deutsche Nationalbibliografie; detailed bibliographic data are available in the Internet at http://dnb.d-nb.de.

Any brand names and product names mentioned in this book are subject to trademark, brand or patent protection and are trademarks or registered trademarks of their respective holders. The use of brand names, product names, common names, trade names, product descriptions etc. even without a particular marking in this works is in no way to be construed to mean that such names may be regarded as unrestricted in respect of trademark and brand protection legislation and could thus be used by anyone.

Cover image: www.ingimage.com

Publisher: Südwestdeutscher Verlag für Hochschulschriften GmbH & Co. KG
Dudweiler Landstr. 99, 66123 Saarbrücken, Germany
Phone +49 681 37 20 271-1, Fax +49 681 37 20 271-0
Email: info@svh-verlag.de

Printed in the U.S.A.
Printed in the U.K. by (see last page)
ISBN: 978-3-8381-2767-5

Copyright © 2011 by the author and Südwestdeutscher Verlag für Hochschulschriften GmbH & Co. KG and licensors
All rights reserved. Saarbrücken 2011

To my family

The work for this thesis was encouraged and supported by a number of people to whom I would like to express my gratitude at this point.

First of all, I would like to appreciate my supervisor Prof. Dr. Hendrik Zipse for giving me the opportunity to do my Ph. D. in his group and his guidance in the course of scientific research presented here. I thank him for all the constructive discussions, especially for the great degree of independence and freedom to explore.

I would like to thank Prof. Dr. Herbert Mayr for acting as my "Zweitgutachter" and assessing this work. I would like to appreciate Hildegard Lipfert for all kind help for my stay in Munich. I acknowledge AK Mayr for the possibility to carry out the low-temperature kinetic measurements in their thermostat.

Furthermore, I would like to thank Johnny Hioe, Raman Tandon and Aliaksei Putau for careful and patient reading and correcting this thesis. My thanks to all the members of our research group: Dr. Ingmar Held, Dr. Yin Wei, Boris "Borix" Maryasin, Florian Achrainer, Christoph Lindner, Dr. Valerio "Wall-E" D'Elia, Dr. Sateesh "DrNa" Patrudu, Johnny "Alter" Hioe, Elija "Slave" Wiedemann, Jowita "Iwota" Humin, Florian Barth, Michael Miserok, Cong "YeYe" Zhang, for their helps and lasting friendships, which have made my time in Germany a pleasant and worthwhile experience. I thank my F-Praktikum student Regina Bleichner for creating nice atmosphere in the lab.

I would especially like to thank my great colleague Dr. Yinghao Liu who I spent the whole Ph. D. time with, for his interesting discussions and helpful suggestions. Additional thanks go to "Russian mafia" Konstantin Troshin and Anna Antipova from AK Mayr, as well as Boris Maryasin, for helpful discussions, nice teatimes and hiking tours.

I acknowledge Ludwig-Maximilians-Universität, München for financial support and Leibniz-Rechenzentrum München for providing some computation facilities.

Most importantly I would like to thank my parents and Ksenia for their love, support, help and encouragement during these years. Thank you very much!

Parts of this book have been published:

1. **Larionov, E.**; Zipse, H.; Organocatalysis: Acylation Catalysts. *WIRES Comp. Mol. Sci.* **2011**, *1*, 601-619.

2. Held, I.; **Larionov, E.**; Bozler, C.; Wagner, F.; Zipse, H.; The Catalytic Potential of 4-Guanidinylpyridines in Acylation Reactions. *Synthesis* **2009**, 2267-2277.

Table of contents

1. General Introduction   1
1.1 Acylation reactions: mechanistic survey   1
1.2 Catalyzed acylation reactions   4
1.2.1 Acid catalysis   4
1.2.2 Base catalysis   6
1.2.3 Nucleophilic mechanism of DMAP-catalyzed acylation   8
1.2.4 Base catalysis mechanism of DMAP-catalyzed transesterification   11
1.3 Objectives   14

2. The Catalytic Potential of Substituted Pyridines in Acylation Reactions: Theoretical Prediction and Experimental Validation   16
2.1 Introduction   16
2.2 Synthesis and catalytic activity of 3,4-diaminopyridines   20
2.2.1 Synthesis of 3,4-diaminopyridines   20
2.2.2 Catalytic activity of 3,4-diaminopyridines   24
2.3 Acetylation enthalpies (ground state model)   26
2.4 Activation enthalpies (transition state model)   31
2.4.1 Relative activation enthalpies   31
2.4.2 Conformational properties of the transition states   35
2.4.3 Influence of the solvation model   36
2.4.4 Discussion   41
2.5 Conclusions   42

3. Applications of the Relative Acylation Enthalpies   44
3.1 Photoswitchable pyridines   44
3.1.1 Introduction   44
3.1.2 Results and Discussion   47
3.1.3 Conclusions and Outlook   54
3.2 Relative acetylation enthalpies for paracyclophane derivatives   55
3.3 Relative isobutyrylation enthalpies for chiral 3,4-diaminopyridines   59
3.4 Relative acetylation enthalpies for ferrocenyl pyridines   63

4. (4-Aminopyridin-3-yl)-(thio)ureas as Acylation Catalysts 68
4.1 Introduction 68
4.2 Achiral (4-aminopyridin-3-yl)-(thio)ureas 70
4.2.1 Acetylation enthalpies of (4-aminopyridin-3-yl)-(thio)ureas 70
4.2.2 Synthesis and catalytic activity of (4-aminopyridin-3-yl)-(thio)ureas 74
4.2.3 Catalysts agregation studied by NMR and kinetic measurements 77
4.3 Chiral (4-aminopyridin-3-yl)-ureas 80
4.3.1 Synthesis of chiral catalysts, derived from (S)-amino acids 80
4.3.2 Derivatization of catalysts by Grignard reagent 81
4.3.3 Acetylation enthalpies and benchmark reaction kinetics 82
4.3.4 Introduction of a linker between the pyridine and urea moieties 85
4.3.5 Potential of (4-aminopyridin-3-yl)-ureas in the kinetic resolution of alcohols 87
4.4 Conclusions 90

5. Theoretical Prediction of Selectivity in KR of Secondary Alcohols 91
5.1 Introduction 91
5.2 Catalytic system with PPY 94
5.2.1 Determination of activation parameters for the PPY-catalyzed acylation reaction 94
5.2.2 Theoretical study of the catalytic cycle with PPY 96
5.3 Catalytic system with Spivey's catalyst 103
*5.3.1 The energy profile of the acylation catalyzed by catalyst **59a*** 103
5.3.2 Reaction barriers and conformational space of TSs 106
*5.3.3 The selectivity rationalization: TS **67** for catalysts **59b**, **59c** and **59e*** 111
*5.3.4 The selectivity prediction for catalysts **59d**, **59f** and **59g*** 114
*5.3.5 Synthesis and selectivity measurements for catalysts **59d**, **59f** and **59g*** 116
5.3.6 Comparison with theoretical predictions 118
5.4 Estimating the stereoinductive potential of the pyridines 120
5.4.1 Chiral 3,4-diaminopyridine derivatives 120
5.4.2 Prochiral probe approach 121
5.4.3 Conformational analysis of transition states 123
5.5 Conclusions 130

6. Summary and general conclusions 131

7. Experimental part 135
Chapter 2. Experimental details 136
Chapter 4. Experimental details 154
Chapter 5. Experimental details 180

8. Appendix (Computational details) 187
Chapter 2. Computational details 187
Chapter 3. Computational details 227
Chapter 4. Computational details 243
Chapter 5. Computational details 255

9. Kinetics of reactions in homogeneous solution: derivation of the kinetic law 283

References 298
Abbreviations 305

Chapter 1. General Introduction

Acyl-transfer reactions are among the most fundamental reactions in organic chemistry and biochemistry. Considering their importance in biochemical and synthetic processes, these reactions have been widely studied both in solution and in the gas phase.

1.1 Acylation reactions: mechanistic survey

Until now a large number of experimental and theoretical[1] studies on ester hydrolysis in aqueous solution have been carried out, resulting in a multitude of possible reaction mechanisms, which are described in many textbooks.[2] Several possible mechanisms of the base-catalyzed ester hydrolysis are shown in Figure 1.1. Early experimental results in aqueous solution showed that acyl-transfer reactions proceed via a stepwise mechanism $B_{AC}2$, which includes tetrahedral intermediates (Figure 1.1a).[3a] Subsequent studies suggested that the reaction can also occur through a one-step, concerted mechanism (Figure 1.1c), when the substrate has a good leaving group.[3b] The two possible mechanisms for ester hydrolysis, $B_{AC}2$ and $B_{AL}2$ (Figure 1.1a,b), which were shown to compete in the gas phase hydrolysis of methyl formate, were studied computationally by Pliego et al. at high *ab initio* level MP4/6–311+G(2df,2p)//MP2/6–31G(d).[1g] The calculated distribution of reaction paths was in excellent agreement with experimental values (85% of $B_{AC}2$). The main challenge in theoretical analysis of aqueous ester hydrolysis is the inclusion of water solvent effects by, for example, a cluster-continuum model[1c] or by explicit inclusion of all solvent molecules.[1f] Biochemically meaningful ester hydrolysis by enzymes was modeled by imidazole-catalyzed hydrolysis in several theoretical studies.[4]

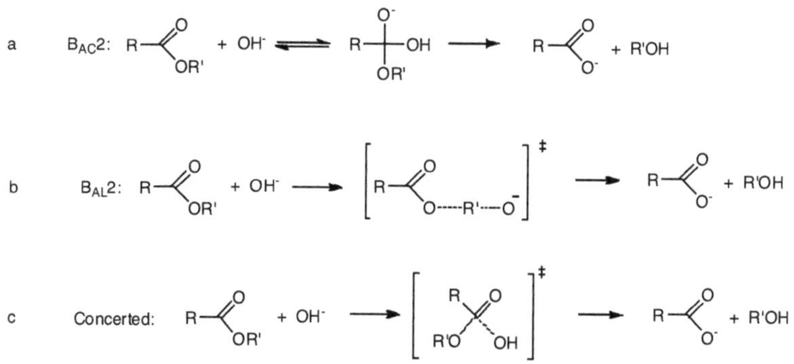

Figure 1.1. Possible mechanisms of the base-catalyzed ester hydrolysis.

The reverse reaction, esterification of alcohols, is also well studied and widely used in organic synthesis.[5] The general mechanisms are well known (Figure 1.2). The nucleophilic species undergoes addition to the carbonyl group, followed by elimination of the halide or carboxylate anion. Kinetic studies of the reaction of alcohols with acid chlorides in polar solvents in absence of basic catalysts generally reveal terms both first-order and second-order in alcohol.[5a,b] The first term is associated with the formation of a tetrahedral intermediate (Figure 1.2a), whose deprotonation is assisted by the solvent molecule (e.g., acetonitrile).[5a] Transition states in which the second alcohol molecule acts as a proton acceptor have been proposed for the second term (Figure 1.2b). The mechanism is concerted when anionic nucleophiles, such as phenoxides, reacted with carboxylic acid derivatives (Figure 1.2c).[5c-e]

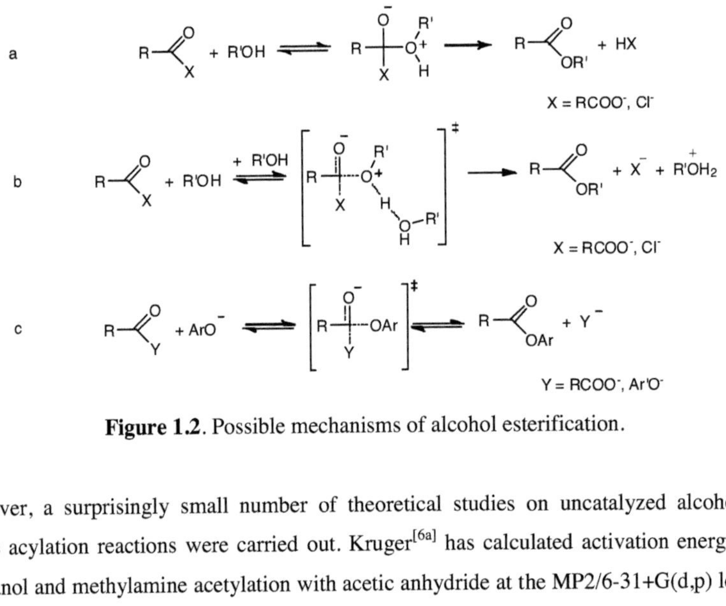

Figure 1.2. Possible mechanisms of alcohol esterification.

However, a surprisingly small number of theoretical studies on uncatalyzed alcohol and amine acylation reactions were carried out. Kruger[6a] has calculated activation energies for methanol and methylamine acetylation with acetic anhydride at the MP2/6-31+G(d,p) level of theory. The activation energy for amide formation was found to be much lower (37 kJ/mol) than that for the corresponding ester (71 kJ/mol) on this occasion. The latter proceeds through a six-membered ring transition state whose structure was found to be similar to that for the hydrolysis of acetic anhydride. It is worth noting that the leaving acetate serves as a base in the transition state to facilitate deprotonation of the attacking alcohol (Figure 1.3a).

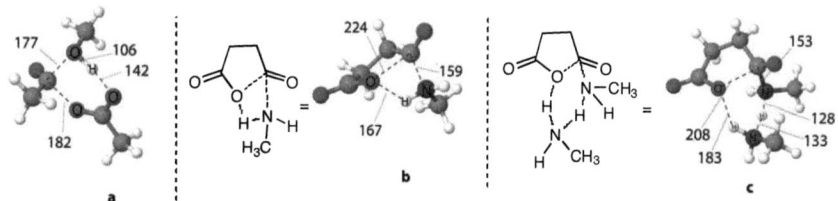

Figure 1.3. Transition state structures for the acylation of methanol by acetic anhydride (a)[6a] and the aminolysis of succinic anhydride: uncatalyzed (b) and catalyzed by methylamine (c)[6b] (distances are given in pm).

A theoretical study of the aminolysis of succinic anhydride,[6b] carried out at high levels of theory (MP2/6-311++G(d,p) and CCSD(T)/6-31G(d) single point calculations using geometries optimized at B3LYP/6-311++G(d,p) level) has shown that inclusion of a second methylamine molecule in the transition state, which facilitates the proton transfer (Figure 1.3b,c), decreases the activation free energy by *ca.* 36 kJ/mol. The calculations furthermore revealed that the concerted mechanism is more favorable than the stepwise pathway for the uncatalyzed and base-catalyzed pathways. The first stage of the stepwise mechanism is rate-determining in both cases.

In summary, the uncatalyzed and reactant catalyzed acylation of alcohols and amines by anhydrides proceeds preferentially through the concerted mechanism. Deprotonation of the alcohol or amine reactants typically dominates the transition state. Amines thus catalyze acyl-transfer reactions by facilitating the proton transfer process. In the following, we summarize which types of acids and bases are used in synthetically successful esterification reactions and how this type of catalysis can be rationalized by appropriate computation.

1.2 Catalyzed acylation reactions

1.2.1 Acid catalysis

Different acidic and basic catalysts and/or activators can be used for the acylation of alcohols by carboxylic acid derivatives.[7a] Commonly used Brønsted acids for the reaction between alcohols and anhydrides are sulfonic, perchloric and sulfuric acids.[7b-d] Significant progress in the use of Lewis acids in the acylation of alcohols with anhydrides was made in the last 15 years, especially owing to the development of metal salts of triflic acid (TfOH = CF_3SO_3H). Commercially available $Sc(OTf)_3$ shows high catalytic activity, even at catalyst loadings as low as 0.1 mol %.[7e,f] The selectivities and reactivities observed for this type of catalyst often contrast those for base-catalyzed reactions, thus providing a synthetically valuable set of tools for ester synthesis. Other metal triflates such as $Bi(OTf)_3$,[8a] $VO(OTf)_2$,[8b] $In(OTf)_3$,[8c] and $Cu(OTf)_2$[8d] also appear to work well in acylation reactions. Finally, it was demonstrated that TMSOTf is a usefull catalyst for the acylation of phenols and alcohols,[8e] providing an economically and chemically attractive alternative to the more established approach employing $Sc(OTf)_3$. In marked contrast to the multitude of synthetic studies only a small number of mechanistic studies on acid-catalyzed acylation reactions have been carried out. Recently Marko et al. have shown that the metal triflate-promoted acylation of alcohols appears to be catalyzed mainly by triflic acid, released by metal triflates at the onset of the reaction.[9a] The active acylation species in the catalytic cycle was shown to be the mixed anhydride acyl triflate (RCOOTf), which is formed in the reaction of anhydride with triflic acid or metal triflates. Theoretical studies of the aminolysis of succinic anhydride **1** revealed that acetic acid can act as a very effective catalyst for this reaction.[6b] Two possible pathways for catalysis by acetic acid were studied computationally, employing the B3LYP/6-311++G(d,p) level of theory, while treating benzene solvent effects at PCM/B3LYP/6-31G(d) level (Figure 1.4). Path A, which involves only the hydroxyl group of acetic acid, shows similar energetics as the base-catalysed pathway, which includes the second methylamine molecule in the transition state. Path B can be termed "bifunctional acid catalysis" as it involves both oxygen atoms of acetic acid in the proton transfer process. This latter option is more favourable than path A by ca. 50 kJ/mol. The calculations also show that both acid-catalyzed pathways proceed in a stepwise manner and are more favourable than the concerted alternatives, in which the first step is rate-determining. These results are in agreement with experimental studies of the aminolysis reaction between benzylamine and succinic anhydride in benzene perfomed by Smagowski and Bratnicka,[9b] which confirm the existence and preferable role of acid catalysis in aminolysis reactions.

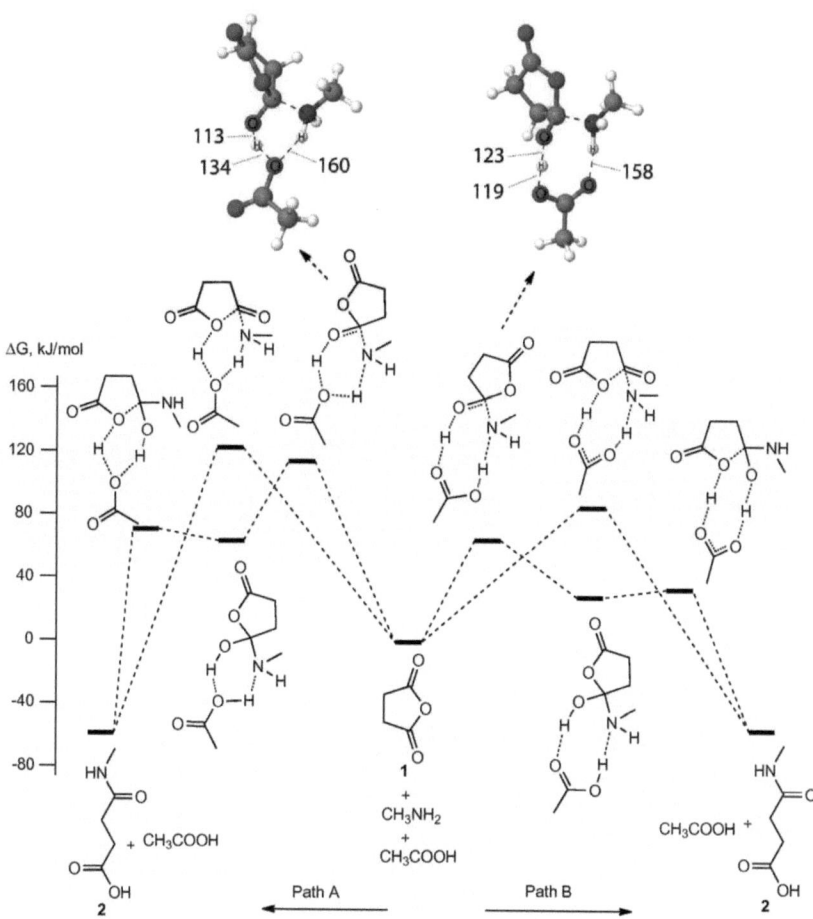

Figure 1.4. Free energy profile for two pathways in the acid-catalyzed reaction of succinic anhydride **1** with methylamine as calculated at PCM/B3LYP/6-31G(d) (benzene) level; structural parameters of the rate-determining transition states are included (distances are given in pm).[6b]

1.2.2 Base catalysis

The acylation of alcohols by acid anhydrides and acyl chlorides can conveniently be carried out in pyridine as a basic solvent or in the presence of tertiary amines such as triethylamine. The latter is the reagent of choice for acylations with alcohols and anhydrides of sufficient (intrinsic) reactivity, or for substrates too labile towards stronger bases.[10a] The tertiary amine/acid chloride procedure is also relatively mild and well suited for a variety of primary and secondary alcohol substrates.[10b] Tertiary alcohols are often unreactive under these reaction conditions, allowing the selective acylation of secondary alcohols. On the basis of results obtained in NMR studies, Oriyama *et al.* have suggested that a benzoyl chloride-tertiary amine complex is the active catalytic species in benzoylation reactions. Chiral tertiary amines derived from (*S*)-proline were subsequently developed for the kinetic resolution of secondary[10c] and primary[10d] alcohols. For the acylation of sterically hindered alcohols electron-rich pyridines such as DMAP (4-dimethylaminopyridine, **3**) provide superior levels of catalytic turnover. This was independently established by the groups of Litvinenko and Steglich.[11a,b] In 1967 Litvinenko *et al.* found that addition of DMAP (**3**) results in a rate-acceleration of ca. 10^4 for the benzoylation of *m*-chloraniline as compared to pyridine.[11a] Steglich and Höfle[11b,c] described in 1969 a very strong catalytic effect of DMAP and 4-pyrrolidinopyridine (PPY, **4**) in acylation reactions of hindered alcohols. Since then DMAP and PPY have been developed as standard catalysts for a wide variety of acylation reactions.[12] Catalysis is usually most efficient in apolar solvents (CH_2Cl_2, $CHCl_3$) and in the presence of 1-3 eq. auxiliary base (NEt_3, i-Pr_2EtN). Zipse *et al.* have shown[13a] that omission of the auxiliary base leads to a dramatic reduction of reaction rate. These can, of course, be compensated by conducting the reaction at higher concentrations and for longer reaction times.[13b]

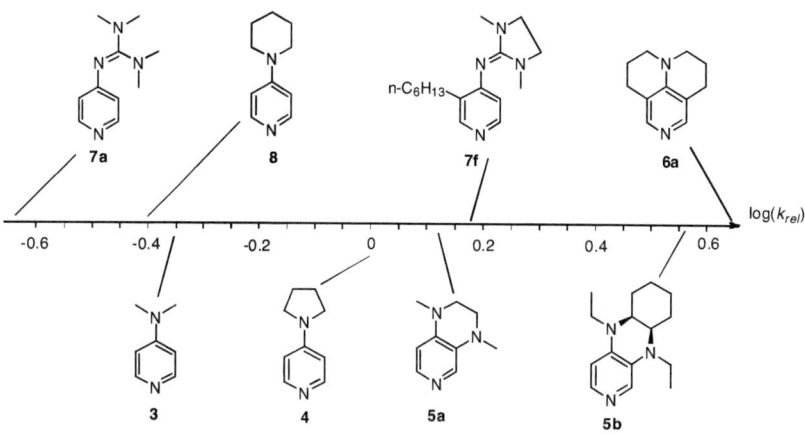

Figure 1.5. Relative catalytic activities of substituted pyridines in acylation reactions.[11c,13a,15]

In experimental studies of the DMAP-catalyzed acylation of cyclohexanol Zipse et al.[14] have shown that the acylation reaction is first order in alcohol, DMAP and anhydride, and zero-order in NEt$_3$. The catalytic efficiency of a variety of DMAP derivatives (Figure 1.5) was studied using the the acetylation of 1-ethynylcyclohexanol with acetic anhydride (2 eq.) in the presence of NEt$_3$ (3 eq.) as the auxiliary base and 0.1 eq. of catalyst in CDCl$_3$ at 23 °C.[26] As shown in Figure 1.5 the most active catalysts currently known are 3,4-diaminopyridine **5b** and the annelated DMAP-derivative **6a**. The catalytic efficiency of some of these compounds has also been studied by Hassner et al. for the acetylation of 1-methylcyclohexanol with acetic anhydride (2.0 eq.) in the presence of NEt$_3$ (2.0 eq.) as solvent and base using a turnover-based assay.[15] Under these conditions derivative **7a** (Figure 1.5) was slightly more catalytically active then DMAP (**3**).

1.2.3 Nucleophilic mechanism of DMAP-catalyzed acylation

The DMAP-catalyzed acylation of alcohols by anhydrides and acyl chlorides is currently believed to proceed via the nucleophilic catalysis mechanism shown in Figure 1.6. This mechanism is supported by the following observations:

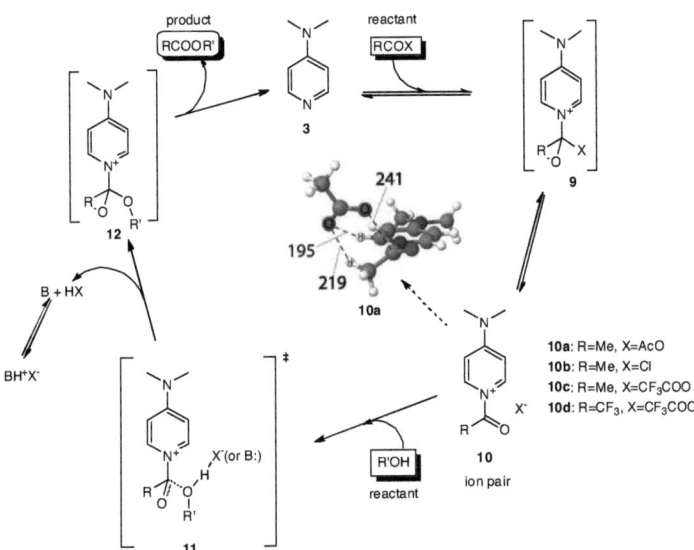

Figure 1.6. Nucleophilic mechanism of alcohol esterification catalyzed by DMAP; the structure shown for ion-pair intermediate **10a** has been obtained at UAHF-PCM B3LYP/6-31G(d) level (distances are given in pm).[18]

1. A dramatic loss of catalytic activity is observed for 2-substituted pyridines, even though this substitution has a small effect on the basicity of these derivatives. Also, DMAP and NEt₃ have similar pK_a values (9.7 and 11.0), but very different catalytic activities. These facts cannot be reconciled easily with a simple base catalysis mechanism.

2. The N-acylpyridinium species **10** formed in the catalytic cycle were detected by NMR spectroscopy. N-acetyl-4-dimethylaminopyridinium chloride **10b** is formed quantitatively upon mixing DMAP and acetyl chloride.[16] This salt is readily soluble in DMSO, but less soluble in CDCl₃.[11c] In mixtures of acetic anhydride and DMAP or PPY in CDCl₃ and CD₂Cl₂ appreciable amounts of N-acetylpyridinium acetate **10a** can be detected by ¹H NMR

spectroscopy at low temperature.[11c] It was shown that at room temperature about 5-10% of PPY is in the form of N-acetyl-4-pyrrolidinopyridinium acetate, and at lower temperatures the relative amount of this intermediate increases. Parallel experiments with pyridine resulted in no detectable formation of N-acetylpyridinium acetate and precipitation of insoluble N-acetylpyridinium chloride, respectively.

3. Steglich *et al.* have shown that the acetylation of 1-ethinylcyclohexanol with Ac_2O (2 eq.) catalysed by DMAP (3 eq.) proceeds three times faster than the reaction with AcCl (2 eq.).[11c] This is in contrast to the equilibrium amounts of acylpyridinium salts in the respective reaction mixtures. Moreover, Wakselman et al. have shown that other N-acetyl-4-dimethylaminopyridinium salts **10** (with X=Cl⁻, TsO⁻ or BF_4^-) do not react with *tert*-butanol in $CHCl_3$, CH_3CN or AcOEt.[17a] A plausible explanation of this fact is that the counterion is responsible for deprotonating the alcohol in the rate-determining transition state. This view is also supported by experiments of Kattnig and Albert,[17b] who found that acetic anhydride acetylates secondary alcohols *ca.* 10 times faster than acetyl chloride when treated with combinations of DMAP and an insoluble carbonate base. Replacing the DMAP/carbonate pair by pyridine as the auxiliary base reverses the relative rates, highlighting the important role of the deprotonation step in the overall reaction mechanism.[17a]

4. The DMAP-catalyzed acetylation of *tert*-butanol with acetic anhydride was studied by Zipse *et al.* at the B3LYP/6-311+G(d,p)//B3LYP/6-31G(d) level of theory (Figure 1.7).[14] The second step of the nucleophilic catalysis mechanism was shown to be rate-determining (enthalpy of transition state **18** relative to reactants ΔH_{298} = +34.8 kJ/mol). One notable feature of the rate-limiting transition state **18** is the concertedness of acetyl and proton transfer. The base-catalyzed mechanism through transition states **20a** or **20b** is less favourable with barriers over 70 kJ/mol, and stepwise alternatives to these pathways are significantly less favourable with barriers over 150 kJ/mol. Since acylation reactions are known to be solvent-dependent, proceeding faster in less polar solvents like hexane and CCl_4,[15] Zipse *et al.* have studied the influence of solvent effects on the reaction barrier employing the PCM/UAHF/B3LYP/6-31G(d) continuum model.[14] For CCl_4, $CHCl_3$, and CH_2Cl_2 as the solvents the base catalysis was still shown to be unfavourable, and the rate-limiting step for nucleophilic catalysis was predicted to be the same as in the gas phase. The activation enthalpy of the nucleophilic catalysis mechanism increases in the order CCl_4 - $CHCl_3$ - CH_2Cl_2, being +55.1, +57.1, and +60.9 kJ/mol respectively. These results are in line with experimental studies by Hassner *et al.*[15]

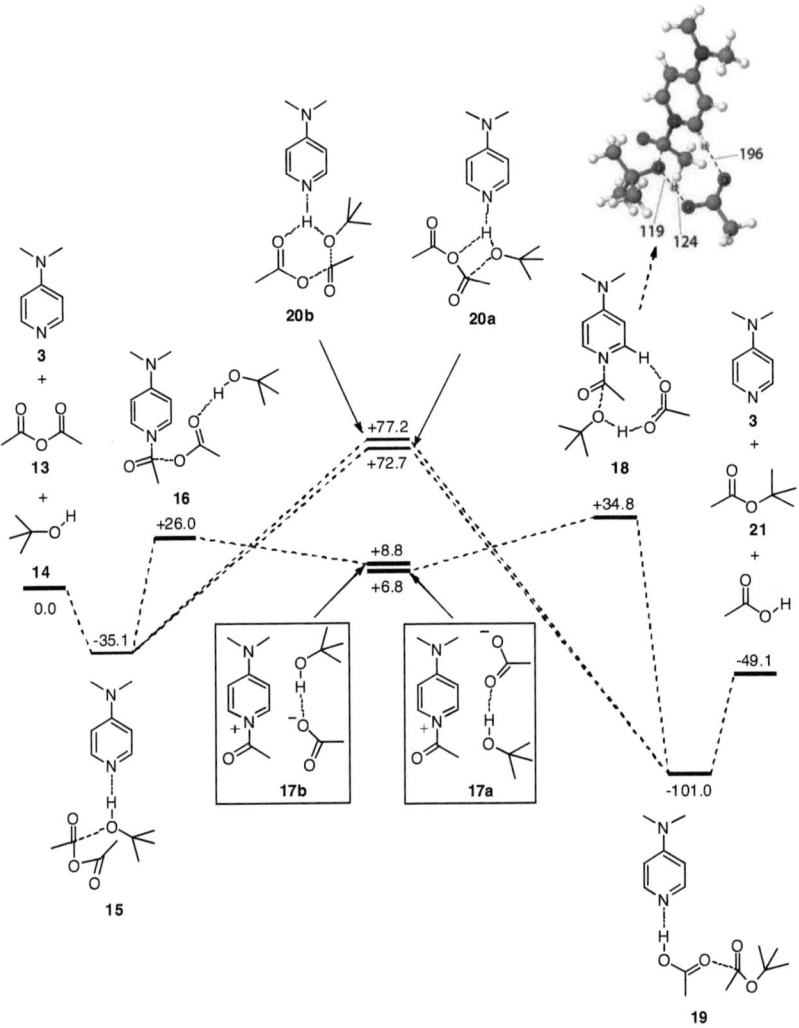

Figure 1.7. Gas-phase enthalpy profile (ΔH_{298}, kJ/mol) for the competing nucleophilic and base catalysis mechanisms in the DMAP-catalyzed reaction of acetic anhydride with *tert*-butanol as calculated at the B3LYP/6-311+G(d,p)//B3LYP/6-31G(d) level of theory (distances are given in pm).[14]

5. Recently Schreiner *et al.* have studied the structures of N-acylated DMAP salts **10a-d** (Figure 1.6) using a variety of different methods.[18] The authors were able to crystallize salts **10b-d**. The analyses of X-ray structures of these salts show that there are weak hydrogen-bonding interactions between anions and the hydrogen at C2 of the pyridine ring, with distances of 2.80 and 2.82 Å in **10c** and **10d**, respectively. Chloride **10b** has much longer distances and the solvent molecules are associated with the chloride anion, therefore **10b** cannot be described as a tight ion pair. NMR coalescence experiments were used to determine the rotational barriers around the N-acyl bond, which are around +38.9±1.3 kJ/mol for all compounds **10a-d**. On the basis of the low-temperature IR- and NMR-measurements, the authors concluded that at room temperature the steady-state concentration of catalytically active **10a** is below 1%. Calculations at DFT-D-B3LYP/6-311+G(d,p)//B3LYP/6-31G(d) level confirm the 'tight' ion pair structure for **10a** with the acetate having hydrogen-bonding interactions with the hydrogen at C2 position of the pyridine ring (1.95 Å) and one of the acetyl group hydrogens (2.19 Å) (Figure 1.6). These distances resemble the corresponding values for the transition state **18** (1.96 and 2.38 Å).

In conclusion, both experimental and theoretical studies support the nucleophilic mechanism of the DMAP-catalyzed acylation of alcohols. The active catalytic species are acylpyridinium cations, whose interactions with the counterion are essential in the rate-limiting deprotonation of alcohol substrates.

1.2.4 Base catalysis mechanism of DMAP-catalyzed transesterification

In 2001 Hedrick *et al.* reported the DMAP-catalyzed ring-opening polymerisation (ROP) of L-lactide **22** as the first example for an organocatalytic living polymerisation.[19a] DMAP and PPY were shown to catalyse the ROP very selectively and without side reactions, providing polymeric products in near-quantitative yield and narrow polydispersities (polydispersity index PDI was 1.1-1.2). The reaction was carried out in CH_2Cl_2 with ethanol or benzyl alcohol as initiator and DMAP or PPY (2-4 equivalents relative to alcohol). Hedrick has suggested a mechanism of ROP, which involves the nucleophilic activation of monomer by DMAP (path A on Figure 1.8), affording acylpyridinium intermediate **24**, which reacts subsequently with the terminal alcohol group of the growing polymer. This mechanism was suggested on the basis of experiments, which do not interfere with the base catalysis mechanism. In 2008 Bonduelle *et al.* have studied computationally the DMAP-catalyzed ROP of L-lactide.[19b] Methanol was chosen as a model for the initiating/propagating alcohol (Figure 1.8). The authors used the B3LYP/6-31G(d) level of theory and included the solvent effects of CH_2Cl_2 through PCM single point calculations at the same level of theory. They have shown that intermediate **24** of the nucleophilic route is 105 kJ/mol higher in energy than the separated reactants, whereas intermediate **27** of the basic route is just 42 kJ/mol higher in energy than the separated reactants. The rate-determining first step of the stepwise basic route was shown to have an activation barrier of 63 kJ/mol in CH_2Cl_2 (transition state **26**). The barrier height for the concerted basic mechanism is comparable to that for the stepwise route, so reaction can occur along both pathways equally well.

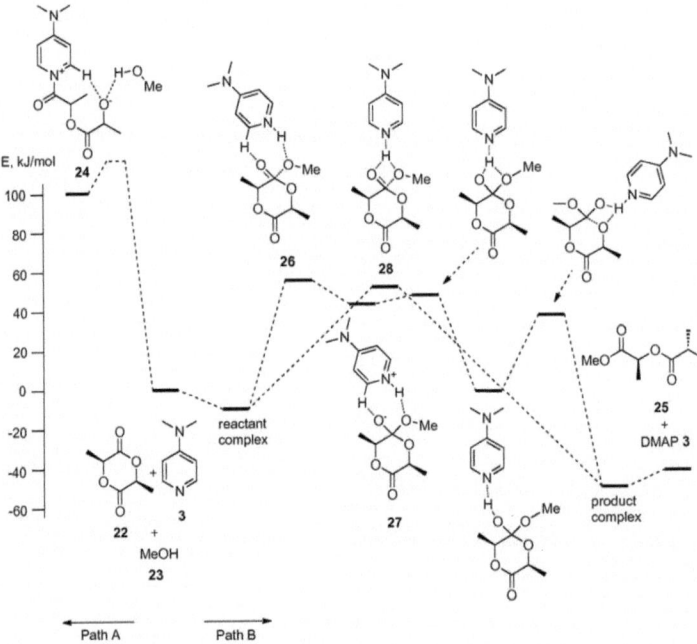

Figure 1.8. Nucleophilic (A) and basic (B) routes of activation in the DMAP-catalyzed ring-opening of L-lactide by methanol as calculated at the B3LYP/6-31G(d) level of theory (PCM/SCRF single-point calculations, including zero-point vibrational energy corrections).[19b]

Why does the basic mechanism of the L-lactide methanolysis notably contrast with the nucleophilic mechanism of the acetylation of *tert*-butanol? Bonduelle et al. tried to answer this question using computational methods.[19b] The authors compared the nucleophilic and basic pathways for the acetylation of methanol and *tert*-butanol. The difference in activation energies between these two pathways amounts to 36 kJ/mol for *tert*-butanol (in accordance with Zipse et al.[14]) and 9 kJ/mol for methanol, suggesting that the two pathways may become competitive for primary alcohols. Comparison of the stabilities of acylpyridinium intermediates formed from acetic and methylsuccinic anhydrides also reveals that the nucleophilic pathway is somewhat disfavoured for cyclic substrates and so the basic route may become more favourable, at least with primary alcohols. Similar to the observations made by Schreiner et al. for acylpyridinium complexes, the authors also note for the DMAP-catalyzed ROP of lactones that DMAP can act as a hydrogen-bond acceptor (through its basic nitrogen center) and a weak hydrogen-bond donor (through one of the *ortho*-hydrogen atoms) as indicated for transition state **26** in Figure 1.9.[19b] This is somewhat reminiscent of the dual role of guanidine base 1,5,7-triazabicyclo[4.4.0]dec-5-ene (TBD) in the catalysed ROP of L-lactide[19c,d] as shown for transition state **29** in Figure 1.9.

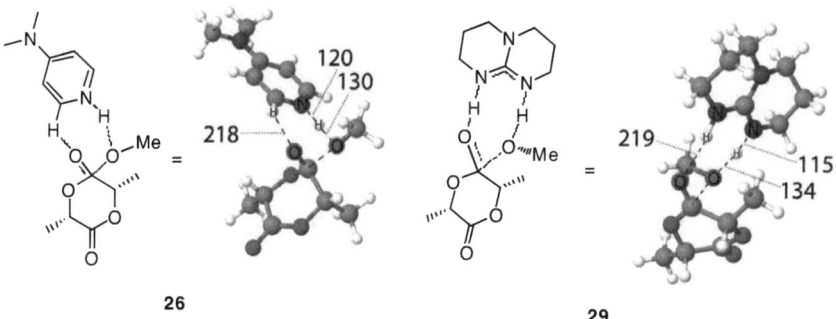

Figure 1.9. Structures of transition state **26** for the DMAP-catalyzed methanolysis of L-lactide **22**, as calculated at B3LYP/6-31G(d) level,[19a] and transition state **29** for the TBD-catalyzed methanolysis of L-lactide, as calculated at MPW1K/6-31+G(d) level[19d] (distances are given in pm).

1.3 Objectives

The calculation of the full reaction profiles shown in Figures 1.5 and 1.6 is a rather time-consuming exercise and the question naturally arises how much of this effort is necessary for the goal of catalyst development. In many of the multi-step catalytic reactions discussed here the rate- and selectivity-determining transition states are accompanied by energetically and structurally related intermediates. The quantitative characterization of the structure and stability of these intermediates represents one of the most efficient approaches for characterizing the properties of a large ensemble of systems (e.g., a group of catalysts). This is demonstrated below using examples from stereoselective catalysis (where the conformational properties of intermediates are relevant) and from the development of more active catalysts (where the relative stability of acyl-intermediates is relevant). Theoretical methods can be used in this context in a variety of ways, starting from the structural optimization of reactants, products, and reaction intermediates up to the full characterization of potential energy surfaces. Whether, in the end, all of this is necessary for the theory-guided development of acylation catalysts is also discussed.

One important aspect of the catalyst design is the development of new active acylation catalysts. In Chapter 2 we compare several ground state and transition state models in their capacity to match kinetic data for the acetylation. In order to base this methodological study on a greater number of measured kinetic data, we have also synthesized a large number of new catalysts based on the 3,4-diaminopyridine framework and evaluated their catalytic activity in the same benchmark reaction. Application of the cheaper ground state model, i.e. acetylation enthalpies, to the design of the photoswitchable pyridines as well as the planar-chiral aminopyridine derivatives, containing paracyclophane or ferrocenyl substituents, is discussed in Chapter 3.

One of the important concepts in the "Organocatalysis" field is activation of an electrophile by (thio)urea-derived catalysts.[21] (Thio)ureas serve as partial Brønsted (weak Lewis) acids, activating the electrophile (carbonyl, epoxide) by N-H double hydrogen bonding. A broad variety of monofunctional and bifunctional chiral (thio)urea organocatalysts have been developed to accelerate various synthetically useful organic transformations employing H-bond accepting substrates, e.g., carbonyl compounds, imines, nitroalkenes as starting materials. The concept of multifunctional catalysis, wherein the catalysts exhibit both Lewis acidity and Brønsted basicity, enables effective transformations, which generally are hard to achieve by a single functional catalyst.[21d] Therefore we envisioned that 3,4-diaminopyridine derivatives bearing a (thio)urea moiety on the 3N-atom could also work as bifunctional catalysts. The design of a new class of acylation catalysts, 3-(thio)urea-4-aminopyridines, is described in Chapter 4. Chiral derivatives of this new class can be prepared via a modular strategy from easily accessible isocyanates. The potential of newly synthesized chiral 3,4-diaminopyridine derivatives will then be explored in the kinetic resolution (KR) of secondary alcohols (Chapter 4).

Finally, some mechanistic insights into the catalytic system for the kinetic resolution, both experimental and theoretical, are described in Chapter 5. The theoretical rationalization and prediction of the experimentally observed stereoselectivities in the KR of *sec*-alcohols is attempted using the key transition state model. Less time-consuming models for selectivity prediction, based on the prochiral probe approach and transition state conformational analysis, will also be attempted.

Chapter 2. The Catalytic Potential of Substituted Pyridines in Acylation Reactions: Theoretical Prediction and Experimental Validation

2.1 Introduction

Since their discovery, N,N-dimethylaminopyridine (DMAP, **3**) and its derivatives such as 4-pyrrolidinopyridine (PPY, **4**) have been employed as versatile and efficient nucleophilic catalysts for a variety of chemical transformations such as the acylation of alcohols.[11,12] Current developments focus on enhancing the selectivity of these catalysts in stereo- and regioselective transformations, in particular in the kinetic resolution of alcohols and amines.[20] Rationalizing the origin of selectivity in these resolution experiments represents an important step in the further development of catalyst. While conformational preferences of critical intermediates in the catalytic cycle determine the stereochemical outcome of substrate turnover, the actual stability of these intermediates with respect to the free catalyst is thought to be of critical importance for turnover rates.

Quantitative approaches for matching catalyst nucleophilicity (a kinetic property) to some sort of affinity number (a thermodynamic property) count among the most traditional methods in physical organic chemistry. The so called "Brønsted plots" (correlating rate constants for catalytic processes with pK_a values) have been widely applied in mechanistic studies.[3] Steglich et al.[11c] have demonstrated a linear correlation between the rates for acyl-transfer reactions in benzene and pK_a values for *para*-substituted pyridines (*e.g.* DMAP (**3**), PPY (**4**), 4-methylpyridine, pyridine) (Figure 2.1). Other nitrogen bases such as DBU or triethylamine do not obey a linear free energy relationship. Remarkably, enantioselectivities of several organocatalyzed transformations can also be correlated with pK_a values of the used catalysts.[22a,b] This approach usually works well for a family of catalysts with comparable steric requirements.[22b,c] For more complicated, stepwise reactions a change in the rate-limiting step or the participation of catalysts in different steps can alter the linearity of the Brønsted plot.[3c]

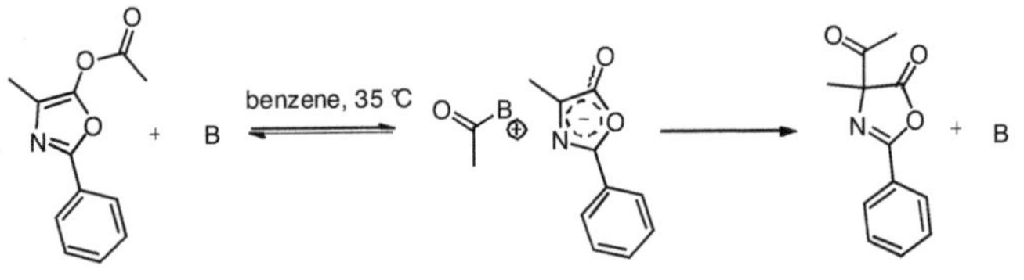

Scheme 2.1. Base-catalyzed acetyl transfer; the reaction was studied in benzene at 35 °C by ^{1}H NMR.

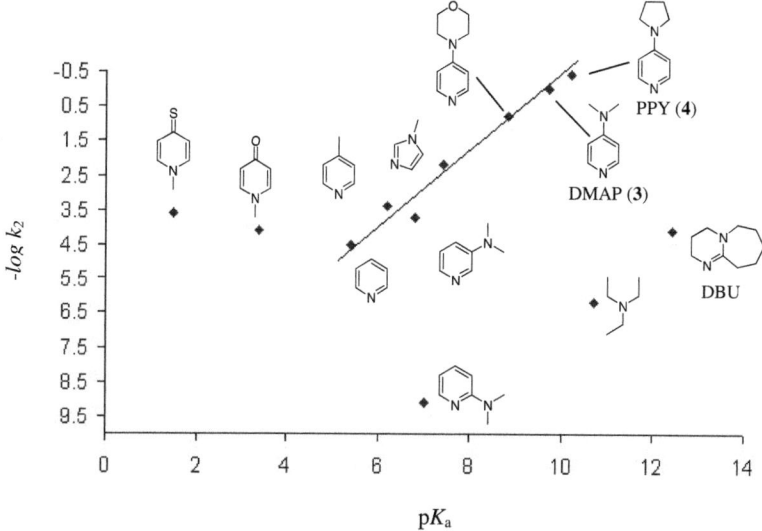

Figure 2.1. Correlation of the relative rates of acyl-transfer with pK_a values for different amines.[11c]

Application of the Brønsted-plot strategy is often restricted by the availability of pK_a values for the solvent used in the actual catalysis experiments. Mayr and coworkers persuasively argued in favor of universal nucleophilicity/electrophilicity scales based on kinetic data recorded for a series of nucleophiles towards benzhydrylium cations as a standard set of electrophiles.[23a] Concerning acylation catalysts, Mayr et al. have observed a good linear correlation between nucleophilicities N of different *para*-substituted pyridines measured in CH_2Cl_2 and their pK_a values in CH_3CN (Figure 2.2).[23b]

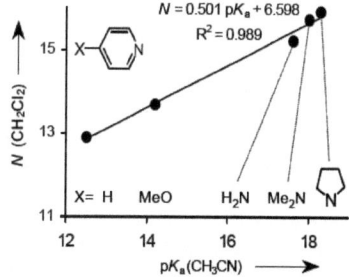

Figure 2.2. Correlation of the nucleophilicities N (in CH_2Cl_2) with pK_a (in CH_3CN) for 4-substituted pyridines.[23b]

All of the above experimental approaches are contingent upon the availability of actual substrate samples, a point not easily reconciled with the realities of catalyst development projects. Theoretically calculated nucleophilicities offer the advantage that predictions can be made for compounds before their respective synthesis, thus avoiding the potential problem of synthesizing compounds of low (catalytic) activity. A number of theoretical studies on enantioselective catalysts have been carried out during the last decade. Most of these concern the rationalization of the stereochemical outcome of these reactions and involve either the full characterization of potential energy surfaces or the localization of the transition states of the selectivity-determining step.[24] In contrast, studies devoted to application of theoretical methods to the prediction of catalyst activity have received comparatively little attention.[22c,25,26] One of the theoretical approaches for predicting nucleophilic activity is based on the minimum values of the molecular electrostatic potential (MEP), which was evaluated by Campodonico *et al.* for a series of substituted pyridines.[25a] Significantly closer to the situation in catalytic processes are approaches involving the calculation of affinity values towards model electrophiles. Since most of the organocatalytic transformations involve nucleophilic attack on carbon, methyl cation affinities (MCA) can serve as the most simple model for a carbon basicity scale.[25b] Zipse *et al.*[25c] have compared experimental rates of the Michael addition reaction of methanol to acrylamide catalyzed by amine and phosphane nucleophiles with pK_a values, proton and methyl cation affinities (calculated at MP2(FC)/6-31+G(2d,p)//B98/6-31G(d) level). The best correlation was obtained by using MCA data ($R^2 = 0.91$), thus illustrating the potential usefulness of this type of data for the development of new nucleophilic catalysts (Figure 2.3).

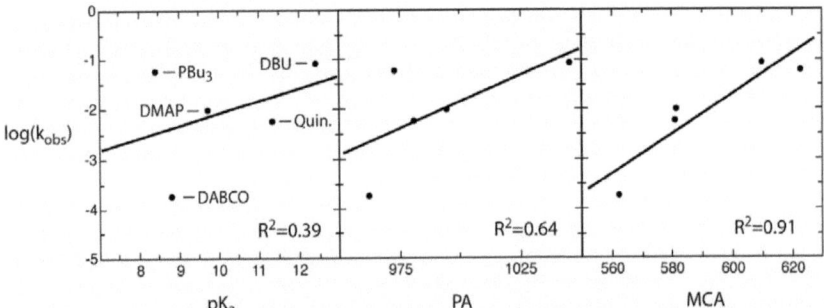

Figure 2.3. Correlation of observed rate constants k_{obs} for the nucleophile-induced addition of methanol to acrylamide with the corresponding pK_a, PA (kJ/mol), and MCA (kJ/mol) values.[25c]

Acetylpyridinium cations are important intermediates in pyridine-catalyzed acylation reactions,[14] and relative acetyl cation affinities were therefore used as the guideline for the development of new pyridine-based nucleophilic catalyst such as aminopyridine **6a** (Figure 2.4), in which the electron-donor properties of the 4-amino substituent were complemented by two annelated ring systems.[26a] The enthalpies of the isodesmic reaction shown in Scheme 2.2 were calculated at the B3LYP/6-311+G(d,p)//B3LYP/6-31G(d) level of theory in the gas phase for a large variety of pyridine catalysts.[26b]

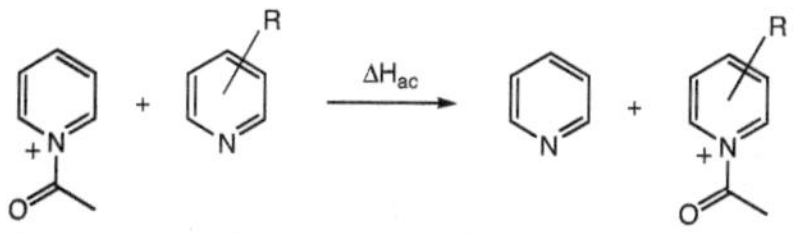

Scheme 2.2. Isodesmic reaction for the calculation of acetylation enthalpies.

This approach was successfully used for the development of highly active catalysts such as 3,4-diaminopyridine derivative **5b**[26c] and 3,4,5-triaminopyridine derivative **30a**,[27] which are similarly active in the acetylation of tertiary alcohol **36a** (Figure 2.7) as aminopyridine **6a**. DMAP-derivatives with smaller (**6c**) or larger (**6d**) annelated rings were found to be similarly active as **3**, despite their higher acylation enthalpies.[28]

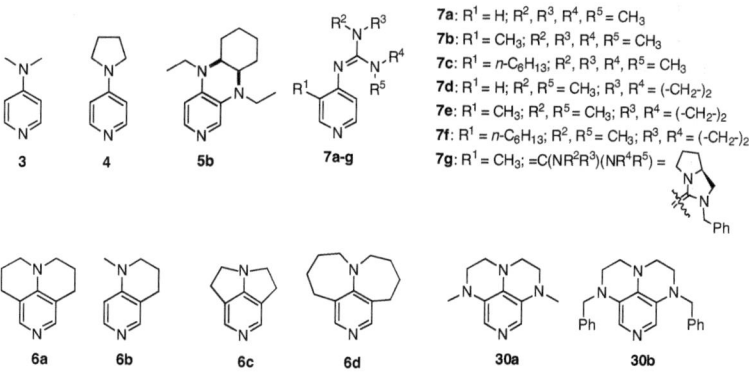

Figure 2.4. Selected nucleophilic catalysts based on the DMAP motif.

Even though the calculated acylation enthalpies correlated very well with reaction half-lives for a narrow class of pyridine catalysts (such as catalysts **3, 4** and **6a**), a less satisfactory correlation is found when including structurally more diverse catalysts, such as 3,4-diamino- and 4-guanidinylpyridines **7a-g**.[13a] This lack of correlation may either be rooted in the

19

fundamental deficits of any free-energy relationship building on the correlation of thermodynamic properties of selected ground state systems with kinetic data for a process involving related species. We would then have to assume that a true transition state model for the benchmark acylation reactions would be clearly superior. Alternatively, the unsatisfactory correlation may be due to technical characteristics of the ground state model used. In order to clarify this point we compare here several ground and transition state models in their capacity to match kinetic data for the acetylation of 1-ethinylcyclohexanol with acetic anhydride in chloroform. In order to base this methodological study on a greater amount of measured rate data, we have also synthesized a larger number of new catalysts based on the 3,4-diaminopyridine framework and measured their catalytic efficiency for the same benchmark reaction (Figure 2.7).

2.2 Synthesis and catalytic activity of 3,4-diaminopyridines

2.2.1 Synthesis of 3,4-diaminopyridines

New 3,4-diaminopyridine catalysts were synthesized from commercially available substances using modifications of previously established procedures.[26c] This involves derivatives of catalysts **5a** and **5b** bearing different substituents (alkyl chains or benzyl groups) on the nitrogen atoms attached to the 3- and 4-position (Figure 2.5).[29] Synthesis of catalysts **5h-j** has already been described in the literature.[30]

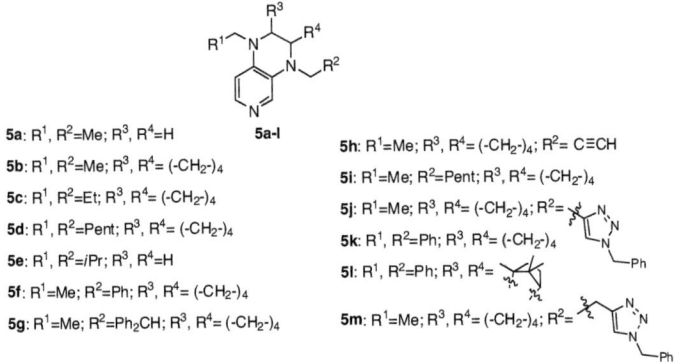

Figure 2.5. Members of the 3,4-diaminopyridine catalyst family.

The intermediates **31a-c** were synthesized from commercially available substances, using procedures described in the literature (Scheme 2.3).[26c] The acetylation of compound **31a** by acetic anhydride has previously been carried out under harsh conditions: 48 h in pyridine as solvent at 100 °C.[26c] Employing a microwave reactor allowed us to reduce the reaction time to 10 min, with an increased reaction temperature of 170 °C and a lower amount of pyridine (1 ml per 1 mmol substrate). Under the optimized conditions acylation by aliphatic anhydrides proceeded smoothly giving up to 86 % yield of products. Acylation of amine **31b** by acyl chlorides employing these optimized conditions afforded the corresponding products in 98 % yield. Subsequent reduction of amide groups by LiAlH$_4$/AlCl$_3$ gives the corresponding catalysts in 66-74 % yield (Scheme 2.3).

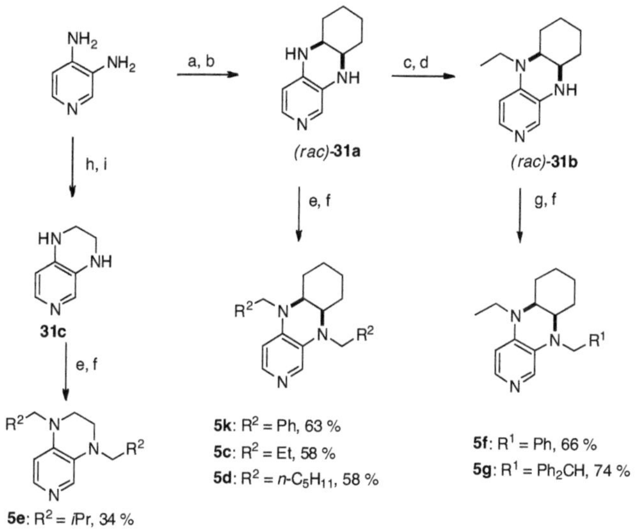

Scheme 2.3. Synthesis of 3,4-diaminopyridine catalysts: a) 1,2-cyclohexanedione, EtOH, 70 °C, 5 h, 90 %; b) LiAlH$_4$, THF, -40 °C, 30 min -> RT, 32 h, 79 %; c) Ac$_2$O, NEt$_3$, CH$_2$Cl$_2$, 5 mol% PPY, 96 %; d) AlCl$_3$, THF, RT, 45 min -> LiAlH$_4$, 0 °C, 1 h -> rf, 8h, 60 %; e) (R^{2}CO)$_2$O, pyridine, MW, 170 °C, 10 min, 60-86 %; f) AlCl$_3$, THF, RT, 45 min -> LiAlH$_4$, 0 °C, 1 h -> rf, 8h, 50-75 %; g) R^{1}COCl, pyridine, MW, 170 °C, 60 min, 98 %; h) glyoxal, EtOH, 70 °C, 2 h, 88 %; i) NaBH$_4$, EtOH, 40 °C, 15 h, 50 %.

All synthesized derivatives of **5b** were obtained as racemates with *cis* configuration of the cyclohexane ring. Incorporating chirality into the cyclohexyl fragment by employing commercially available *(S)*-(+)-camphorquinone **32** for the condensation with 3,4-diaminopyridine allowed us to synthesize catalyst **5l** as a single diastereomer (Scheme 2.4).

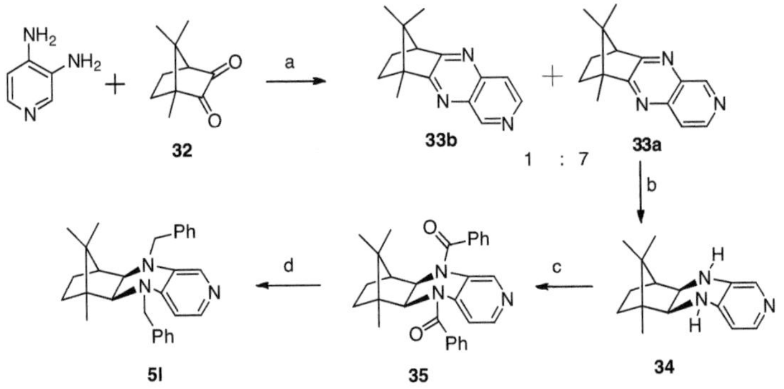

Scheme 2.4. Synthesis of camphor derivative **5l**: a) AcOH, rf, 6 h, 85 %; b) NaBH$_4$, BH$_3$, THF, 40 °C, 15 h, 55 %; c) PhCOCl, pyridine, MW, 150 °C, 60 min, 64 %; d) AlCl$_3$, THF, RT, 45 min -> LiAlH$_4$, 0 °C, 1 h -> rf, 8 h, 55 %.

Condensation with **32** proceeds in acetic acid under refluxing conditions and gives a mixture of regioisomers **33a** and **33b** in a ratio of 7:1 with 85% yield. The major product **33a** was purified by recrystallization from a cyclohexane/diethyl ether mixture (1:1). The regioselectivity of condensation was determined through detection of NOE enhancements between H$_2$ and H$_5$ protons of the pyridine ring and protons of the camphor fragment in compound **33a** (Scheme 2.5). The regioselectivity could be unambiguously assigned through the X-ray analysis of this compound (see Experimental Part).

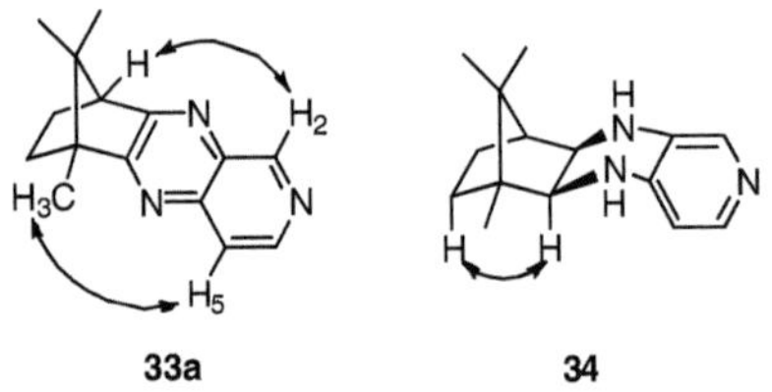

Scheme 2.5. Observed NOE enhancements in compounds **33a** and **34**.

Reduction of compound **33a** by NaBH$_4$/BH$_3$ in THF gave diamine **34** with 55% yield and 87% diastereomeric excess, as determined by chiral HPLC (Scheme 2.4). The stereochemistry of the major diastereomer was determined through detection of NOE enhancements between the protons of the camphor-ring (Scheme 2.5). Benzoylation of diamine **34** in the microwave reactor gave compound **35** as a mixture of diastereomers with de = 70 %. It was possible to isolate the major diastereomer of **35** by simple recrystallization from diethyl ether. The stereochemistry of the previous reduction step was unambiguously determined by X-ray analysis of compound **35** (Figure 2.6). Subsequent reduction of the amide groups by LiAlH$_4$/AlCl$_3$ gives the final product **51** in 55 % yield as a single diastereomer (determined by chiral HPLC de > 99 %). This catalyst was used for kinetic and theoretical studies.

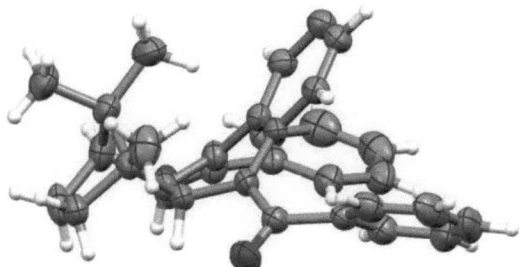

Figure 2.6. X-ray structure of compound **35**.

2.2.2 Catalytic activity of 3,4-diaminopyridines

The catalytic potential of newly synthesized 3,4-diaminopyridines has been explored in the acetylation of tertiary alcohol **36a** (Figure 2.7). The reaction proceeds to full conversion with all studied catalysts. The rate of reaction was characterized by its half-life time $t_{1/2}$, which was extracted from the conversion-time plot through fitting to a second-order kinetic rate law. The obtained data is shown in Table 2.1. For the sake of comparison we also include kinetic data for the known catalysts DMAP (**3**), PPY (**4**), **5a** and **5b**, as well as for recently published 3,4-diaminopyridine derivatives **5i** and **5j**.[30]

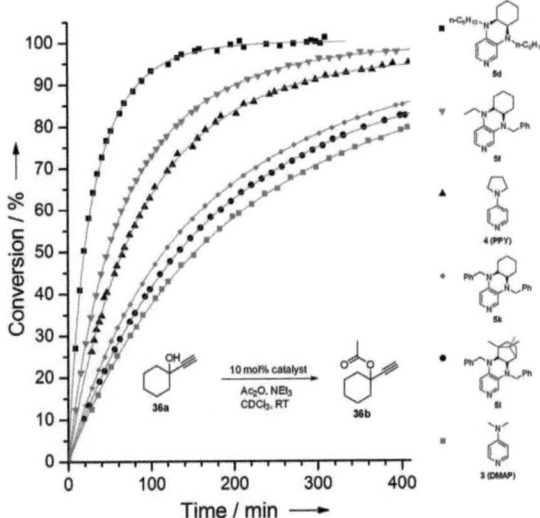

Figure 2.7. Kinetic profiles of the acetylation reaction, catalyzed by pyridine derivatives (determined by ^{1}H NMR spectroscopy of the reaction mixture).

Table 2.1. Catalytic activity of the 3,4-diamino and 4-aminopyridines.[a]

Entry	Catalyst	R^1	R^2	R^3, R^4	$t_{1/2}$ [min]
1	DMAP (3)				151 ± 1.7 [b]
2	5l	Ph	Ph	camphor[d]	138 ± 2.0
3	5k	Ph	Ph	(-CH$_2$-)$_4$	116 ± 2.0
4	PPY (4)				67 ± 1.0 [b]
5	5a	Me	Me	H	51 ± 1.0 [b]
6	5f	Me	Ph	(-CH$_2$-)$_4$	49 ± 1.0
7	5g	Me	Ph$_2$CH	(-CH$_2$-)$_4$	44 ± 1.0
8	5e	*i*Pr	*i*Pr	H	43 ± 0.5
9	5h	Me	C≡CH	(-CH$_2$-)$_4$	35 ± 1.0
10	5i	Me	Pent	(-CH$_2$-)$_4$	21 ± 1.0 [c]
11	5c	Et	Et	(-CH$_2$-)$_4$	21 ± 1.0
12	5d	Pent	Pent	(-CH$_2$-)$_4$	19 ± 0.5
13	5b	Me	Me	(-CH$_2$-)$_4$	18 ± 0.5 [b]
14	5j	Me	triazolyl[d]	(-CH$_2$-)$_4$	16 ± 0.5 [c]

[a] Conditions: 0.2 M alcohol **36a**, 2.0 equiv of Ac$_2$O, 3.0 equiv of NEt$_3$, 0.1 equiv catalyst, CDCl$_3$, 23.0±1.0 °C. [b] Data from ref. 26c. [c] Data from ref. 30. [d] See Figure 2.5.

All studied 3,4-diaminopyridine derivatives promote this reaction effectively, being faster than DMAP (**3**). Catalysts **5c**, **5d** and **5i** show almost the same half-life times as the parent derivative **5b** (entries 10-12, Table 2.1). This indicates that the variation of the alkyl substituents on the nitrogen atoms of the 3,4-diaminopyridine motif did not lead to larger changes in catalytic activity. For catalysts **5f** and **5g**, however, we note that introduction of a benzyl or diphenylethyl group on the N3 position results in a 2-fold drop of catalytic activity compared to **5b**. Introduction of two benzyl groups on both 3- and 4-nitrogen centers as in **5k** (entry 3, Table 2.1) decreases the catalytic activity even more (6-fold relative to **5b**). This observation is in remarkable contrast to recent findings of David *et al.*, who have shown that dibenzyl derivative **30b** is much more nucleophilic than the parent substance **30a** (Figure 2.4).[31] The integration of other π-systems such as the triazolyl fragment in catalyst **5j** is not accompanied by any loss in activity and provides even better catalytic efficiency than the parent system **5b** (entries 13 and 14, Table 2.1). The chiral derivative **5l** is less catalytically active in acetylation reactions than the corresponding racemic dibenzyl analogue **5k** (entries 2 and 3, Table 2.1), but still more effective than DMAP (**3**).

2.3 Acetylation enthalpies (ground state model)

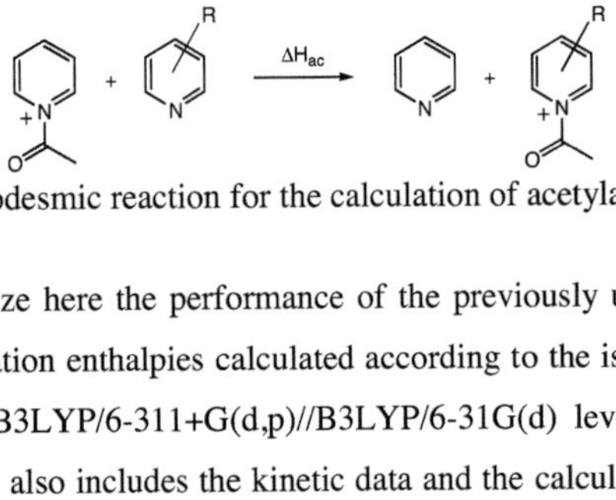

Scheme 2.2. Isodesmic reaction for the calculation of acetylation enthalpies.

As a first step we analyze here the performance of the previously used ground state model based on relative acetylation enthalpies calculated according to the isodesmic reaction shown in Scheme 2.2 at the B3LYP/6-311+G(d,p)//B3LYP/6-31G(d) level of theory in the gas phase. The analysis now also includes the kinetic data and the calculated acylation enthalpies for the newly synthesized catalysts (Table 2.2). Under the condition that these acetylation enthalpies have a direct relation to the respective activation free energies, a linear correlation of acetylation enthalpies with $\ln(1/t_{1/2})$, which is proportional to $\ln(k_2)$ should be expected (Figure 2.8). The quality of this correlation is expressed by correlation coefficient R^2, which is therefore assumed to estimate the performance of a given model. Beside analysis of the correlation coefficients, the individual examination of the data is also performed.

Inspection of the obtained results shows that the B3LYP/6-311+G(d,p)//B3LYP/6-31G(d)[32] level of theory for the calculation of acetylation enthalpies fails to correlate with experimental rates for a wide range of catalysts (correlation coefficient $R^2 = 0.1239$). As a first refinement of the ground state model in Scheme 2.2, gas phase energies were calculated at the

MP2(FC)/6-31+G(2d,p)//B98/6-31G(d) level of theory used recently for the calculation of methyl cation affinities.[25b] This includes initial geometry optimization with the B98[33] hybrid functional in combination with the 6-31G(d) basis set and subsequent calculation of thermochemical corrections at 298.15 K at the same level (no scaling factors were applied). The thermochemical corrections have then been combined with single point energies calculated at MP2(FC)/6-31+G(2d,p) level to yield enthalpies cited as "H_{298} (MP2-5)" in the text. In order to verify that important conformations had not been missed in previous studies the conformational space of all catalysts has been searched using the OPLS force field and the systematic search routine implemented in MACROMODEL 9.7.[34] All stationary points located at force field level have then been reoptimized at B98/6-31G(d) level. Solvent effects for chloroform have been accounted for by additional single point calculations with the polarizable continuum model (PCM) at RHF/6-31G(d) level with UAHF radii. The resulting enthalpies in solution were then Boltzmann-averaged over all available conformers. The most stable conformations of free catalysts (in the energy window of 20 kJ/mol) were then acetylated on the pyridine nitrogen (two orientations are possible here) and the resulting structures were treated in the same way as described for the free catalysts before. All quantum mechanical calculations have been performed with Gaussian 03.[35] The acetylation enthalpies calculated at different levels of theory are collected in Table 2.2.

Table 2.2. Relative acetylation enthalpies calculated according to the isodesmic reaction shown in Scheme 2.2 at four levels of theory.

Entry	Catalyst	$\ln(1/t_{1/2})$ ($t_{1/2}$ [min])	ΔH_{ac} (B3LYP)[a] [kJ mol^{-1}]	ΔH_{ac} (B98)[a] [kJ mol^{-1}]	ΔH_{ac} (MP2-5)[a] [kJ mol^{-1}]	ΔH_{ac} (MP2-5/solv)[a] [kJ mol^{-1}]
1	py	-	0.0	0.0	0.0	0.0
2	7a[d]	-5.69	-113.1	-114.6	-98.9	-67.5
3	7d[d]	-5.56	-118.9	-120.6	-102.0	-70.1
4	3[c]	-5.02	-82.1	-82.1	-77.2	-61.3
5	5l	-4.93	-124.3	-125.9	-123.0	-75.4
6	5k	-4.75	-124.3	-125.5	-124.1	-79.9
7	7b[d]	-4.64	-120.5	-121.0	-105.6	-73.0
8	7g[d]	-4.62	-133.1	-133.8	-116.4	-76.7
9	4[c]	-4.20	-93.1	-93.0	-87.5	-67.6
10	6b[b]	-4.14	-96.0	-95.6	-90.3	-71.7
11	7c[d]	-4.14	-123.1	-123.8	-109.0	-73.8
12	7e[d]	-4.04	-126.7	-126.8	-107.0	-75.4
13	5a[c]	-3.93	-115.5	-113.7	-108.6	-81.4
14	5f	-3.89	-125.3	-127.0	-127.4	-84.1
15	5g	-3.78	-126.5	-124.8	-121.8	-80.1
16	7f[d]	-3.78	-130.1	-131.8	-109.1	-75.5
17	5e	-3.76	-121.1	-120.6	-117.2	-82.7
18	5h	-3.56	-122.4	-120.3	-116.3	-81.8
19	5i	-3.04[e]	-128.8	-128.7	-121.9	-84.8
20	5c	-3.04	-129.4	-126.6	-121.8	-85.1
21	5b[d]	-2.89	-127.1	-126.0	-119.6	-85.2
22	5j	-2.77[e]	-142.3	-138.5	-122.1	-82.1
23	6a[b]	-2.71	-108.9	-106.9	-102.3	-82.3
25	5m	-	-	-131.0	-132.7	-83.2
	Correlation coefficient R^2 [f]		0.1239	0.0732	0.1628	0.5829

[a] Levels of theory: "B3LYP": B3LYP/6-311+G(d,p)//B3LYP/6-31G(d); "B98": B98/6-31G(d); "MP2-5": MP2(FC)/6-31+G(2d,p)//B98/6-31G(d); "MP2-5/solv": MP2(FC)/6-31+G(2d,p)//B98/6-31G(d) with PCM/UAHF/RHF/6-31G(d) solvation energies for chloroform. [b] Data from ref. 26a. [c] Data from ref. 26c. [d] Data from ref. 13a. [e] Data from ref. 30. [f] Correlation coefficient R^2 of the acetylation enthalpies with relative reaction rates $\ln(1/t_{1/2})$.

Analysis of the obtained data, i.e. correlation coefficients R^2, shows that the combination of MP2(FC)/6-31+G(2d,p) single point calculations with thermochemical corrections to 298 K at B98/6-31G(d) level gives a systematically better correlation with relative rates than the 'old' B3LYP/6-311+G(d,p)//B3LYP/6-31G(d) method. Since the actual acylation reaction is carried out in chloroform, the application of the appropriate solvation model for the calculation is necessary to get closer to experiment. Indeed employment of PCM single point calculation at HF/6-31G(d) level with UAHF radii dramatically improves the correlation (R^2 = 0.5829), which can quantitatively be expressed by the equation: $\Delta H_{ac} = -5.9703 \cdot \ln(1/t_{1/2}) - 101.28$. In contrast, the use of more expensive methods for single point calculations, such as MP2/6-311+G(2d,p) or MP2/6-311+G(3df,2p), does not improve the correlation between acetylation enthalpies and relative rates (Table 2.3).

Table 2.3. Relative acetylation enthalpies ΔH_{ac} (in kJ mol^{-1}) calculated according to the isodesmic reaction shown in Scheme 2.2 using different basis sets for the MP2 single point computations (only the best conformers were taken into account).

Catalyst	ln(1/$t_{1/2}$)	MP2-5[a]	MP2-6[b]	MP2-7[c]	MP2-5[a]	MP2-6[b]	MP2-7[c]
		in the gas phase			in chloroform[d]		
py		0.0	0.0	0.0	0.0	0.0	0.0
3	-5.02	-77.2	-77.4	-77.3	-61.3	-61.5	-61.4
4	-4.20	-87.5	-87.8	-87.7	-67.6	-67.9	-67.8
7a	-5.69	-98.9	-100.9	-101.1	-67.9	-69.9	-70.1
7b	-4.64	-105.6	-107.6	-107.7	-73.2	-75.2	-75.3
6a	-2.71	-102.3	-102.2	-102.1	-82.3	-82.1	-82.0
5k	-4.75	-124.1	-125.7	-125.0	-81.4	-82.9	-82.2
5a	-3.93	-108.6	-110.0	-109.6	-81.6	-82.9	-82.6
5b	-2.89	-119.6	-121.1	-121.7	-86.3	-87.7	-88.4
Correlation coefficient R^2 [e]		0.1083	0.0877	0.0920	0.5227	0.4472	0.4603

[a] "MP2-5" = MP2/6-31+G(2d,p)//B98/6-31G(d). [b] "MP2-6 " = MP2/6-311+G(2d,p)//B98/6-31G(d).
[c] "MP2-7" = MP2/6-311+G(3df,2p)//B98/6-31G(d).
[d] With solvation energies calculated at PCM/UAHF/RHF/6-31G(d) level. [e] Correlation coefficient of the acetylation enthalpies with relative reaction rates ln(1/$t_{1/2}$).

Even though the quality of the correlation is moderate and does not allow very precise predictions, a closer inspection of the data indicates that separate correlations of better fidelity exist for each of the catalyst families: 4-guanidinylpyridines **7a-g**, 4-dialkylaminopyridines and 3,4-diaminopyridines **5a-l** (Figure 2.8). The best correlation is obtained for the 4-

dialkylaminopyridines ($R^2 = 0.9712$), although the correlation is built only on four data points. The correlation for the 4-guanidinylpyridines is moderate ($R^2 = 0.7331$), but becomes much better ($R^2 = 0.9336$) when one does not include into the correlation catalyst **7g**, which contains a benzyl group in the guanidinyl substituent and differs thus from other catalysts **7a-f** (Figure 2.8). The 3,4-diaminopyridine derivatives **5a-l** show the worst correlation among these families ($R^2 = 0.6441$). It should be noted that considerable residuals between the actual and fitted values have been obtained for pyridines **5f**, **5g**, **5j**, **5k** and **5l**, which contain benzyl groups in the 3-amino substituents. Furthermore, considering only the best conformer enthalpies instead of Boltzmann-averaging over all the conformations is very desirable, since it would dramatically reduce the computational time. Indeed, the best conformer acetylation enthalpies are very close to the Boltzmann-averaged values (up to 0.6 kJ/mol) and can be correlated with relative reaction rates. However, a full conformational search is necessary in order to find the most stable conformations, which can then be used for calculations of acetylation enthalpy.

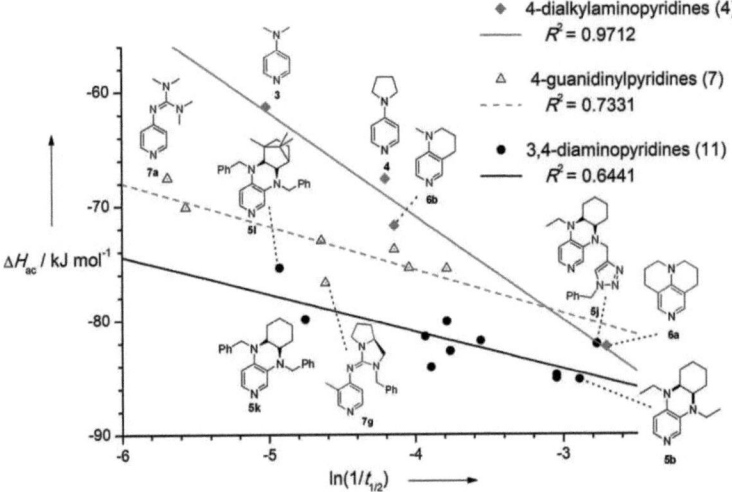

Figure 2.8. Correlation between the acetylation enthalpies (calculated at MP2/6-31+G(2d,p)//B98/6-31G(d) level with solvation energies at PCM/UAHF/RHF/6-31G(d) level) and relative reaction rates. The number of data points in each family is shown in brackets.

From the results obtained so far, we can conclude that calculation of relative acetylation enthalpies can be used for the optimization of the substitution pattern for a given catalyst family. However the correlation in the 3,4-diaminopyridine family is still not sufficient to make more exact predictions. The most inaccurate predictions are obtained for catalysts containing benzyl groups (**7g, 5f, 5j** and **5k**). This implies that some interactions with the phenyl groups play an important role in determining the activity of catalysts in the acetylation reaction. In order to study the nature of this phenomenon and find more accurate models for the activity prediction, we have studied the relative activation enthalpies for a selection of catalysts.

2.4 Activation enthalpies (transition state model)

2.4.1 Relative activation enthalpies
Theoretical[14] and experimental[12] studies of the complete catalytic cycle for the DMAP-catalyzed acetylation of alcohols have shown that the second step (attack of alcohol on the acetylpyridinium cation) is rate-limiting (Figure 1.7). The most obvious model for the prediction of relative acetylation rates would thus be the calculation of the rate-limiting transition state (Scheme 2.6). Since a search of the conformational space of the corresponding transition states for the experimentally used alcohol **36a** (1-ethynylcyclohexanol) would be too time consuming, we have chosen *t*-butanol as a model alcohol. Furthermore, the conformational space for the transition state with DMAP has already been explored, but at a different level of theory (B3LYP/6-31G(d)).[14]

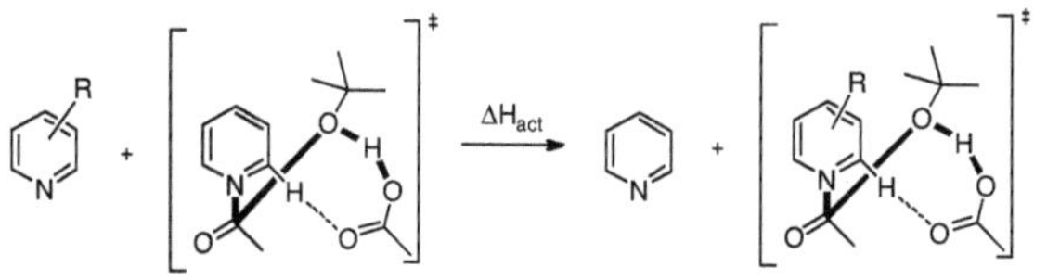

Scheme 2.6. Isodesmic reaction for the calculation of relative activation enthalpies ("frozen transition states" model). The bonds marked bold were frozen; bond lengths, Å: r (N-C) = 1.577, r (C-O) = 1.933, r (O-H) = 1.190, r (H-O) = 1.243.[14]

In the first model, called "frozen transition states", we explore the conformational space by geometry optimization of different transition state conformations with selected frozen bonds (Scheme 2.6, distances were taken from ref. 14) to energy minima at B98/6-31G(d) level, followed by frequency analysis at the same level and single point calculations at MP2(FC)/6-31+G(2d,p) level. We have chosen a series of pyridine derivatives with different substitution pattern (Figure 2.9), including well studied catalysts **3, 4, 6a, 7a-b** and **5a-b**, as well as the new 3,4-diaminopyridine derivatives **5j**, **5k** and **5l**, which have the largest deviations from the correlation line in Figure 2.8. There are generally four possible orientations of the alcohol/anhydride part of the TS for each catalyst conformer: two orientations of the acetyl group and two variants of the alcohol attack on the reaction center (from the front face or the back face of the pyridine ring). For symmetrical aminopyridines such as DMAP (**3**) and PPY (**4**) this number reduces to one possible orientation. Up to four best conformations of the free catalyst were used for chiral 3,4-diaminopyridines to obtain the initial geometries of TSs, which were then optimized to energy minima with selected frozen bonds. The actual numbers of TS conformers are shown in the Appendix (Table A2.11).

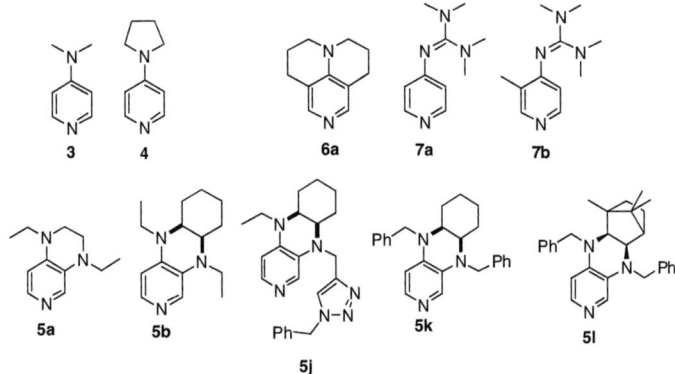

Figure 2.9. Selected DMAP derivatives studied in transition state model.

Table 2.4. Relative activation enthalpies ΔH_{act} and Gibbs free energies ΔG_{act} (in kJ mol^{-1}) as calculated at the MP2(FC)/6-31+G(2d,p)//B98/6-31G(d) level of theory in the gas phase and with inclusion of solvent effects in chloroform at different levels.[c]

Catalyst	ln(1/$t_{1/2}$) ($t_{1/2}$ [min])	"frozen transition states"		"optimized transition states"				
		ΔH_{act} [a] gas phase	ΔH_{act} [a] "solv1"	ΔH_{act} [a] "solv1"	ΔH_{act} [a] "solv2"	ΔH_{act} [a] "solv3"	ΔH_{act} [b] "solv1"	ΔG_{act} [a] "solv1"
py		0.00	0.00	0.00	0.00	0.00	0.00	0.00
7a	-5.69	-24.40	-20.50	-23.89	0.03	-7.84	-24.18	-25.07
3	-5.02	-21.94	-21.86	-25.71	-8.77	-10.82	-25.71	-26.53
7b	-4.64	-27.84	-23.01	-27.87	-6.05	-8.29	-28.68	-26.76
4	-4.20	-24.79	-24.67	-28.49	-16.72	-18.52	-28.48	-28.48
6a	-2.71	-33.94	-29.65	-34.57	-13.71	-16.17	-33.90	-35.28
5l	-4.93	-62.25	-40.69	-41.41	-12.29	-14.55	-41.45	-40.28
5k	-4.75	-55.75	-39.09	-43.47	-15.99	-18.00	-42.64	-41.36
5a	-3.93	-51.23	-41.46	-47.20	-18.51	-20.72	-45.98	-44.12
5b	-2.89	-54.86	-44.40	-49.99	-22.36	-24.84	-49.63	-46.71
5j	-2.77	-64.23	-49.28	-53.07	-18.03	-22.64	-52.66	-49.76
Correlation coefficient R^2 [d]								
All catalysts		0.1653	0.3361	0.4024	0.5434	0.5791	0.3974	0.4208
4-Aminopyridines		0.7176	0.9911	0.9895	0.5628	0.5186	0.9652	0.9516
3,4-Diaminopyridines		0.0061	0.7760	0.9510	0.6657	0.8709	0.9636	0.9419

[a] Boltzmann-averaged over the maximum available number of conformers. [b] Based on the best conformers. [c] Methods for calculating PCM single point energies in chloroform: "solv1": RHF/6-31G(d) with UAHF radii; "solv2": RHF/6-31G(d) with Pauling radii; "solv3": B98/6-31G(d) with Pauling radii. [d] Correlation coefficients R^2 of the activation enthalpies with relative reaction rates ln(1/$t_{1/2}$).

Relative activation enthalpies ΔH_{act} calculated at the combined MP2(FC)/6-31+G(2d,p)//B98/6-31G(d) level of theory in the gas phase, do not correlate well with relative reaction rates (Table 2.4). Inclusion of solvent effects of chloroform at PCM/RHF/6-31G(d) level with UAHF radii significantly improves the correlation, as was already observed for the acetylation enthalpies. The studied catalysts can now be divided into two families: 3,4-diaminopyridines and 4-aminopyridines, the latter family includes 4-dialkylaminopyridines and 4-guanidinylpyridines. The overall correlation coefficient is still modest ($R^2 = 0.3361$), but correlation in each family is much better compared to acetylation enthalpies. However, the quality of correlation for 3,4-diaminopyridines still does not allow precise predictions of reaction rates ($R^2 = 0.7760$).

The modest correlation in the "frozen transition states" model may be due to the fact that the structures used for the enthalpy calculation have constrained bonds and are not the real transition states. We have subsequently optimized the most stable conformations, obtained in the "frozen transition states" model, followed by frequency analysis at B98/6-31G(d) level and single point calculation at MP2 level as well as PCM solvation energy calculations. The energies of the transition states have changed negligibly after the optimization process (up to 2 kJ mol^{-1} at B98 level), showing thus that the previously obtained constrained structures are very close to "real" transition states. This fact allowed us to take only several best conformations of "frozen transition states" for the TS optimization (generally 3-4 conformations for 3,4-diaminopyridines). The actual numbers of optimized TS conformers are listed in the Appendix (Table A2.11).

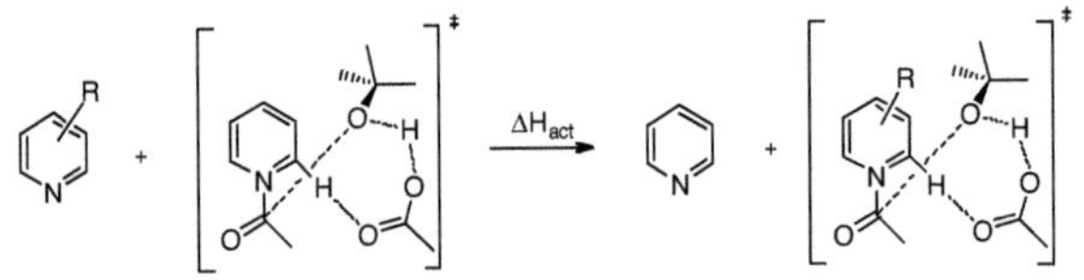

Scheme 2.7. Isodesmic reaction for the calculation of relative activation enthalpies ("optimized transition states" model).

This model, called "optimized transition states", gives much better correlation with relative reaction rates, especially in the 3,4-diaminopyridine family (Table 2.4). Employing relative activation free energies instead of enthalpies does not improve correlation significantly. From the other side, considering only the best conformer enthalpies instead of Boltzmann-averaging over all the conformations is very desirable, since it would dramatically reduce the computational time. Indeed, the best conformer activation enthalpies also correlate with relative reaction rates (Figure 2.10). The correlation coefficients obtained for each family of catalysts ($R^2 = 0.96$) allow very precise prediction of the reaction rates.

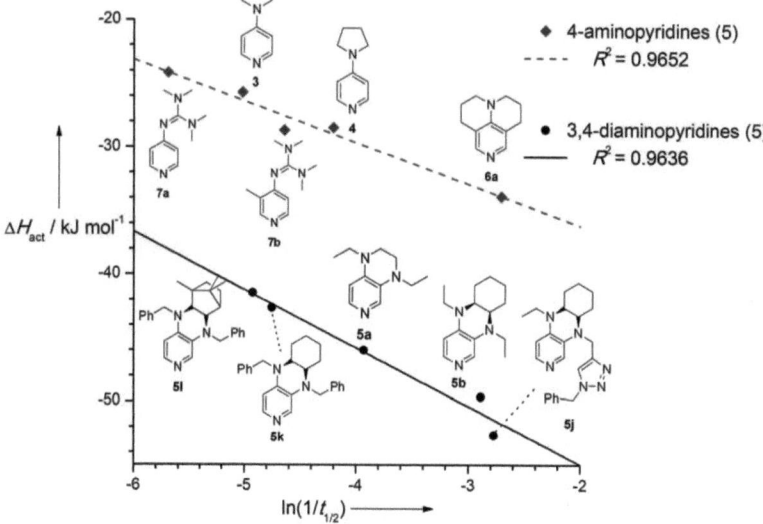

Figure 2.10. Correlation between relative activation enthalpies ("optimized transition states" model, based on the best conformers) and relative reaction rates; enthalpies were calculated at MP2/6-31+G(2d,p)//B98/6-31G(d) level with solvation energies at PCM/UAHF/RHF/6-31G(d) level). The number of data points in each family is shown in brackets.

2.4.2 Conformational properties of the transition states

A thorough analysis of the transition state structures for the pyridine derivatives studied here reveals that all transition states have an interaction between the acetate oxygen and the *o*-hydrogen of the pyridine ring, as was already mentioned in the case of DMAP.[14] This interaction has been recently proposed for the acetylated DMAP species by combined experimental and theoretical studies.[18] For 3,4-diaminopyridines we have found that additional interactions between the acetate oxygen and H^a hydrogens of the methylene group do affect the relative stability of the transition state (Figure 2.11). For all studied 3,4-diaminopyridines the transition states with acetate pointing towards the hydrogens H^2 and H^a are systematically more stable than the transition states with acetate groups pointing towards only one *ortho*-hydrogen H^6 (see Chapter 5 for thorough conformational analysis). This observation is very helpful for the transition state conformational search, since the number of conformers can be reduced by a factor of two. Furthermore, for catalysts **5j**, **5k** and **5l** with aromatic substituents one additional interaction of the acetate oxygen with the aromatic *ortho*-C-H hydrogen (like H^b) was observed in the transition states. NBO analysis of the best conformer of the TS with **5j** shows a weak hydrogen bond between acetate and H^2 hydrogen via overlap of the oxygen lone pair and the σ* orbital of C-H bond (Figure 2.11). Other interactions between the acetate oxygen and C-H bonds have very small orbital overlap and are mainly of electrostatic nature.

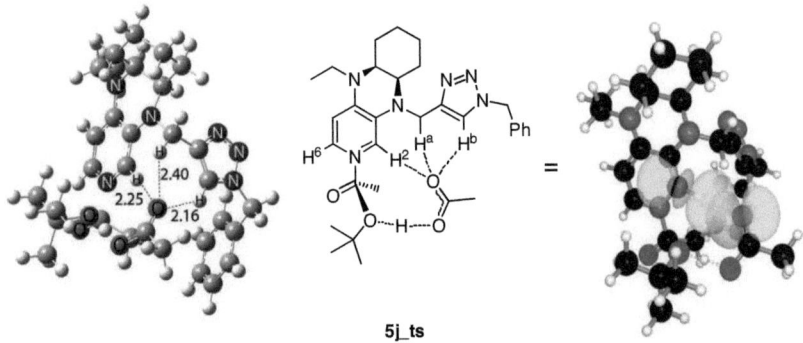

Figure 2.11. Structure of the best conformer of the TS for **5j** (left), distances between acetate oxygen and neighbouring hydrogens are given in Å; NBO analysis shows overlap of LP (O) and σ* (C-H^2) (right).

2.4.3 Influence of the solvation model

As was mentioned above the relative activation enthalpies calculated in the gas phase do not correlate with relative rates, whereas combination with PCM solvation energies gives an excellent correlation (see Table 2.4). In order to get more insight into the solvent effect and probably improve the correlation, we have briefly studied the influence of the solvation model on the correlation of activation enthalpies with reaction rates. First, we varied the PCM solvation model by changing the radii and the level of theory (Table 2.4). Results show that using Pauling radii gives slightly better overall correlation than UAHF radii (Figure 2.12). However, correlation in the catalyst families becomes much worse when employing Pauling radii. The level of theory used for PCM single point calculations does not have a large influence on the correlation coefficient.

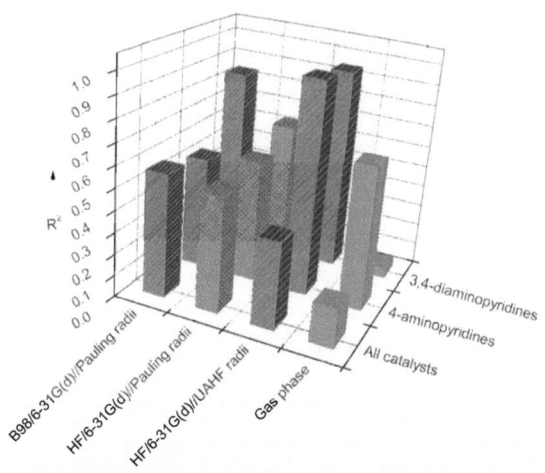

Figure 2.12. Influence of the PCM solvation model on the correlation coefficient between activation enthalpies and reaction rates.

In addition, we have briefly studied the influence of explicit solvation on the relative activation enthalpies and their correlation with relative rates. First, we propose that the highly basic pyridines could form a complex with the alcohol, affecting thus the activation enthalpy. In order to study this effect, we have calculated relative activation enthalpies of the isodesmic reaction shown in Scheme 2.8 for a selection of catalysts (4-aminopyridines: **3**, **4**, **7a** and **6a**; 3,4-diaminopyridines: **5a**, **5b**, **5k** and **5l**). The results reveal that the catalyst – alcohol complex formation has a small influence on the relative activation enthalpies, even deteriorating their correlation with relative rates (Table 2.5, Figure 2.13).

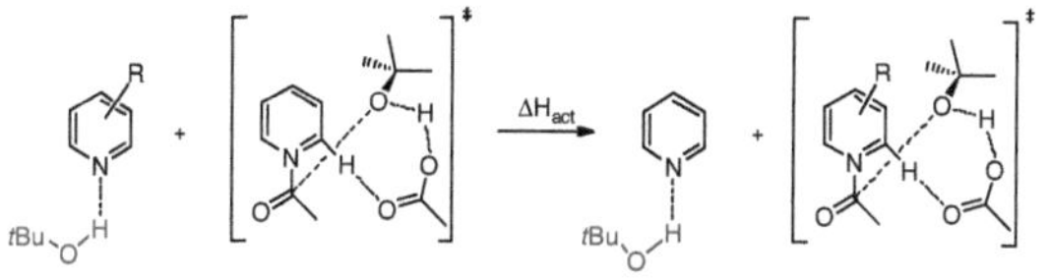

Scheme 2.8. Isodesmic reaction for the calculation of relative activation enthalpies with included formation of catalyst – alcohol complex.

Table 2.5. Relative activation enthalpies ΔH_{act} with included formation of catalyst – alcohol complex as calculated at MP2(FC)/6-31+G(2d,p)//B98/6-31G(d) level of theory in the gas phase and with inclusion of solvent effects in chloroform at different levels.

Catalyst	Without explicit solvation "solv1" [a]	With formation of catalyst – alcohol complex		
		gas phase	"solv1" [a]	"solv2" [a]
py	0.00	0.00	0.00	0.00
3 (DMAP)	-25.71	-20.33	-22.50	-16.27
4 (PPY)	-28.48	-22.26	-24.56	-23.43
6a	-33.90	-29.95	-29.41	-19.49
7a	-24.18	-15.59	-17.72	-11.11
5a	-45.99	-47.95	-47.07	-32.09
5b	-49.63	-53.85	-51.55	-39.17
5k	-42.64	-47.54	-42.56	-33.86
5l	-41.46	-48.11	-38.74	-23.21
Correlation coefficient R^2 [b]				
All 8 catalysts	**0.3460**	**0.1648**	**0.2458**	**0.2707**
4-Aminopyridines	**0.9937**	**0.9810**	**0.9587**	**0.4538**
3,4-Diaminopyridines	**0.9981**	**0.7833**	**0.9424**	**0.6337**

[a] Methods for calculation PCM single point energies in chloroform: "solv1": RHF/6-31G(d) with UAHF radii; "solv2": RHF/6-31G(d) with Pauling radii. [b] Correlation coefficients R^2 of the activation enthalpies with relative reaction rates $\ln(1/t_{1/2})$.

Second, we have briefly studied the influence of explicit solvation by chloroform on the relative activation enthalpies for the same selection of catalysts (4-aminopyridines: **3**, **4**, **7a** and **6a**; 3,4-diaminopyridines: **5a**, **5b**, **5k** and **5l**). The results show that solvation of catalysts slightly improves the overall correlation (Table 2.6). We have observed computationally that chloroform preferably solvates the oxygen atom of the acetate in the transition states.

Table 2.6. Relative activation enthalpies ΔH_{act} (in kJ mol^{-1}) calculated according to the isodesmic reaction shown in Scheme 2.9 at the MP2(FC)/6-31+G(2d,p)//B98/6-31G(d) level of theory in the gas phase and with inclusion of solvent effects in chloroform by PCM.

Catalyst	without explicit solvation by CHCl$_3$		with explicit solvation of catalyst by CHCl$_3$		with explicit solvation of TS by CHCl$_3$		with explicit solvation of TS and catalyst by CHCl$_3$	
	gas phase	"solv1" [a]	gas phase	"solv1" [a]	gas phase	"solv1" [a]	gas phase	"solv1" [a]
Py	0.00	0.00	0.00	0.00	0.00	0.00	0.00	0.00
3 (DMAP)	-24.50	-25.71	-20.65	-23.16	-28.05	-23.57	-24.20	-21.01
4 (PPY)	-27.35	-28.48	-22.28	-24.96	-30.96	-25.64	-25.88	-22.12
6a	-37.30	-33.90	-30.27	-30.06	-37.92	-32.06	-31.57	-28.22
7a	-27.70	-24.18	-7.50 [c]	-16.12	-22.42	-19.08	-6.37 [c]	-11.02
5a	-54.71	-45.99	-46.29	-43.32	-50.58	-40.92	-43.31	-38.25
5b	-59.17	-49.63	-51.79	-47.40	-63.86	-47.96	-56.78	-45.73
5k	-59.16	-42.64	-48.21	-40.17	-65.16	-40.80	-53.73	-38.33
5l	-62.88	-41.46	-50.33	-37.70	-66.26	-38.90	-54.09	-35.14
Correlation coefficient R^2 [b]								
All 8 catalysts	0.0640	0.3460	0.3367	0.3705	0.2124	0.4014	0.3224	0.4125
4-Aminopyridines	0.7165	0.9937	0.8511	0.9038	0.9775	0.9817	0.7258	0.8485
3,4-Diaminopyridines	0.1932	0.9981	0.4798	1.0000	0.0001	0.8172	0.0791	0.7988

[a] Method for calculating PCM single point energies in chloroform; "solv1": RHF/6-31G(d) with UAHF radii.
[b] Correlation coefficients R^2 of the activation enthalpies with relative reaction rates $\ln(1/t_{1/2})$.
[c] Catalyst **7a** was shown to be explicitly solvated by chloroform preferentially on the guanidinyl nitrogen (at MP2(FC)/6-31+G(2d,p)//B98/6-31G(d) level in the gas phase).

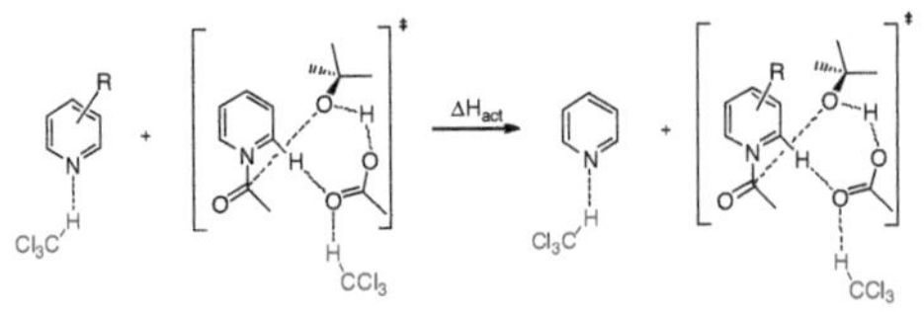

Scheme 2.9. Isodesmic reaction for the calculation of relative activation enthalpies with included explicit solvation by chloroform.

Relative activation enthalpies of the isodesmic reaction shown in Scheme 2.9, with included explicit solvation of both catalyst and TS by chloroform, have a moderate correlation with relative rates, when calculated in the gas phase ($R^2 = 0.3224$). Combination of explicit solvation and UAHF-PCM single point calculations at RHF/6-31G(d) level shows slightly improved correlation with all catalysts included ($R^2 = 0.4125$). However, the correlation in each catalyst family is worse than without explicit solvation (Figure 2.13).

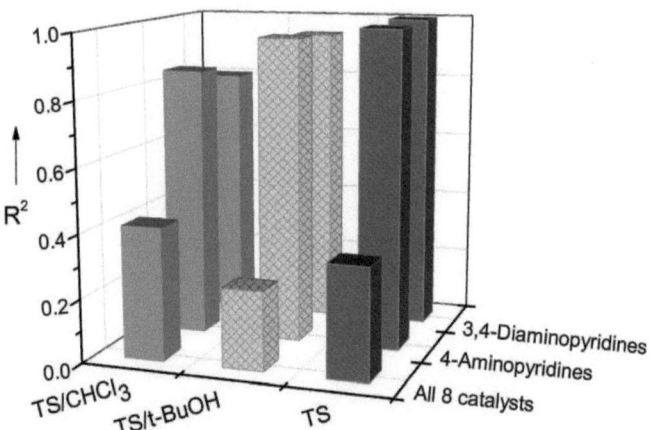

Figure 2.13. Influence of the explicit solvation by alcohol ("TS/t-BuOH") and chloroform ("TS/CHCl$_3$") on the correlation coefficient between reaction rates and activation enthalpies calculated at MP2/6-31+G(2d,p)//B98/6-31G(d) level with PCM/UAHF/RHF/6-31G(d) solvation energies. 4-Aminopyridines: **3**, **4**, **6a** and **7a**; 3,4-diaminopyridines: **5a**, **5b**, **5k** and **5l**.

2.4.4 Discussion

The variation of the solvation model (employing Pauling radii for PCM or explicit solvation by alcohol/chloroform) has a relatively small influence on the overall correlation of the activation enthalpies with relative acetylation rates. However, irrespective of the used solvation model all catalysts can be divided into two families that can be described by separate correlation lines with similar slopes. The distances between these lines are usually larger than the activation enthalpy differences inside the catalyst families (*e.g.* 14 kJ mol^{-1} on Figure 2.10). One possible explanation of this phenomenon would be a systematic overestimation of dispersion interactions at the MP2 level of theory.[65] From other side, the DFT functional with dispersion correction B3LYP-D was shown to predict this type of interactions more accurately.[76] Therefore we have briefly studied whether using the B3LYP-D/6-311+G(d,p) level of theory for single point calculations (instead of MP2) would decrease the gap between the correlation lines. The quantum chemical calculations at B3LYP-D level have been performed with ORCA 2.6.4.[81]

Table 2.7. Relative activation enthalpies ΔH_{act} (in kJ mol^{-1})[a] calculated according to the isodesmic reaction shown in Scheme 2.7 at the B3LYP-D/6-311+G(d,p)//B98/6-31G(d) and MP2(FC)/6-31+G(2d,p)//B98/6-31G(d) levels of theory in the gas phase and with inclusion of solvent effects in chloroform at different levels.[b]

Catalyst	ln(1/$t_{1/2}$)	B3LYP-D/6-311+G(d,p) //B98/6-31G(d) [b]				MP2(FC)/6-31+G(2d,p) //B98/6-31G(d) [b]			
		gas phase	solv1	solv2	solv3	gas phase	solv1	solv2	solv3
6a	-2.71	-38.49	-35.70	-14.91	-17.35	-37.30	-34.57	-13.71	-16.17
5b	-2.89	-55.40	-46.05	-17.84	-20.18	-59.17	-49.99	-22.36	-24.84
$\Delta\Delta H_{act}$ [c]		16.91	10.35	2.92	2.83	21.87	15.42	8.65	8.67

[a] Activation enthalpies are Boltzmann-averaged at 298 K. [b] Methods for calculation PCM single point energies in chloroform: "solv1": RHF/6-31G(d) with UAHF radii; "solv2": RHF/6-31G(d) with Pauling radii; "solv3": B98/6-31G(d) with Pauling radii. [c] The enthalpy differences $\Delta\Delta H_{act} = \Delta H_{act}$ (**6a**) – ΔH_{act} (**5b**).

The well-studied catalysts **6a** and **5b** were chosen as the reference points, since they have similar catalytic activities but very different activation enthalpies (the difference at "MP2-5" level is 21.9 kJ mol^{-1}). When the B3LYP-D/6-311+G(d,p) level of theory is used for single point calculations instead of MP2(FC)/6-31+G(2d,p) level (in "optimized transition states" model), the activation enthalpy difference between **6a** and **5b** slightly decreases (Table 2.7). The same effect is observed when the PCM/UAHF/RHF/6-31G(d) solvation energies are included (15.4 kJ mol^{-1} with MP2(FC)/6-31+G(2d,p) and 10.4 kJ mol^{-1} with B3LYP-D/6-311+G(d,p) single points). This implies that dispersion interactions play some role for the stability of transition states. The lowest activation enthalpy differences between **6a** and **5b** (2.8 kJ mol^{-1}) are observed when Pauling radii are used for calculations of PCM solvation energies. Therefore, using B3LYP-D level for single point calculations can bring together the correlation lines of different catalyst families. The geometry optimization at B3LYP-D level could probably yield more accurate structures than at B98 level.

2.5 Conclusions

In summary, the results show that relative acetylation enthalpies, calculated at MP2(FC)/6-31+G(2d,p)//B98/6-31G(d) level with inclusion of solvent effects in chloroform using the PCM approach at HF/6-31G(d) level with UAHF radii, can be used for the approximate optimization of the substitution pattern of a given catalyst family (such as 3,4-diaminopyridines or 4-guanidinylpyridines). Activation enthalpies (calculated at the same level of theory) give much better correlation with relative acylation rates (R^2 = 0.96 in each family) and can be used for the precise prediction of catalyst activity. However, the latter model needs much more effort: first, the transition states have generally twice as many conformers as acetylated catalysts; second, optimization and frequency analysis for the transition states are also more time-consuming than for acetylated species (*e.g.* roughly 6 times more for catalyst **5k**).

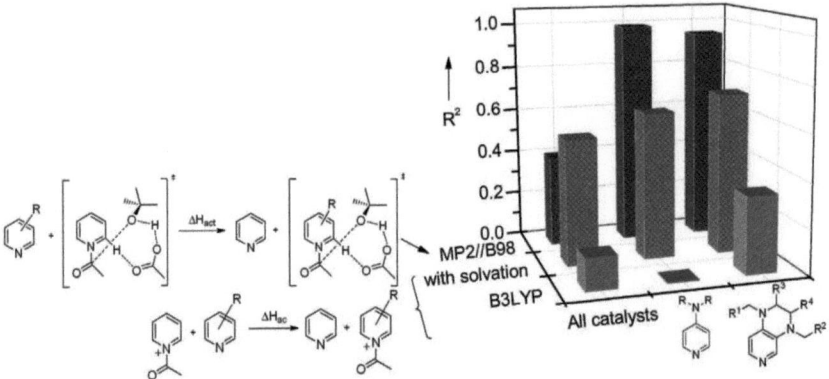

Figure 2.14. Comparison of different models for the prediction of catalytic activity.

Additionally, we show that the variation of the solvation model (employing Pauling radii for PCM calculations or explicit solvation by alcohol/chloroform) has a small influence on the overall correlation, but worsens the correlation in each catalyst family. Existence of the two catalyst families with separate correlation lines persists by using different solvation models. It implies that rather chemical than computational reasons are responsible for the splitting of correlation lines. One possible reason could be a substantial stabilization of the acetylpyridinium intermediate in the case of 3,4-diaminopyridines, which generally have more negative acetylation enthalpies than 4-aminopyridines. It would then increase the reaction barrier of the second step (see Figure 1.6). However, this reason can be disproved since the formation of the acetylated catalyst during the kinetic measurements was not observed. Another possibility can be the difference in the rate-limiting steps for different classes of pyridine derivatives, *i.e.* the second step for 4-aminopyridines and the first step for 3,4-diaminopyridines (Figure 1.6). This hypothesis can be examined by computational study of the full reaction profile (experimental measuring the reaction rate order for alcohol would not be informative in this case since the alcohol can participate in the first step).[14] Finally, the base catalysis mechanism, which has higher activation energy for DMAP,[14] can be imagined for 3,4-diaminopyridines.

Chapter 3. Applications of the Relative Acylation Enthalpies
3.1 Photoswitchable pyridines

3.1.1 Introduction

Photoswitchability of molecular properties for the development of "smart" devices and materials has been investigated quite intensively in recent years.[36] The use of light as a trigger offers different advantages, since it is a non-contact stimulus that can be manipulated perfectly by modern optics. The functional response of the molecular system to light is mediated by photochromic moieties, which enable reversible activation and deactivation (switching). While a number of molecular properties have successively been rendered photoswitchable in recent years, the reversible photomodulation of catalytic activity has so far been poorly explored. Most reported examples suffer from a lack of generality and low ON/OFF ratios.[36] Integrating an external stimulus to regulate catalytic activity offers a means to control catalyzed chemical processes and modulate the behavior of functional materials containing them.

In order to realize a reversible photoreaction in photoswitchable catalysts, photochromic moieties have to be incorporated into the catalyst system. The reactivity of the photoswitchable catalyst can be regulated by light of different wavelengths and the interface between system and stimulus is provided by a suitable photochrome. The state of higher reactivity is commonly referred to as the ON state, whereas the state of lower reactivity is referred to as the OFF state. A switchable catalyst system is characterized by the ON/ OFF ratio k_{rel}, which is given by the ratio of rate constants for substrate conversion by the ON and OFF states: $k_{rel} = k_{ON}/k_{OFF}$. A high activity ratio not only implies high reactivity associated with the ON state, but also a low OFF state reactivity to avoid undesirable background reactivity, that is, substrate conversion by the OFF state, over long periods of time.

The first photoswitchable nucleophilic catalyst has been reported by Hecht et al.[37] The design was based on the reversible steric shielding of the lone pair of a nitrogen atom, thus deteriorating both its basicity and nucleophilicity (Figure 3.1). Several piperidine derivatives were synthesized, in which a sterically bulky photochromic azobenzene shield was attached through a spiro junction. The conformation of the catalyst in the OFF state (E isomer) is largely restricted, thus efficiently shielding the lone pair of the piperidine. Upon irradiation, E→Z isomerization is induced, leading to a relative movement of the 3,5-disubstituted azobenzene fragment away from the piperidine moiety and therefore opening access to the basic/nucleophilic nitrogen atom (Figure 3.1). Acid–base titration experiments in acetonitrile reveal a difference in pK_a values of approximately one unit. Furthermore, the photoswitchable piperidines were exploited as general base catalysts in a nitroaldol (Henry) addition of nitroethane to 4-nitrobenzaldehyde. Optimizing the structure of the catalyst by introducing suitable substituents at the piperidine N atom to prevent N-inversion and at the 3,5-positions of the azobenzene moiety to more efficiently block the pyridine nitrogen atom led to an enhancement of the ON/OFF ratio to more than 35. Later on, Hecht at al. have immobilized the photoswitchable piperidine catalyst on silica gel.[38] Several successful and fully reversible switching cycles at the surface were demonstrated by UV/Vis absorption spectroscopy of suspensions in dichloromethane. Remarkably, the pH of a suspension of these functionalized silica particles in water could be photomodulated by approximately 0.3 units.

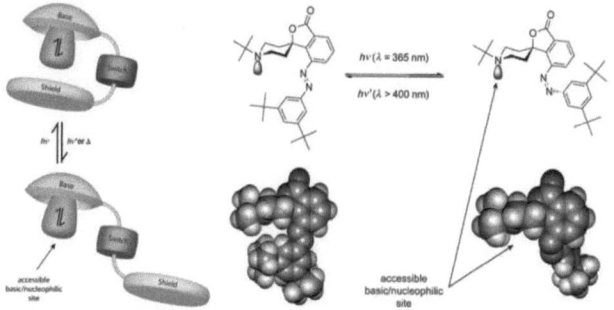

Figure 3.1. Concept of a photoswitchable base on the basis of reversible steric shielding (left); chemical structures of investigated azobenzene derivatives and single crystal X-ray structures (right). Reproduced from ref. 37.

It is surprising that only a few examples of successful reactivity switching by a photochrome-mediated electronic modulation of a catalyst's active site have been described to date.[39] Obviously pyridine derivatives are suitable model systems for electronic photomodulation since their nucleophilicity/basicity depends on the electronic properties of the substituents. Derivatives of DMAP and 3,4-diaminopyridine are particularly attractive, since they display a high efficiency in different catalyzed reactions, which can be controlled reversibly by light. Thus several 3,4-diaminopyridine derivatives bearing the azobenzene substituent were proposed as potential photoswitchable nucleophilic catalysts, which could be applied in acylation reactions controlled reversibly by light. Since theoretical predictions can be made for compounds before their respective synthesis, we decided first to study computationally the acetylation enthalpies for the proposed derivatives (calculated at MP2(FC)/6-31+G(2d,p)//B98/6-31G(d) level with solvation energies at PCM/UAHF//HF/6-31G(d) level).

Chapter 3

Table 3.1. Relative acetylation enthalpies for potential photoswitchable catalysts. Data for known catalysts **DMAP**, **PPY** and **5b** are also shown.

Catalyst	diaza1		diaza2		diaza3		DMAP	PPY	5b
	ΔH_{ac} (kJ/mol)	ΔH_{ac} (kJ/mol)	ΔH_{ac} (kJ/mol)	ΔH_{ac} (kJ/mol)	ΔH_{ac} (kJ/mol)	ΔH_{ac} (kJ/mol)	ΔH_{ac} (kJ/mol)		
R =	cis	trans	cis	trans	cis	trans			
			with solvation (CHCl$_3$)[a,c]						
CN	-69.8	-64.0	-58.2	-61.1	-	-	-61.3	-67.6	-85.2
H	-67.0	**-67.6**	-60.9	**-63.0**	**-69.4**	-67.1			
OMe	-67.6	**-69.5**	-61.9	**-65.1**	-	-			
			without solvation (gas phase)[b,c]						
CN	**-120.6**	-92.0	**-110.4**	-99.3	-	-	-77.2	-87.5	-119.6
H	-99.7	**-104.6**	-103.2	**-109.6**	**-100.2**	**-101.1**			
OMe	-102.8	**-109.7**	-105.8	**-114.1**	-	-			

[a] Acetylation enthalpies calculated at MP2(FC)/6-31+G(2d,p)//B98/6-31G(d) level with solvation model PCM$_{chloroform}$ at HF/6-31G(d) level with UAHF radii, and then Boltzmann averaged over all conformations.
[b] Acetylation enthalpies calculated at MP2(FC)/6-31+G(2d,p)//B98/6-31G(d) level, and then Boltzmann averaged over all conformations.
[c] More negative acetylation enthalpies in pairs *cis-trans* are marked bold.

3.1.2 Results and Discussion

Relative acetylation enthalpies for all studied azobenzene derivatives (calculated with inclusion of solvent effects in chloroform) are in the region between -58 and -70 kJ/mol, *i.e.* are comparable with DMAP and PPY (Table 3.1). The **diaza1** derivatives with the ethylene bridge show acetylation enthalpies, which are lower than for the corresponding **diaza2** derivatives with the diphenylethylene bridge. The **diaza1** derivatives are therefore expected to be more catalytically active in the acylation reactions. Changing the substitution pattern from 3N-methyl-4N-azobenzene in **diaza1** to 3N-azobenzene-4N-methyl in **diaza3** has a small effect on the stability of the acetyl intermediate. Substitution of the *para*-hydrogen in *trans*-derivatives for the electron withdrawing CN group decreases the stability of the acetyl intermediate and substitution for the electron donating OCH$_3$ group – increases the stability. Moreover, acetylation enthalpies of *trans*-substituted derivatives can be correlated with σ-constants of *para*-substituents (Figure 3.2).

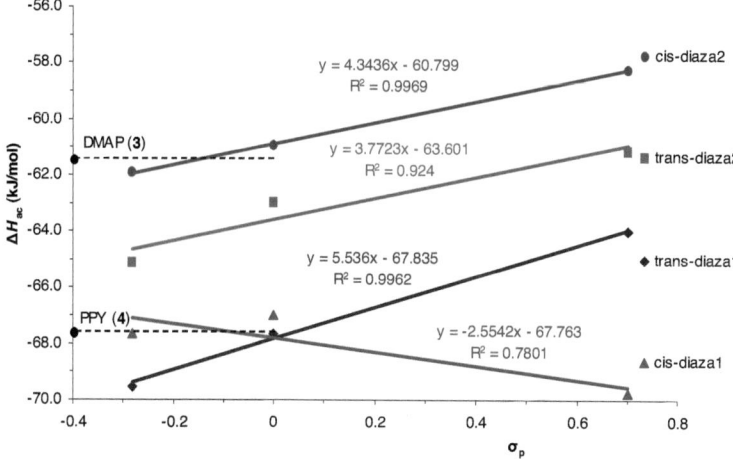

Figure 3.2. Correlation of acetylation enthalpies for *trans* and *cis* isomers (calculated at MP2(FC)/6-31+G(2d,p)//B98/6-31G(d) level with solvation model PCM at HF/6-31G(d) level with UAHF radii) with σ-constants of *para*-substituents; σ$_p$(OMe) = -0.28; σ$_p$(H) = 0; σ$_p$(CN) = +0.70.[40]

The photoswitchability of catalysts can be characterized by the ON/OFF ratio $k_{rel} = k_{ON}/k_{OFF}$, which is directly connected with the difference $\Delta\Delta H_{ct}$ between acetylation enthalpies for the *cis* and *trans* isomers (Figure 3.3). This difference is relatively low for **diaza2** derivatives (up to 3.2 kJ/mol). For **diaza1** derivatives the difference $\Delta\Delta H_{ct}$ between acetylation enthalpies is smaller, except for *p*-cyano catalyst *p*-CN-**diaza1**, which has a larger negative difference of -5.8 kJ/mol. The largest difference between *cis* and *trans* isomers together with the most negative acetylation enthalpy for the *trans* isomer are observed in the case of the *p*-methoxy **diaza2** (Figure 3.3, dotted green circle), as well as the *p*-cyano **diaza1** (Figure 3.3, green circle). This suggests that these derivatives can be potent photoswitchable catalysts. However, the comparatively moderate acetylation enthalpies (which are in the range of DMAP and PPY) for all studied derivatives can result in low absolute catalytic activities of these compounds (*cf.* Chapter 2, Figure 2.8).

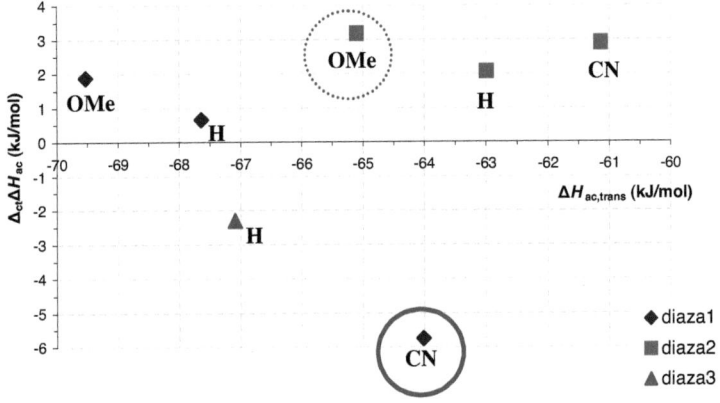

Figure 3.3. Acetylation enthalpy differences $\Delta_{ct}\Delta H_{ac} = \Delta H_{ac,cis} - \Delta H_{ac,trans}$ for **diaza1** and **diaza2** derivatives with different *para* substituents *vs* acetylation enthalpies of *trans* derivative $\Delta H_{ac,trans}$, as calculated at MP2(FC)/6-31+G(2d,p)//B98/6-31G(d) level with solvation model PCM at HF/6-31G(d) level with UAHF radii.

While the acetylation enthalpy differences $\Delta_{ct}\Delta H_{ac}$ determine the ON/OFF ratio,[36] the differences between enthalpies of *cis* and *trans* azobenzene isomers ΔH_{ct} also plays an important role for the kinetics of the thermal isomerisation *cis* –> *trans*.[41] The obtained results (Table 3.2) show that *p*-methoxy substituted **diaza1** and **diaza2** derivatives have the largest enthalpy differences ΔH_{ct}.

Table 3.2. Relative stabilities of *cis* and *trans* isomers of catalysts **diaza1**, **diaza2** and **diaza3**, as expressed by enthalpy differences ΔH_{ct} (positive values imply more stable *trans* isomers).

$$\Delta H_{ct} = H_{cis} - H_{trans} \text{ (kJ/mol)}^a$$

R=	diaza1	diaza2	diaza3
CN	+48.6	+41.0	
H	+50.7	+43.5	+48.6
OMe	+51.5	+46.6	

[a] Enthalpies H_{298} were calculated at MP2(FC)/6-31+G(2d,p)//B98/6-31G(d) level with solvation model PCM at HF/6-31G(d) level with UAHF radii, and then Boltzmann averaged at 298 K over all conformations.

Acetylation enthalpy differences $\Delta_{ct}\Delta H_{ac}$ for **diaza1** and **diaza2** derivatives, calculated at MP2(FC)/6-31+G(2d,p)//B98/6-31G(d) level *in the gas phase*, have generally larger absolute values than the corresponding differences calculated *with solvation* (Figure 3.4). The largest values of the differences $\Delta_{ct}\Delta H_{ac}$ are observed in the case of *p*-cyano substitution (up to −28.6 kJ/mol for **diaza1** derivatives), thus showing that solvent plays an important role for the photo-switchability of these compounds. Examination of the data in Table 3.1 shows that *cis*-derivatives of *p*-cyano-**diaza1** and *p*-cyano-**diaza2** have surprisingly low (more negative) acetylation enthalpies, calculated at MP2(FC)/6-31+G(2d,p)//B98/6-31G(d) level in gas phase.

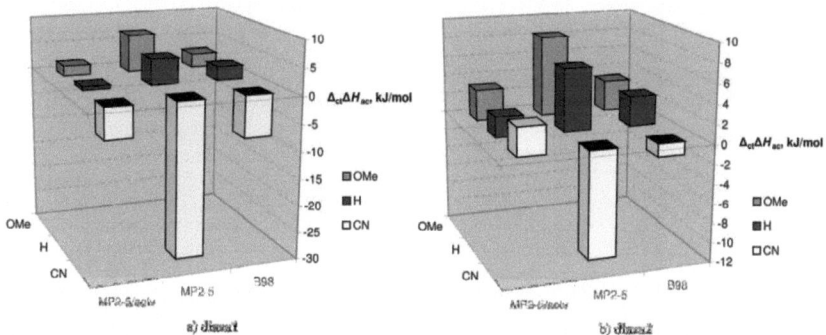

Figure 3.4. Acetylation enthalpy differences $\Delta_{ct}\Delta H_{ac} = \Delta H_{ac,cis} - \Delta H_{ac,trans}$ (in kJ/mol) for **diaza1** (a) and **diaza2** (b) derivatives with different *para* substituents, as calculated at the different levels of theory: "B98" = B98/6-31G(d)//B98/6-31G(d); "MP2-5" = MP2(FC)/6-31+G(2d,p)//B98/6-31G(d); "MP2-5/solv" = MP2(FC)/6-31+G(2d,p)//B98/6-31G(d) with solvation model PCM at HF/6-31G(d) level with UAHF radii.

A thorough analysis of the conformational space for studied pyridines and their acetylated forms reveals an interesting structural feature of the acetylated catalysts *cis*-*p*-cyano-**diaza2** and *cis*-*p*-cyano-**diaza1**. The most stable gas phase conformation **4_ac1** of the acetylated *cis*-*p*-cyano-**diaza2** has surprisingly low energies at both B98 and MP2 levels of theory (comparing with conformation **4_ac2**, which has the opposite orientation of the acetyl group, Table 3.3). The cyano group in conformation **4_ac1** is in close proximity to the acetyl group (Figure 3.5, right). A smaller molecular dipole moment of conformation **4_ac1** comparing with **4_ac2** (Table 3.3) leads to less negative solvation energy of this conformer (−63.76 kJ/mol for **4_ac1** and −80.08 kJ/mol for **4_ac2**), which compensates a larger stability of **4_ac1** in gas phase. As a result, this conformation does not affect the acetylation enthalpy at the combined level of theory including solvent effects. This type of conformation for acetylated *cis*-*p*-cyano-**diaza1** (Figure 3.5, left) has the lowest energies both in the gas phase and in chloroform. Replacing the CN group by methoxy or hydrogen in these conformations does not give stable species, which after optimization give "open" conformers with larger distances between the phenyl and pyridine rings.

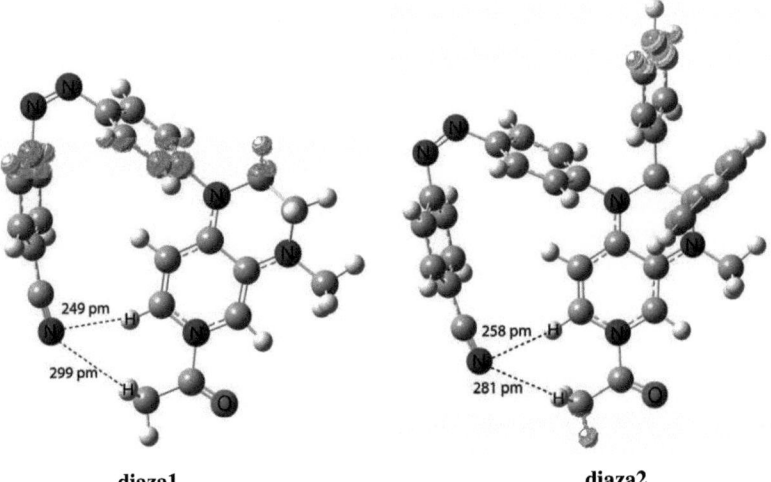

diaza1   diaza2

Figure 3.5. Structures of the most stable gas phase conformers of the acetylated catalysts *cis*-*p*-cyano-**diaza1** and *cis*-*p*-cyano-**diaza2** (optimized at B98/6-31G(d) level).

Table 3.3. Comparison of **4_ac1** and **4_ac2** conformers of acetylated *cis-p*-cyano-**diaza2**.

Parameter	4_ac1	4_ac2
r_1, pm	281	998
r_2, pm	454	606
Dihedral angle φ_1, °	115.9	135.6
Molecule dipole moment, D	3.43	11.98
q_N [a]	−0.345	−0.280
q_H [a]	+0.287	+0.282 [c]
q_{ac} [a]	+0.292	+0.282
Relative enthalpies ΔH_{298} at different levels of theory[b] (kJ/mol)		
"B98"	0	12.7
"MP2-5"	0	30.0
"MP2-5/solv"	1.8	15.5

[a] NPA charges are given in units of elemental charge e.
[b] Relative to the most stable conformer of acetylated *cis-p*-cyano-**diaza2** at the corresponding level of theory.
[c] Charge is averaged over three hydrogens of acetyl group.

The described extraordinary stability can originate from electrostatic interactions between the negatively charged nitrogen of the CN group and the positively charged hydrogens of the COCH$_3$ group or dispersion interactions between phenyl rings of azobenzene. Natural population analysis (NPA) shows that the CN nitrogen atom is more negatively charged in the **4_ac1** conformer of acetylated *cis-p*-cyano-**diaza2** as compared to conformer **4_ac2**; the hydrogen atoms of the acetyl group are also more positively charged (Table 3.3). This implies that electrostatic interactions between the CN nitrogen and the acetyl hydrogens could play an important role for the exceptional stability of the **4_ac1** conformer in the gas phase. In order to clarify this point, we have briefly studied the relative stabilities of the acetylated *cis-p*-

cyano-**diaza1'** conformers, which contain a propylene bridge instead of the phenylene group in the azobenzene moiety (Table 3.4). The propylene bridge was chosen since it is geometrically more similar to the phenylene group than the $(CH_2)_2$ or $(CH_2)_4$ fragments. The results show that the conformer **ac1**, analogue of the "closed" conformation *cis-p*-cyano-**diaza1_4_ac1**, also has lower energies at both B98 and MP2 levels of theory, comparing with conformation **ac2**, which has the opposite orientation of the acetyl group (Table 3.4). The "opened" conformer **ac3** has much higher energy in the gas phase, but becomes more stable than **ac1** and **ac2**, when the solvation energy is added. This implies that the dispersion interactions between phenyl rings of the azobenzene substituent in **diaza1** and **diaza2** derivatives have a relatively small influence on the described extraordinary stability of acetylated species.

Table **3.4**. Relative enthalpies of acetylated *cis-p*-cyano-**diaza1'** at different levels of theory.

ΔH_{298} (kJ/mol)[a]	**ac1**	**ac2**	**ac3**
"B98"	0.0	18.3	9.3
"MP2-5"	0.0	20.2	29.1
"MP2-5/solv"	1.3	10.3	0.0

[a] Relative enthalpies ΔH_{298} of conformers of acetylated *cis-p*-cyano-**diaza1'** at three levels of theory.

In all studied model systems **diaza1** and **diaza2** the aryl substituent in 4N position is not coplanar to the pyridine ring, decreasing significantly the conjugation between the *para*-substituent of the azobenzene group and the pyridine nitrogen. The derivatives **diaza4** of **cat11un**, bearing the phenyldiazenyl group in the *para*-position, were suggested as potential photoswitchable catalysts, since the pyridine nitrogen and the azobenzene substituent are directly conjugated in these systems. A brief study of the respective acetylation enthalpies (Table 3.5) shows that the suggested derivatives have low stabilities of acetyl intermediates, which would result in low absolute catalytic activities of these compounds. The enthalpy difference $\Delta_{ct}\Delta H_{ac}$ between acetylation enthalpies for the *cis* and *trans* isomers is also moderate (Table 3.5). It implies that the proposed substitution pattern is insufficient for the development of photoswitchable catalysts.

Table 3.5. Relative acetylation enthalpies for catalysts **5b**, **cat11un** and its azobenzene derivatives, calculated according to the isodesmic reaction.

Catalyst	ΔH_{ac} (kJ/mol)	
	With solvation[a]	Without solvation[b]
5b	−85.2	−119.6
cat11un	−60.1	−91.5
trans-diaza4	−52.9	−87.5
cis-diaza4	−49.1	−79.9
$\Delta_{ct}\Delta H_{ac}$	+3.8	+7.6

[a] Calculated at MP2(FC)/6-31+G(2d,p)//B98/6-31G(d) level with solvation model PCM at HF/6-31G(d) level with UAHF radii.
[b] Calculated at MP2(FC)/6-31+G(2d,p)//B98/6-31G(d) level in thr gas phase.

3.1.3 Conclusions and Outlook

The studied model systems **diaza1** and **diaza2** are shown computationally to be potentially photoswitchable, since they have different acetylation enthalpies in the *cis* and *trans* states. The largest effects on the acetylation enthalpies are observed for the *p*-cyano substituted derivatives, which have some interactions between CN and COCH$_3$ groups in *cis*-derivatives. However, these effects are mainly electrostatic in nature. In order to further increase these effects, the azobenzene moiety should be directly introduced into the 5-position of the pyridine ring (Scheme 3.1). The suggested structures have also the advantage that the pyridine nitrogen would be more nucleophilic in the ON state. Alternatively, incorporating bulky substituents into the azobenzene part could further increase the ON/OFF ratio by steric shielding of the pyridine nitrogen in the OFF state, as was successfully used in the design of the photoswitchable piperidine catalyst.[37]

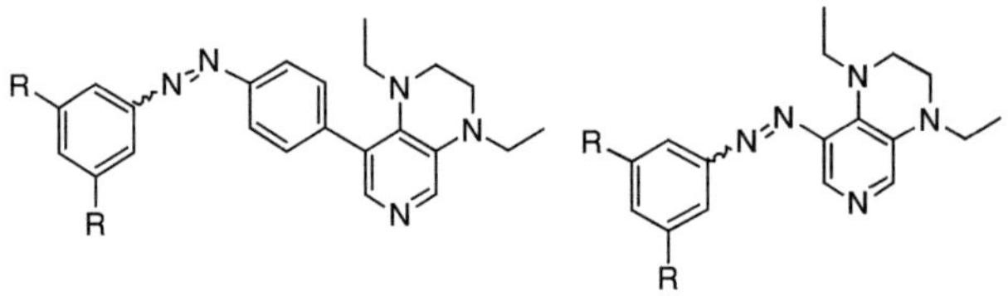

Scheme 3.1. Structures of potentially photoswitchable pyridines.

3.2 Relative acetylation enthalpies for paracyclophane derivatives

Planar-chiral motifs are often used as typical design elements of nucleophilic catalysts or chiral ligands for transition-metal catalysis.[30,42a] These structures are usually derived from metallocenes, including ferrocene,[43] or from the [2.2]paracyclophane framework. Compounds with the [2.2]paracyclophane scaffold[44] have already been applied in enantioselective synthesis, for example, titanium complexes of salen-type derivatives of [2.2]paracyclophane in the enantioselective formation of cyanohydrins of aromatic aldehydes and in the enantioselective addition of diethylzinc to aromatic aldehydes.[42] A number of [2.2]paracyclophane derivatives have been employed as organocatalysts.[45] Surprisingly, there are no reports to date of planar-chiral DMAP derivatives, containing the [2.2]paracyclophane moiety. Here we study computationally the influence of the paracyclophane substituent on the catalytic activity and conformational properties of DMAP derivatives.

Relative acetylation enthalpies ΔH_{ac} for a series of 3-paracyclophane-4-aminopyridines were calculated according to the isodesmic reaction shown in Scheme 2.2 at MP2/6-31+G(2d,p)//B98/6-31G(d) level with inclusion of solvent effects at PCM/UAHF/RHF/6-31G(d) level (Table 3.6). The obtained data show that a 3-paracyclophane substituent decreases the relative stability of the pyridinium cation by 5.9 kJ/mol for the derivative **para1** as compared to DMAP (**3**), and 0.9 kJ/mol for **para2**, as compared to PPY (**4**). While a methyl group in the paracyclophane moiety at pseudo-*ortho* position does not almost affect the acetylation enthalpy (**para3**), the amide functional group has a large effect on the acetylation enthalpy (**para4**): it stabilizes the pyridinium cation by ca. 14 kJ/mol. The catalyst **para4** would be expected to be as active as **5b** or **6a** in acylation reactions, combination of this feature with the planar chirality makes this compound potentially useful for the KR experiments. The distances between the acetyl group and the pyridine nitrogen ($r_{(C-N)}$, pm) and the natural charges of the acetyl group ($q_{NPA}(ac)$) were also calculated and are collected in Table 3.6. In contrast to 3,4-diaminopyridines there is no correlation between the acetylation enthalpies of 3-paracyclophane-4-aminopyridines and the charge or distance parameters.

Table 3.6. Relative acetylation enthalpies and structural parameters for 3-paracyclophane-4-amino and 3,4,5-trialkylpyridines, as calculated at MP2/6-31+G(2d,p)//B98/6-31G(d) level with inclusion of solvent effects in chloroform at PCM/UAHF/RHF/6-31G(d) level. Data for catalysts DMAP (**3**), PPY (**4**), **5b** and **6a** are also shown.

Catalyst	Structure	On the basis of the energetically lowest conformer			Averaged
		q_{NPA} (ac)[a,b]	r(C-N)(pm)[a]	ΔH_{ac} (MP2-5/solv) (kJ/mol)	ΔH_{ac} (MP2-5/solv) (kJ/mol)
py		0.368	153.3	0.0	0.0
para1		0.284	147.4	-56.1	-55.4
DMAP (3)		0.302	148.3	-61.3	-61.3
para2		0.278	147.0	-67.3	-66.7
para3		0.276	146.9	-67.1	-66.9
PPY (4)		0.295	147.9	-67.6	-67.6
para4		0.279	148.1	-81.1	-80.5

Catalyst	Structure	On the basis of the energetically lowest conformer			Averaged
		q_{NPA} (ac)[a,b]	r(C-N)(pm)[a]	ΔH_{ac} (MP2-5/solv) (kJ/mol)	ΔH_{ac} (MP2-5/solv) (kJ/mol)
6a		0.283	147.3	-82.3	-82.3
5b		0.276	147.1	-86.3	-85.2
37a		0.341	151.4	-26.3	-26.3
37b		0.336	151.0	-28.3	-27.7
37c		0.333	150.8	-28.1	-27.4

[a] Charge and distance parameters of the most favorable conformer
[b] In units of elemental charge e

The 4-aminopyridine derivatives DMAP (**3**) and **6a** differ structurally mainly by two alkyl substituents in 3- and 5- positions of the pyridine ring. This causes a large change in the acetylation enthalpy (by ca. 20 kJ/mol) and a 10-fold acceleration of the acetylation reaction rate (see Chapter 2). In order to check the influence of the alkyl substituents we have calculated the acetylation enthalpies for 3,4,5-trialkylpyridines **37a-c**. The obtained data show (Table 3.6) that alkyl groups have a moderate donating effect, much smaller than the dialkylamino group. Eventually the studied trialkylpyridines seem not to be active enough to be potential catalysts for acylation reactions.

The conformational space of the neutral derivative **para4** and its acetylated form has been analyzed. Figure 3.6 shows a pictorial representation of the relative enthalpies of conformers in the range 0–25 kJ/mol. Analysis of the obtained data shows that the acetylation of compound **para4** leads to a reduction of the conformational space of this molecule. Furthermore, examination of the best conformer structures for compound **para4** and its acetylated form (Figure 3.6) reveals that a conformational change occurs in the amide group of the catalyst upon acylation. The oxygen atom of the amide group becomes directed towards the acetyl group, probably due to electrostatic interactions between oxygen and neighbouring hydrogens (*ortho* hydrogen of pyridine ring and acetyl hydrogen). These interactions are probably responsible for the exceptional stability of this acetylpyridinium cation mentioned above.

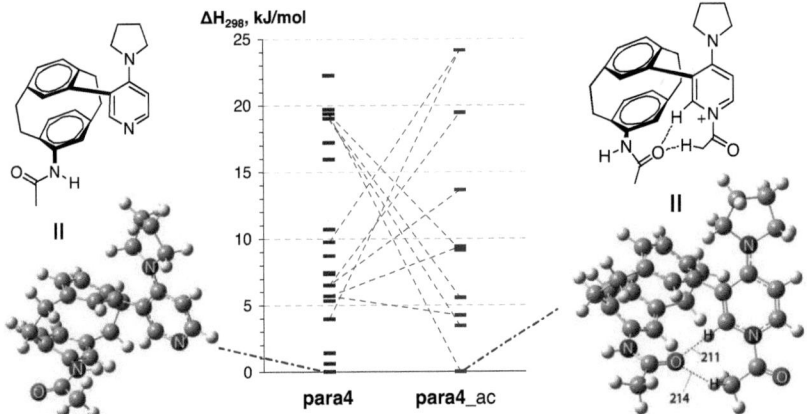

Figure 3.6. Relative enthalpies of conformers of compound **para4** and its acetylated form **para4_ac** at MP2/6-31+G(2d,p)//B98/6-31G(d) level within an energy window of 25 kJ/mol. The structurally related conformers of **para4** and **para4_ac** are connected with dotted lines. Structures of the most stable conformers are also shown (distances are given in pm).

3.3 Relative isobutyrylation enthalpies for chiral 3,4-diaminopyridines

The kinetic resolution of *sec*-alcohols, catalyzed by chiral pyridines, often utilizes isobutyric anhydride as the resolution reagent. It was shown that this anhydride has the best chances for successful kinetic resolution experiments with secondary alcohols.[46] Since the isobutyryl group is more sterically bulky than the acetyl group, it is not clear, whether the acetylation enthalpies are suitable for the prediction of the catalytic activity for chiral pyridine derivatives. For this reason we decided to study whether isodesmic reaction of isobutyryl-transfer, shown in Scheme 3.2, can be used for the prediction of the catalytic activity of chiral pyridine derivatives.

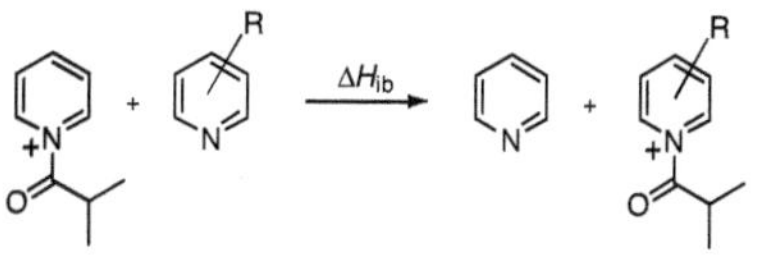

Scheme 3.2. Isodesmic reaction for the calculation of isobutyrylation enthalpies.

The enthalpies of model isodesmic isobutyryl-transfer reaction, shown in Scheme 3.2, were calculated at MP2(FC)/6-31+G(2d,p)//B98/6-31G(d) level in the gas phase and with inclusion of solvent effects in chloroform using PCM single point computations at RHF/6-31G(d) level with UAHF radii (Table 3.7). The distance between the isobutyryl group and the pyridine nitrogen ($r_{(C-N)}$, pm) and the natural charge of the isobutyryl group ($q_{NPA}(ac)$) are also included in Table 3.7.

Table 3.7. Relative isobutyrylation enthalpies at 298.15 K for pyridine derivatives calculated according to the isodesmic reaction shown in Scheme 3.2 at the MP2(FC)/6-31+G(2d,p)//B98/6-31G(d) level of theory (in the gas phase and in CHCl$_3$).

Catalyst	ΔH_{ib} (kJ/mol)		r (C–N)^a (pm)	q$_{NPA}$ (iPrCO)^{a,b}
	in the gas phase	in chloroform^c		
Py	0	0.0	155.0	0.379
DMAP (3)	-73.5	-59.5	149.5	0.311
PPY (4)	-84.0	-66.4	149.0	0.303
6a	-98.2	-80.0	148.4	0.291
5b	-114.6	-84.1	148.1	0.284
5k	-123.7	-81.4	148.4	0.286
38a	-108.9	-77.6	148.4	0.288
38b	-115.3	-82.3	148.1	0.284
5l	-123.5	-76.7	148.7	0.290
5l'	-123.5	-75.3	148.6	0.286

^{a} Charge and distance parameters of the most favorable conformer were calculated at B98/6-31G(d) level.
^{b} NPA charges in units of elemental charge e.
^{c} With solvation model PCM$_{chloroform}$ at HF/6-31G(d) level with UAHF radii.

First, the isobutyrylation enthalpies were calculated for achiral 4-aminopyridines DMAP (**3**), PPY (**4**) and **6a** as well as the known 3,4-diaminopyridine derivatives **5b**, **5k** and **5l**. Figure 3.8 shows that the isobutyrylation enthalpies correlate well with the corresponding acetylation enthalpies. However, correlation with relative acetylation rates becomes worse when using isobutyrylation enthalpies ($R^2 = 0.3034$) instead of acetylation enthalpies ($R^2 = 0.4592$).

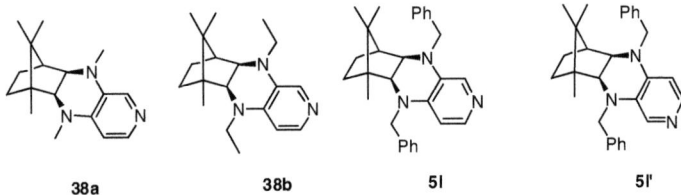

Figure 3.7. Structures of chiral camphor-derived catalysts.

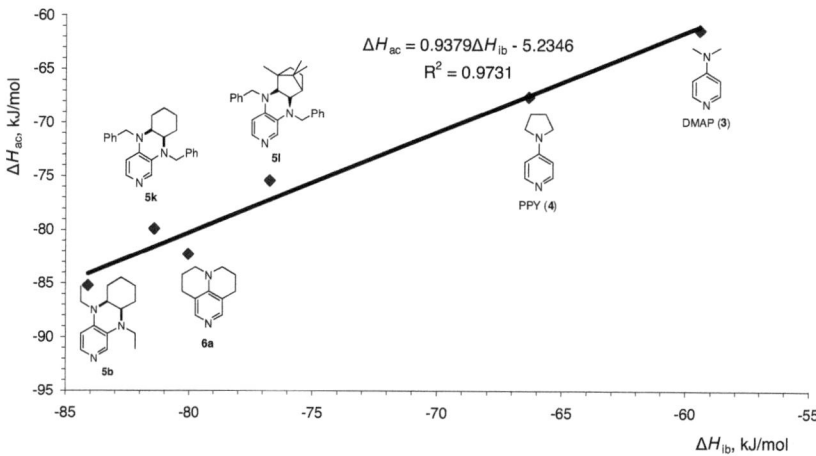

Figure 3.8. Correlation between isobutyrylation and acetylation enthalpies, as calculated at MP2(FC)/6-31+G(2d,p)//B98/6-31G(d) level with inclusion of solvent effects at PCM/UAHF/RHF/6-31G(d) level.

The chiral 3,4-diaminopyridines **38a-b**, **5l** and **5l'**, which contain the chiral fragment of camphor, were then studied computationally (Figure 3.7). Examination of the data in Table 3.7 shows that replacement of the cyclohexyl ring in **5b** by the camphor fragment in **38b** slightly lowers the isobutyrylation enthalpy. Replacement of ethyl groups in **5b** and **38b** by benzyl groups as in **5k** and **5l** enhances the stability of the isobutyryl intermediate in the gas phase by 8-9 kJ/mol, but lowers it when the solvation energies are included. The similar values obtained for the isomers **5l** and **5l'** indicate that the stereochemistry of the camphor fragment has little influence on the stability of isobutyryl species.

The structural parameters, such as C–N distance and the charge of isobutyryl group of acylated species, are also included in Table 3.7. Figure 3.9 shows that greater thermochemical stability correlates with shorter C–N bond distances and smaller isobutyryl group charges.

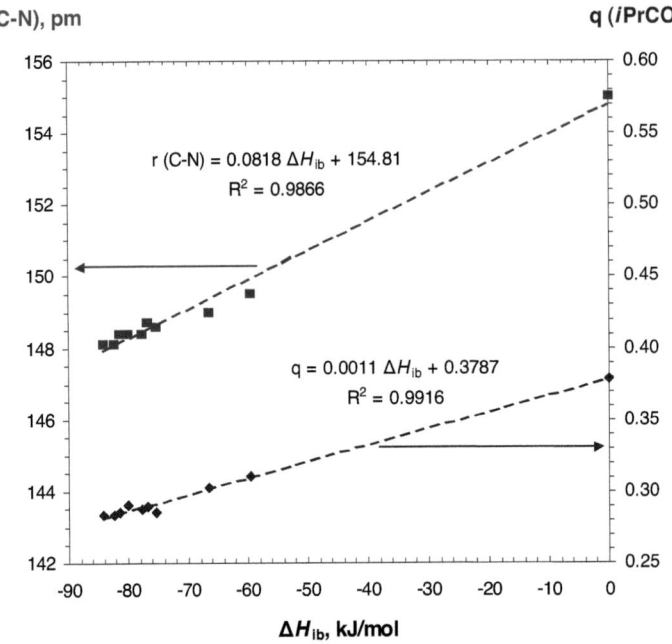

Figure 3.9. Correlation of the isobutyryl group charges (right axes) and C–N bond distances (left axes) calculated at B98/6-31G(d) level with isobutyrylation enthalpies calculated at MP2(FC)/6-31+G(2d,p)//B98/6-31G(d) level with inclusion of solvent effects at PCM/UAHF/RHF/6-31G(d) level.

3.4 Relative acetylation enthalpies for ferrocenyl pyridines

Since its discovery in 1951, ferrocene has increasingly established itself as an efficient and generally applicable backbone in chiral ligands and catalysts due to the possibility of introducing and exploiting both central and planar chirality. A main advantage of planar chirality as a control element is that it does not racemize as compared to, for example, axial chirality seen in the classic biarylic systems, and today several efficient methods exist for introducing planar chirality onto the ferrocene backbone.[47] One of the effective methods is *ortho*-lithiation of chiral sulfoxides, which can be obtained from available asymmetric compounds. This method has an advantage in avoiding the racemate resolution step and was successfully applied to the synthesis of different planar chiral *P,N*-ligands.[48]

The development of asymmetric DMAP derivatives **39** based on planar chiral ferrocene has been reported by Fu *et al*.[43] These catalysts were shown to perform very well for a variety of reactions,[49a,b] such as kinetic resolutions of alcohols[49c] and amines[49d-f] (Figure 3.10). Later attempts in designing ferrocene-based planar chiral DMAP analogues resulted in less efficient catalysts. A C_2-symmetric catalyst **40** gives moderate levels of selectivity in KR of secondary alcohols.[50] The catalyst **41**, which has one ferrocenyl substituent in 3-position of DMAP, gave modest enantioselectivities in dynamic KR and rearrangement reaction of azlactones.[51] On the other hand the catalysts **40** and **41** exhibit low to moderate catalytic activity. The analogue of **41** which has a ferrocenyl substituent in *ortho*-position is completely inactive as nucleophilic catalyst.[51]

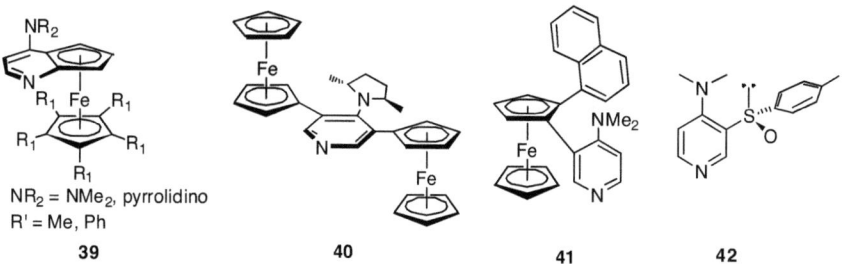

Figure 3.10. Structures of chiral DMAP derivatives **39-42**.

Thus the planar chiral ferrocene moiety seems to be promising in the design of new asymmetric DMAP derivatives. The activity issue is also important for the modeling of new catalysts. For instance, the chiral catalyst **42**, which contains a sulfoxide group, shows moderate enantioselectivities but very low activity in the KR of alcohols.[52] The observed rate difference has been explained by the electron-withdrawing character of the sulfoxide group. In order to shed light on the influence of the sulfoxide group on the activity of ferrocenyl pyridines, we have performed a computational study of the relative acetylation enthalpies for a series of differently designed chiral catalysts, which can potentially be synthesized employing the *ortho*-lithiation of chiral ferrocene sulfoxides. Most of the studied pyridine derivatives contain the remaining sulfoxide fragment. Another important goal was to investigate how different structural features influence the activity of ferrocene substituted pyridines.

Since the HF level of theory (and correspondingly MP2 which contains HF) is known to have convergence problems when applied to ferrocene system, the "MP2-5/solv" method discussed above for the calculation of acetylation enthalpies was not applicable in this case. For this reason the DFT computations at the B3LYP/6-311+G(d,p)//B3LYP/6-31G(d) level of theory were used to evaluate relative acetylation enthalpies for the studied ferrocenylpyridines, which were then compared with the data for "standard" 4-aminopyridines DMAP (**3**), PPY (**4**) and **6a**. The obtained data are shown in Table 3.8.

Table 3.8. Relative acetylation enthalpies for ferrocenylpyridines, as calculated at the B3LYP/6-311+G(d,p)//B3LYP/6-31G(d) level of theory. Data for 4-aminopyridines DMAP (**3**), PPY (**4**) and **6a** are also shown.

Catalyst	Structure	ΔH_{ac} B3LYP/6-311+G(d,p)//B3LYP/6-31G(d) (kJ/mol)	
		best conformer	averaged
py		0.0	0.0
DMAP (**3**)		−82.1	−82.1
PPY (**4**)		−93.0	−93.1
6a		−108.9	−108.9
o-cpstol		−18.2	−18.8
m-cpstol		−59.5	−59.5
p-cpp		−64.1	−63.4
p-cpstol		−68.0	−67.4

Table 3.8 (continued)

Catalyst	Structure	ΔH_{ac} B3LYP/6-311+G(d,p)//B3LYP/6-31G(d) (kJ/mol)	
		best conformer	averaged
p-cpstol_en		-68.9	-68.4
m-cpstolDMAP		-118.8	-118.4
m-cpstolDMAP_en		-113.5	-114.2
m-cpstolPPY		-123.4	-123.5
m-cpstolPPY_en		-127.3	-127.9

Analysis of the obtained results shows, that the 2'-sulfoxidoferrocenyl group acts as electron-donating group (compare **py** with **o-cpstol**, **m-cpstol** and **p-cpstol**), but is a much weaker donor than the dialkylamino substituent (compare **p-cpstol** and **DMAP**). The donating ability of this group increases when going from the *ortho* to the *para* position. The sulfoxide group, which is expected to be electron withdrawing, has a small influence on the acetylation enthalpy (compare **p-cpp** and **p-cpstol**), even displaying weak donating properties. The more potent catalysts **m-cpstolDMAP** and **m-cpstolPPY**, containing a 2'-sulfoxidoferrocenyl group in *meta* and a dialkylamino group in *para* position, have acetylation enthalpies in the range of **6a** and **5b**, consequently they are expected to be active catalysts in acylation reactions. The studied catalysts have two chiral elements: the ferrocene plane and the sulphur centre; changing the configuration of the sulphur chiral center from (*S*) to (*R*) gives the diastereomeric species. However these changes to diastereomers do not have a systematic influence on the acetylation enthalpies of the corresponding catalysts (compare **p-cpstol**, **m-cpstolDMAP** and **m-cpstolPPY** with **p-cpstol_en**, **m-cpstolDMAP_en** and **m-cpstolPPY_en** respectively).

Examination of the best conformer structures for the 4-dialkylamino-3-ferrocenylpyridines and their acylated forms (Figure 3.11) reveals that a conformational change occurs in the sulfoxide group of catalysts upon acylation. The oxygen atom of the sulfoxide group becomes directed towards the acetyl group, probably due to electrostatic interactions between oxygen and neighbouring hydrogens (*ortho* hydrogen of pyridine ring and acetyl hydrogen). This feature appears in catalysts **m-cpstolDMAP** and **m-cpstolPPY**, as well as in their diastereomeric analogues **m-cpstolDMAP_en** and **m-cpstolPPY_en**. It can be potentially advantageous for the deprotonation of alcohol molecule in the rate-limiting step (see Figure 1.7), and probably for distinguishing two alcohol enantiomers, what would make these derivatives useful for the KR of alcohols.

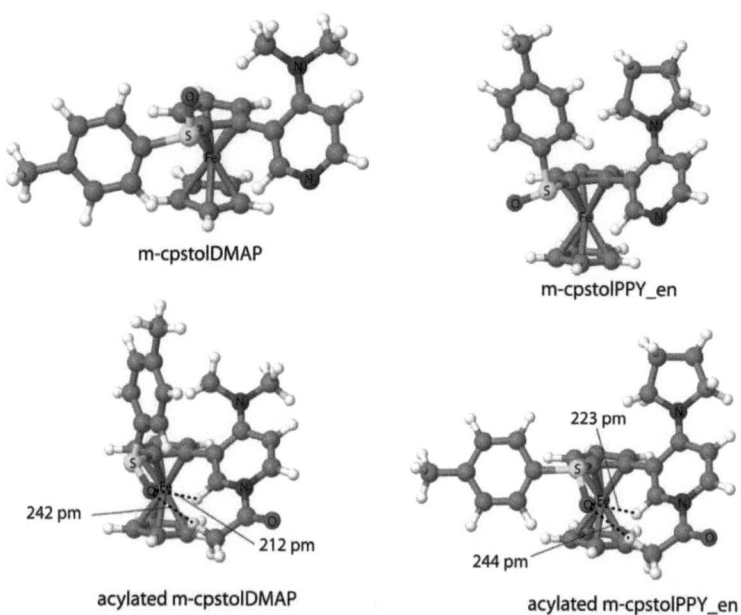

Figure 3.11. Structures of the most stable conformers for the catalysts **m-cpstolDMAP** and **m-cpstolPPY_en** and their acetylated forms.

Chapter 4. (4-Aminopyridin-3-yl)-(thio)ureas as Acylation Catalysts

4.1 Introduction

The kinetic resolution (KR) of racemic secondary alcohols serves nowadays as the primary testing reaction for the design and development of chiral DMAP derivatives.[20] Enantiopure secondary alcohols are important targets in organic chemistry (*e.g.* natural products, bioactive non-natural products, and chiral ligands), and a diverse array of approaches has been described for their synthesis. Kinetic resolution via enzymatic acylation/deacylation is one widely used method, although this strategy can suffer from drawbacks such as low volumetric throughput, high cost, and poor generality.[53] The selectivity factor s = [rate of fast-reacting enantiomer]/[rate of slow-reacting enantiomer] provides a measure of the efficiency of a kinetic resolution.[54] For calibration, a kinetic resolution that proceeds with a selectivity factor greater than 10 gives unreacted starting material with >90% ee at 62% conversion. This value (s = 10) is often employed as the threshold for a synthetically useful kinetic resolution. Of course, even higher selectivity factors are desirable, e.g., a process with s > 50 affords starting material with >99% ee at 55% conversion.[54] Figure 4.1 shows a selection of chiral DMAP derivatives.

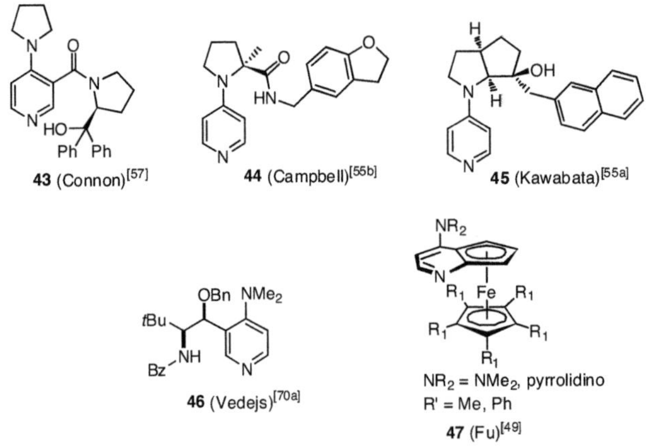

Figure 4.1. Structures of several chiral DMAP derivatives employed in the KR of alcohols.

Several attempts of rationalizing the origins of the selectivity in KR experiments catalyzed by nucleophilic catalysts can be found in the literature. Kawabata *et al.* have proposed that the 'closed' conformation of acylpyridinium cation derived from **45** is necessary for controlling the π-facial reactivity of the N-acylpyridinium intermediate, which directs the enantioselectivity of the subsequent acylation of alcohols.[55a] Theoretical studies by Zipse et al. are in full support of this hypothesis.[56] Recently, computational work of Houk et al. has supported the π-interaction hypothesis of chiral recognition in the kinetic resolution of secondary benzylic alcohols.[24b] Another important structural feature of several chiral pyridine derivatives is the presence of the group able to form hydrogen bonds with reactants: tertiary OH in Kawabata's and Connon's and amide NH in Vedejs's and Campbell's catalysts. For instance, Connon et al. have shown that the ability of **43** to serve as an active and enantioselective acyl-transfer catalyst is due to a combination of aryl-pyridinium π-π interactions and substrate-catalyst H-bonding.[57] This implies that some kind of bifunctional activation by basic pyridine nitrogen and acidic hydrogen can play an important role for the chiral recognition by these catalysts.

An excellent example of merging nucleophilic and hydrogen bonding catalysis for the KR of amines was demonstrated by Seidel and co-workers (Figure 4.2).[58] The authors employed a combination of DMAP (nucleophilic catalyst) and chiral thiourea (hydrogen bonding catalyst) to achieve selectivities s up to 56 for the KR of propargylic and benzylic amines. Only benzoic anhydride worked well under these conditions.[58a] The authors proposed that the co-catalyst binds a benzoate anion via hydrogen bonding and forms thus a chiral ion-pair with benzoylpyridinium cation, which is then responsible for high enantioselectivities in this reaction.[58a] The simultaneous use of DMAP und chiral thiourea have also been successfully applied to the Baylis-Hillman reaction[59a] and conjugate additions to nitroalkenes.[59b,c]

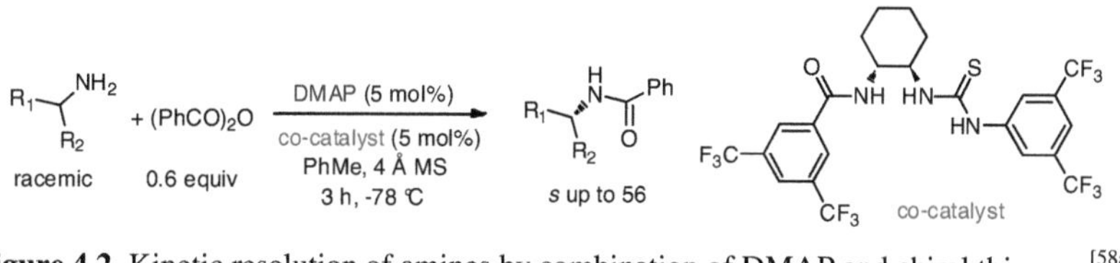

Figure 4.2. Kinetic resolution of amines by combination of DMAP and chiral thiourea.[58a]

The concept of multifunctional catalysis, wherein the catalysts exhibit both Lewis acidity and Brønsted basicity, has been first developed by Shibasaki et al.[60] The synergistic cooperation of two functional groups in the active site helps to improve the reactivity as well as the stereodiscrimination. Later on, a variety of asymmetric transformations have been realized by this powerful concept.[61] An ideal set of multifunctional catalysts should contain two or more Lewis- or Brønsted active sites, which act in several different activation modes. The bi/multifunctional catalysts enable effective transformations, which generally are hard to achieve by a single functional catalyst.[62] Multifunctional catalysts have been successfully applied to Michael addition,[63a] Henry reaction, [63b] Strecker reaction, [63c] Morita-Baylis-Hillman reaction, [63d] and a wide range of enantioselective carbonyl α-functionalization processes.[63e-g] Several bifunctional catalysts based on the DMAP motif were applied to aza-Morita-Baylis-Hillman reaction[64a] and Michael addition.[64b]

Therefore we envisioned that 3,4-diaminopyridine derivatives bearing a (thio)urea moiety on the 3N-atom could also work as bifunctional catalysts. The 3,4-diaminopyridine motif would make these derivatives more active than DMAP, and using a chiral (thio)urea backbone would make them stereoselective. The synthesis of derivatives from readily available precursors **31b** and **31d** and different iso(thio)cyanates would allow screening of the substitution pattern (Scheme 4.1).

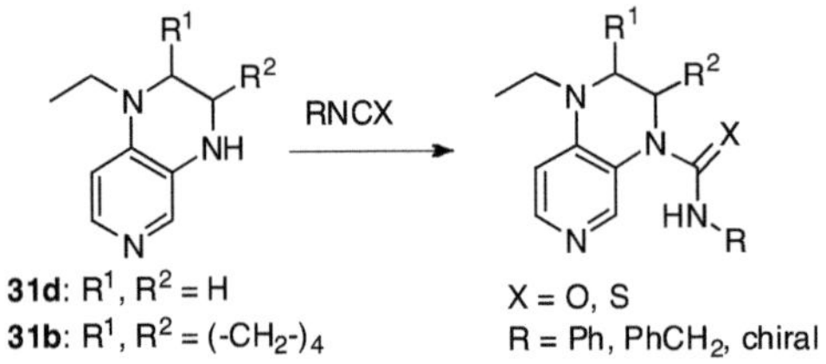

Scheme 4.1. Synthesis of (4-aminopyridin-3-yl)-(thio)ureas.

4.2 Achiral (4-aminopyridin-3-yl)-(thio)ureas

One important issue in developing new catalysts is their activity. As a first step, the achiral variants of new catalysts were therefore studied experimentally and computationally in order to investigate which factors influence their catalytic activity in acylation reactions.

4.2.1 Acetylation enthalpies of (4-aminopyridin-3-yl)-(thio)ureas

As shown in Chapter 2, the stability of acetylpyridinium cations as expressed through the isodesmic equation shown in Scheme 2.2 can be used to assess the activity of substituted pyridines in acylation reactions. Acetylation enthalpies for a variety of (4-aminopyridin-3-yl)-(thio)ureas were calculated at the MP2(FC)/6-31+G(2d,p)//B98/6-31G(d) level of theory with solvation energies calculated at PCM/UAHF/RHF/6-31G(d) level (Table 4.1).

Table 4.1. Relative acetylation enthalpies and stability parameters for (4-aminopyridin-3-yl)-(thio)ureas calculated at MP2/6-31+G(2d,p)//B98/6-31G(d) level with inclusion of solvent effects at PCM/UAHF/RHF/6-31G(d) level.

DMAP   5a   5b
ΔH_{ac} = -61.3   -81.4   -85.2 kJ/mol

X	R^1, R^2	R^3	$\Delta H_{ac}{}^a$ (kJ/mol)	$\Delta H_{form}{}^b$ (kJ/mol)	$\Delta\Delta E$ (kJ/mol)^c free catalyst	$\Delta\Delta E$ (kJ/mol)^c acetylated catalyst
cat8lur1	O	H	Ph	-64.3	-78.4	20.6	28.8
cat8lur2	S	H	Ph	-63.2	-59.3	16.4	20.8
cat8lur1f	O	H	3,5-(CF$_3$)$_2$C$_6$H$_3$	-51.2	-88.6	30.6	39.8
cat8lur3	O	H	CH$_2$Ph	-64.3	-85.3	28.0	31.7
cat8lur4	S	H	CH$_2$Ph	-57.9	-77.2	29.2	31.3
cat8lur5	O	H	(S)-PhMeCH	-59.1	-80.1	33.3	29.2
cat11ur1	O	(-CH$_2$-)$_4$	Ph	-63.0	-72.3	24.2	29.3
cat11ur2	O	(-CH$_2$-)$_4$	CH$_2$Ph	-61.2	-84.6	31.4	34.4
cat11ur3	O	(-CH$_2$-)$_4$	3,5-(CF$_3$)$_2$C$_6$H$_3$	-63.7	-81.0	25.2	39.4
cat11ur4	S	(-CH$_2$-)$_4$	3,5-(CF$_3$)$_2$C$_6$H$_3$	-59.5	-69.3	21.0	34.4

^{a} Boltzmann averaged values.
^{b} Enthalpies of reaction, shown in Scheme 4.2, were calculated at MP2/6-31+G(2d,p)//B98/6-31G(d) level with inclusion of solvent effects at PCM/UAHF/RHF/6-31G(d) level.
^{c} $\Delta\Delta E = [E_{stacked}(MP2) - E_{non\text{-}stacked}(MP2)] - [E_{stacked}(B98) - E_{non\text{-}stacked}(B98)]$

The acetylation enthalpies of the majority of (4-aminopyridin-3-yl)-(thio)ureas are in the range between –58 and –65 kJ/mol, except catalyst **cat8lur1f**, which has a less negative enthalpy of –50.9 kJ/mol. The thiourea derivatives generally have less negative acetylation enthalpies than the corresponding urea derivatives. In general all theoretically studied pyridine derivatives are expected to be active enough to promote acylation of alcohols. A very important issue is the stability of these derivatives with respect to dissociation to the building blocks **31b/31d** and iso(thio)cyanate. This is included in Table 4.1 as the enthalpy of the corresponding forward reaction shown in Scheme 4.2. The obtained data show that the thiourea derivatives should be less stable than the corresponding urea-pyridines.

Scheme 4.2. Calculated enthalpies of (thio)ureas formation from amines and iso(thio)cyanates.

By more careful analysis of the conformational space of the studied systems it was found that relative energies of conformers at different levels of theory depend on the presence of stacking interactions in the system. The largest effect was found for the acetylated catalysts which contain the 3,5-bis(trifluoromethyl)phenyl substituent (Table 4.1). Analysis of the structures of these conformers reveals that the pyridine and benzene ring of the urea unit are close to each other (*ca.* 400 pm) in "stacked" conformers, whereas the rings are far away from each other in "non-stacked" conformers (Table 4.2).

Table 4.2. Comparison of the most stable "stacked" and "non-stacked" conformers for the acetylated catalyst **cat81ur1f**.

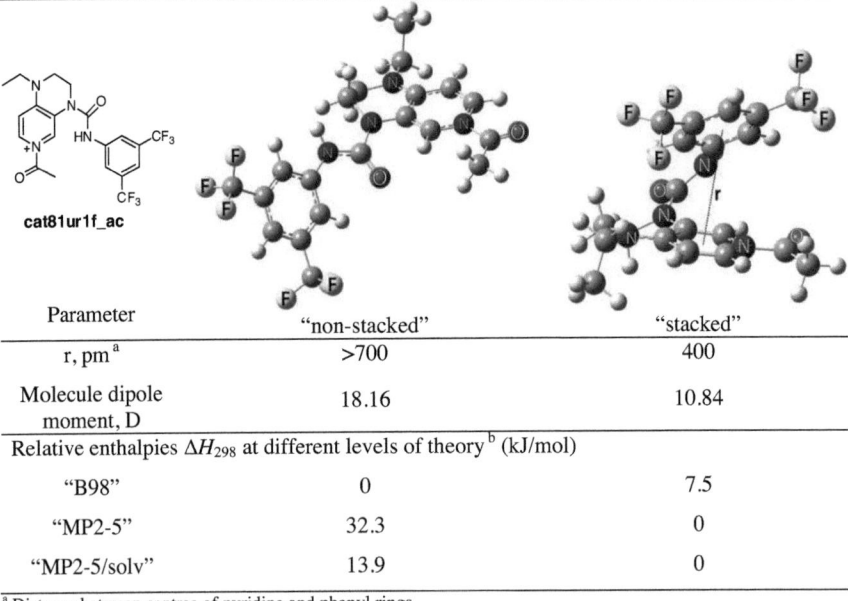

Parameter	"non-stacked"	"stacked"
r, pm [a]	>700	400
Molecule dipole moment, D	18.16	10.84
Relative enthalpies ΔH_{298} at different levels of theory [b] (kJ/mol)		
"B98"	0	7.5
"MP2-5"	32.3	0
"MP2-5/solv"	13.9	0

[a] Distance between centres of pyridine and phenyl rings.
[b] Levels of theory: "B98" = B98/6-31G(d)//B98/6-31G(d); "MP2-5" = MP2(FC)/6-31+G(2d,p)//B98/6-31G(d); "MP2-5/solv" = MP2(FC)/6-31+G(2d,p)//B98/6-31G(d) with solvation energies at PCM/HF/6-31G(d) level with UAHF radii.

It is known that DFT functionals, particularly the B98 functional, cannot predict dispersion interactions accurately.[56] When using the MP2 method for single point calculations, "stacked" conformations become much more stable than "non-stacked" by *ca.* 20–30 kJ/mol (Table 4.2). In order to quantify these interactions, the energy differences between the most stable "stacked" and "non-stacked" conformers at MP2 and B98 levels were calculated:

$$\Delta\Delta E = [E_{stacked}(MP2) - E_{non\text{-}stacked}(MP2)] - [E_{stacked}(B98) - E_{non\text{-}stacked}(B98)]$$

Analysis of the obtained data shows that acetylated species have larger $\Delta\Delta E$ than free catalysts, probably because of the partially charged pyridine ring in the former (Table 4.1). The maximum values are observed for the acetylated catalysts **cat8lur1f** (39.8 kJ/mol) and **cat11ur3** (39.4 kJ/mol). Such big differences in relative energies of conformers imply that a careful conformational search is necessary to find the most stable conformers and get therefore more accurate values of acetylation enthalpies. Remarkably, the energy differences $\Delta\Delta E$ are smaller for the catalyst **cat11ur4**, which is the sulphur-containing analogue of **cat11ur3**. A smaller molecular dipole moment of "stacked" conformations comparing with "non-stacked" (Table 4.2) leads to less negative solvation energy of the former, decreasing thus the energy differences between these types of conformations (13.9 kJ/mol for acylated **cat8lur1f**).

However, one of the serious shortcomings of MP2 theory is a noticeable overestimation of the dispersion interaction energy.[65] In order to verify whether such "stacked" conformations exist in experiment, the X-ray structure of catalyst **cat11ur3** was compared with computations (Table 4.3). The conformation of the ethyl group cannot be resolved by X-ray analysis, which is in a full accordance with computations, which predict very similar energies for the two most stable "non-stacked" conformations. When the MP2(FC)/6-31+G(2d,p) single point calculations are combined with thermochemical corrections at B98/6-31G(d) level, the "stacked" conformation becomes more stable than "non-stacked" by 3.7 kJ/mol (Table 4.3). The "stacked" conformation is therefore predicted to be the lowest in energy in the gas phase. However, the difference in dipole moments leads to better solvation of "non-stacked" conformers in chloroform (at PCM/UAHF//HF/6-31G(d) level). Consequently both types of conformations are almost equally populated in solution (Table 4.3). In summary, the "non-stacked" conformations of **cat11ur3** observed by the X-ray analysis have the lowest energies at B98/6-31G(d) level in the gas phase, whereas "non-stacked" and "stacked" conformations are energetically close in solution (at "MP2-5/solv" level).

Table 4.3. Comparison of the most stable "stacked" and "non-stacked" conformers for the catalyst **cat11ur3**.

Parameter	"non-stacked"	"non-stacked"	"stacked"
r, pm [a]	>700	>700	413
Molecule dipole moment, D	8.18	8.09	5.08
Relative enthalpies ΔH_{298} at different levels of theory [b] (kJ/mol)			
"B98"	1.2	0.0	21.4
"MP2-5"	3.6	3.7	0.0
"MP2-5/solv"	0.9	0.0	1.0

[a] Distance between centres of the pyridine and phenyl rings.
[b] Levels of theory: "B98" = B98/6-31G(d)//B98/6-31G(d); "MP2-5" = MP2(FC)/6-31+G(2d,p)//B98/6-31G(d); "MP2-5/solv" = MP2(FC)/6-31+G(2d,p)//B98/6-31G(d) with solvation energies at PCM/HF/6-31G(d) level with UAHF radii.

4.2.2 Synthesis and catalytic activity of (4-aminopyridin-3-yl)-(thio)ureas

Synthesis of (4-aminopyridin-3-yl)-(thio)ureas was carried out by stirring the corresponding 3,4-diaminopyridine precursors and iso(thio)cyanates in CH_2Cl_2 for 8 h at room temperature, followed by column chromatography on SiO_2 (Scheme 4.1). The yields are up to 90 %. Several derivatives are unstable under the column conditions, e.g. thioureas **cat8lur2** and **cat8lur4**, which were obtained in low yields (20 %).

The catalytic activity of the synthesized derivatives was determined as the reaction half-life for the acetylation of 1-ethinylcyclohexanol with acetic anhydride in $CDCl_3$ at 23 °C in the presence of 10 mol % of the respective catalyst. The reactions with all studied catalysts proceed to full conversion. The rate of reaction was characterized by the half-life time $t_{1/2}$, which was extracted from the conversion-time plot through fitting with a second-order kinetic rate law.[11] The obtained data are shown in Table 4.4.

Table 4.4. Catalytic activity and acetylation enthalpies of the (4-aminopyridin-3-yl)-(thio)ureas.

Pyridine part	Ethylene bridge		Cyclohexane bridge	
Isocyanate	ΔH_{ac} (kJ/mol) [b]	$t_{1/2}$ (min) [a]	ΔH_{ac} (kJ/mol) [b]	$t_{1/2}$ (min) [a]
PhNCO	-64.2	395[c]	-62.9	412
PhCH$_2$NCO	-64.2	723	-61.3	395
3,5-(CF$_3$)$_2$C$_6$H$_3$-NCO	-50.9	1200[c]	-63.1	249
PhNCS	-62.6	[c]	-	-
PhCH$_2$NCS	-57.7	[c]	-	-
3,5-(CF$_3$)$_2$C$_6$H$_3$-NCS	-	-	-59.5	339
Ph-CH(-)-NCO	-59.8	725	-	-

[a] Conditions: 0.2 M alcohol, 2.0 equiv of Ac$_2$O, 3.0 equiv of NEt$_3$, 0.1 equiv catalyst, CDCl$_3$, 23.0±1.0 °C.
[b] Calculated at MP2/6-31+G(2d,p)//B98/6-31G(d) level with solvation energies in chloroform calculated at PCM/UAHF//RHF/6-31G(d) level.
[c] Catalysts are unstable under conditions of the benchmark reaction.

A graphical representation of the measured catalytic activities is shown in Figure 4.3. Examination of the obtained results shows that the derivatives based on the motif of catalyst **5b** with cyclohexane bridge are generally more catalytically active than the derivatives with ethylene bridge. They are also more stable under the reaction conditions (*e.g.* 10 % of catalyst **cat81ur1** is acylated by acetic anhydride under the conditions of NMR-kinetics). No correlation between acetylaton enthalpies and experimental half-lives is observed.

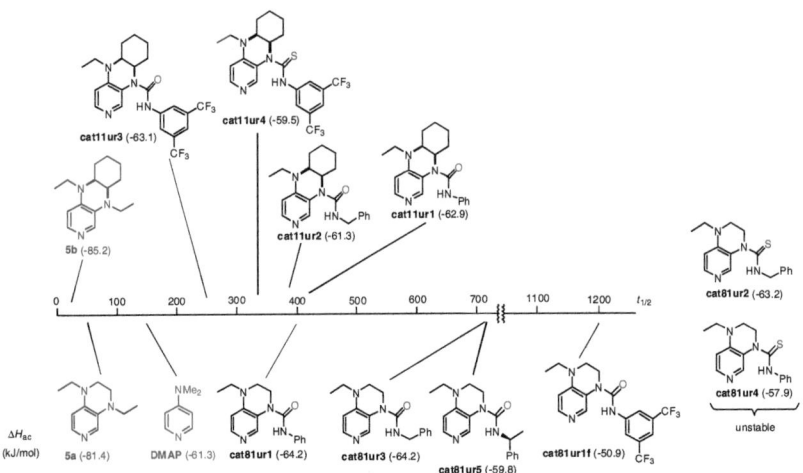

Figure 4.3. Half-lives of the benchmark acetylation reaction catalyzed by (4-aminopyridin-3-yl)-(thio)ureas. The relative acetylation enthalpies ΔH_{ac} (kJ/mol) are shown in brackets.

Introduction of electron withdrawing substituents into the phenyl ring of catalyst **cat11ur1** increases the catalytic activity in acylation reaction: the 3,5-bis-(trifluoromethyl)phenyl derivatives **cat11ur3** and **cat11ur4** are the most active catalyst among the studied pyridines. The enhanced activity can be explained by the increased acidity of the NH hydrogen of the urea group due to the electron withdrawing character of the 3,5-bis(trifluoromethyl)phenyl group. The X-ray single crystal analysis of catalyst **cat11ur3** indicates a hydrogen bonding interaction between the NH hydrogen and pyridine nitrogen of the second catalyst molecule (Figure 4.4).

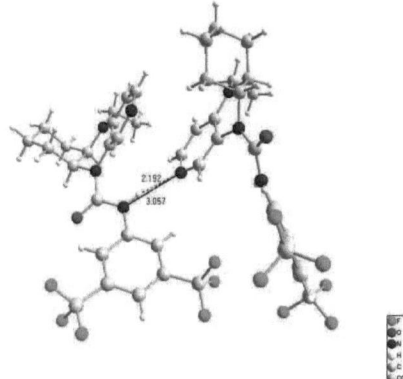

Figure 4.4. Crystal structure of **cat11ur3**: a hydrogen bonding interaction between two catalyst molecules. Distances are given in Å.

4.2.3 Catalysts aggregation studied by NMR and kinetic measurements

In order to check whether such hydrogen bonding could play some role in solution (and consequently during the kinetic measurements), we studied the dependence of ^{1}H NMR spectra on the concentration for derivatives **cat8lur1f** and **cat11ur3**, as well as for **cat11ur1** and **cat11ur2**. Results show that the chemical shifts do depend on concentration (Figures 4.5 and 4.6). The largest changes are in the case of NH hydrogen (up to +1.2 ppm for **cat8lur1f**) and *ortho*-hydrogen of the pyridine ring (up to –0.35 ppm for **cat8lur1f**). The chemical shifts of other protons vary less significantly with varying concentration (up to 0.2 ppm, see Figure 4.5).

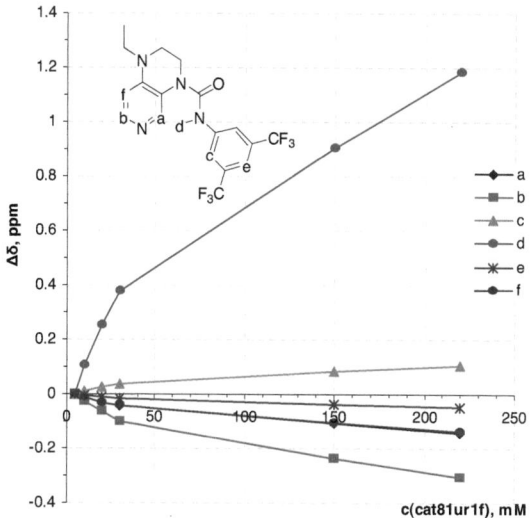

Figure 4.5. Concentration dependence of the ^{1}H NMR chemical shifts of different hydrogens in **cat8lur1f**.

Among the different catalysts 3,5-bis(trifluoromethyl)phenyl urea derivatives **cat8lur1f** and **cat11ur3** show the largest effect of concentration on chemical shifts (Figure 4.6). The large variations of the NH hydrogen chemical shift for derivatives **cat8lur1f** and **cat11ur3** prove that the hydrogen bonding between the NH hydrogen and pyridine nitrogen is also relevant in solution.

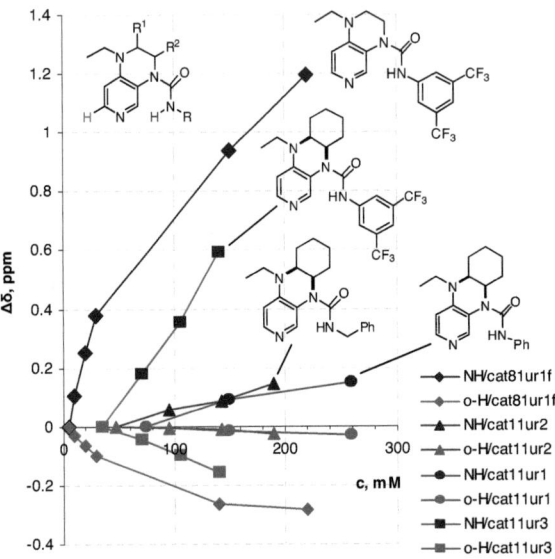

Figure 4.6. Concentration dependence of the NMR chemical shifts of NH and *ortho*-hydrogens for catalysts **cat11ur1**, **cat11ur2**, **cat81ur1f** and **cat11ur3**.

In order to check, whether these interactions play some role in the catalyzed acylation of alcohols, concentration dependent kinetic measurements were carried out. The initial rate method was used to evaluate the influence of the catalyst loading on the rate of the benchmark reaction (Table 4.5). The rates of acetylation of alcohol **36a** were measured in the initial interval of the reaction (up to 10 % conversion).

Table 4.5. Initial rates of catalyzed acetylation of alcohol **36a**, measured at different concentrations of catalysts **cat11ur2** and **cat11ur3**.[a]

Catalyst loading, mol %	Initial rate, 10^{-6} M^{-1} s^{-1}	
	cat11ur2	**cat11ur3**
2.5	1.92	5.44
5	4.44	8.76
7.5	-	14.04
10	7.80	19.56
15	12.54	28.36

[a] Conditions: 0.2 M 1-(ethinyl)cyclohexanol, 2.0 equiv of Ac_2O, 3.0 equiv of NEt_3, 2.5-10 % catalyst, $CDCl_3$, 23.0±1.0 °C.

The initial rates of acetylation, measured for two catalysts **cat11ur2** and **cat11ur3**, show a linear dependence on the catalyst concentration (Figure 4.7). Both lines intercept the rate axis approximately at zero, thus showing only a small influence of the background reaction on the reaction rates. In conclusion, possible catalyst aggregation due to hydrogen bonding, which was proposed on the basis of concentration-dependent NMR spectra of catalysts (Figure 4.6), does not play any role in the kinetics of the catalyzed acetylation of alcohols. This fact can be explained through the presence of triethylamine as an auxiliary base in the benchmark acylation reaction, which can compete with pyridine in hydrogen bonding with urea NH hydrogen atoms.

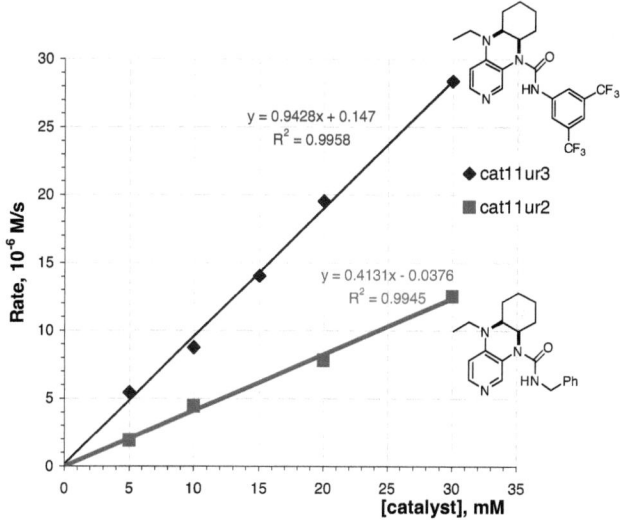

Figure 4.7. Dependence of the initial rate of acetylation on catalyst concentration.

Chapter 4

4.3 Chiral (4-aminopyridin-3-yl)-ureas

4.3.1 Synthesis of chiral catalysts, derived from (S)-amino acids

The achiral (4-aminopyridin-3-yl)-ureas were shown to have only a moderate activity in alcohol acylation. Studying chiral derivatives, which can be obtained in a straightforward way by coupling aminopyridine precursors with chiral isocyanates, would be more attractive. These chiral pyridine derivatives can potentially be used for the kinetic resolution of alcohols or catalyzed enantioselective Steglich rearrangement.[20,55] First, we employed commercially available (S)-1-phenylethylisocyanate **48** as a coupling partner for aminopyridine precursors. An enantiomerically pure catalyst **PhEt** was obtained in 69 % yield (Scheme 4.3). The cyclohexane-bridged analogue **CyPhEt** was obtained in 80 % yield as 1:1 diastereomeric mixture (determined by ^{1}H NMR). The diastereomers of **CyPhEt** could not be resolved by column chromatography or by recrystallization.

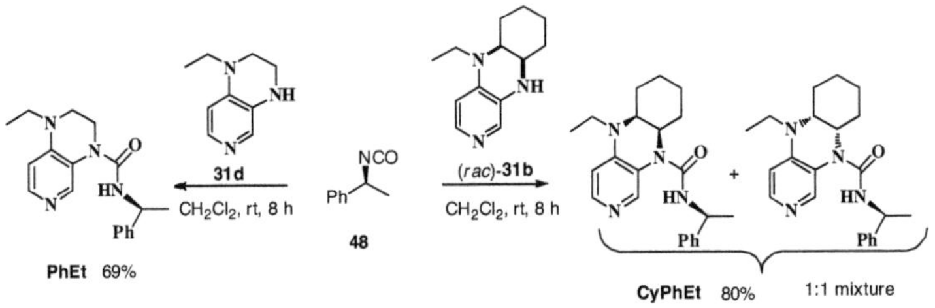

Scheme 4.3. Synthesis of chiral 3-(1-phenylethylurea)-4-aminopyridines.

Chiral aminoacids are versatile starting materials for the preparation of chiral isocyanates. It was shown that the reaction of aminoacid esters **49** with Boc$_2$O in the presence of DMAP gives the corresponding isocyanates **50** after only 15 min stirring at room temperature.[66] Purification of these compounds requires low temperature (–30 °C) column chromatography.[66] However, addition of the aminopyridine precursor directly to the reaction mixture allows carrying out a one-pot transformation to urea-derivatives (Scheme 4.4). Catalysts **ValOMe**, **TleOMe** and **PheOMe**, obtained from (S)-valine, (S)-tert-leucine and (S)-phenylalanine, respectively, were isolated in good yields (34–68 % after 2 steps) as enantiomerically pure compounds (as was shown by chiral HPLC analysis and optical rotation measurements).

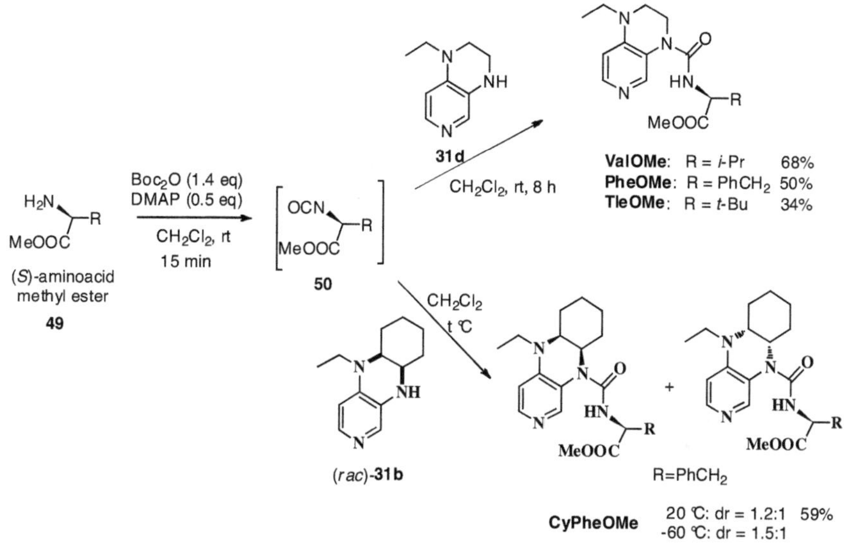

Scheme 4.4. Synthesis of chiral (4-aminopyridin-3-yl)-ureas from (S)-aminoacids.

The reaction of phenylalanine-derived isocyanate with the cyclohexane-bridged precursor **31b**, which has been obtained as a racemic mixture of *cis*-isomers, gives a mixture of diastereomeric compounds **CyPheOMe** (Scheme 4.4). The diastereomeric ratio of **CyPheOMe** depends on the reaction temperature: dr = 1.2:1 at 20 °C and 1.5:1 at –60 °C (when an excess of amine is used). This implies that the reaction of racemic amine **31b** with chiral isocyanates can potentially be used for the kinetic resolution to enantiomers. The diastereomers of **CyPhOMe** could not be separated by column chromatography or by recrystallization.

4.3.2 Derivatization of catalysts by Grignard reagent

The diphenylcarbinol moiety has recently been shown to be responsible for the chiral recognition in the KR of alcohols, catalyzed by chiral DMAP derivatives.[57] Derivatization of the ester group in catalysts **PheOMe** and **ValOMe** was carried out by the reaction with excess PhMgBr in THF, giving catalysts **PhePh$_2$OH** and **ValPh$_2$OH**, respectively (Scheme 4.5). The reaction of PheOMe with 1-naphthylmagnesium bromide yielded catalyst **PheNph$_2$OH**. All these derivatives were obtained as enantiomerically pure compounds (as was shown by chiral HPLC analysis and optical rotation measurements).

81

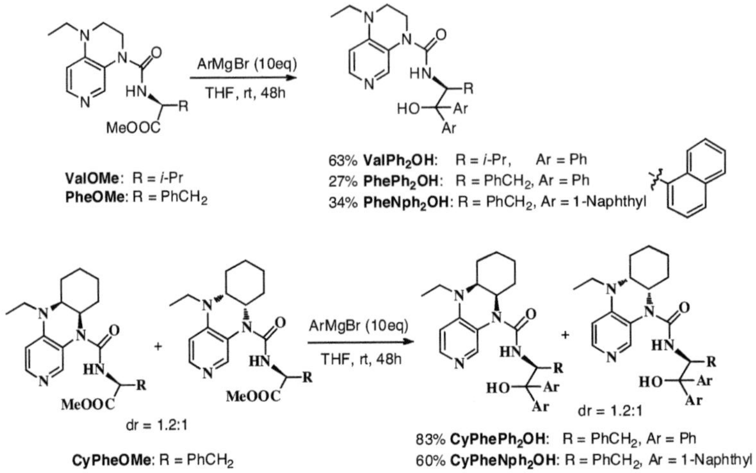

Scheme 4.5. Derivatization of chiral pyridine derivatives by Grignard reaction.

Grignard reaction, carried out with the diastereomeric mixture of **CyPheOMe** obtained at room temperature (dr = 1.2:1), gave catalysts **CyPhePh$_2$OH** and **CyPheNph$_2$OH** as a 1.2:1 mixture of diastereomers (Scheme 4.5). The diastereomeric ratio can be determined by NMR, as well as by chiral HPLC, giving the same results. Catalyst **CyPhePh$_2$OH** can be resolved by recrystallization from EtOH/ethylacetate (1:1), giving an enantioenriched mixture with dr = 96:4% (determined by chiral HPLC).

4.3.3 Acetylation enthalpies and benchmark reaction kinetics

Several chiral derivatives were tested in the benchmark acetylation reaction using 10% catalyst loading (Table 4.6). The reactions with all studied catalysts proceeded to full conversion. Derivatives **PhEt** and **ValPh$_2$OH** had almost the same half-lives as the achiral analogues (*cf.* Figure 4.3). Catalyst **PheOMe** was eventually less active than the other studied chiral catalysts. The derivative **CyPhePh$_2$OH**, which contains a cyclohexane bridge, was shown to be the most catalytically active among the chiral urea pyridines.

Table 4.6. Catalytic activity and acetylation enthalpies of the chiral (4-aminopyridin-3-yl)-ureas.

Catalyst	Structure	ΔH_{ac}, kJ/mol [b]	$t_{1/2}$, min [a]
PhEt		-59.1	725 ± 20
PheOMe		-70.6	1192 ± 18
ValPh$_2$OH		-	776 ± 7
CyPhePh$_2$OH		-	457 ± 2
PhePh$_2$OH		-55.0	-
cat81ur6		-88.9	-

[a] Conditions: 0.2 M alcohol, 2.0 equiv of Ac$_2$O, 3.0 equiv of NEt$_3$, 0.1 equiv catalyst, CDCl$_3$, 23.0±1.0 °C.
[b] Calculated at MP2/6-31+G(2d,p)//B98/6-31G(d) level with solvation energies at PCM/UAHF/RHF/6-31G(d) level.

The acetylation enthalpies, calculated for several chiral catalysts (Table 4.6), failed to correlate with catalytic activity, as was observed for achiral urea pyridine derivatives. However, calculations helped to shed light on the conformational preferences of these catalysts (Figure 4.8). As mentioned above, the intramolecular π-π-interactions in the N-acylpyridinium intermediate can be responsible for the chiral recognition in the kinetic resolution of secondary alcohols. Therefore we analyzed, which conformational changes occur in the catalyst structure upon acetylation and whether any "stacking" interactions are relevant for the studied chiral catalysts.[56]

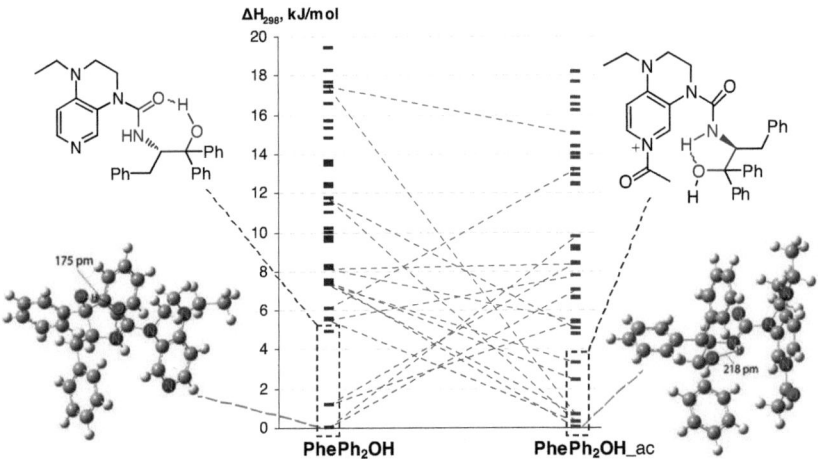

Figure 4.8. Relative enthalpies of conformers of catalyst **PhePh$_2$OH** and its acetylated form within an energy window of 20 kJ/mol (calculated at MP2/6-31+G(2d,p)//B98/6-31G(d) level with solvation energies at PCM/UAHF/RHF/6-31G(d) level). The structurally related conformers of **PhePh$_2$OH** and **PhePh$_2$OH_ac** are connected with dotted lines. Structures of the most stable conformers are also shown.

In contrast to related achiral derivatives, the energy differences between "stacked" and "non-stacked" conformers at MP2 and B98 levels (ΔΔE) for the chiral catalysts **PheOMe** and **PhePh$_2$OH** are smaller (around 15–18 kJ/mol). This implies that dispersion interactions are less significant for these derivatives. Figure 4.8 shows a pictorial representation of the relative enthalpies of conformers of the neutral derivative **PhePh$_2$OH** and its acetylated form in the range 0–20 kJ/mol. Analysis of the obtained data shows that the acetylation of compound **PhePh$_2$OH** does not narrow its conformational space. Noticeably, the most stable conformations of the catalyst **PhPh₂OH** have a hydrogen bond between OH hydrogen and C=O oxygen, forming thus a 7-membered ring (Figure 4.8). Furthermore, examination of the structures of the most stable conformers for compound **PhPh₂OH** and its acetylated form (within an energy window of 4 kJ/mol) reveals that a conformational change occurs upon acetylation: the hydrogen bond is getting broken, but a new hydrogen bond between NH hydrogen and OH oxygen appears with formation of a 5-membered ring. Therefore, the OH-group would be able to form hydrogen bonds with reactants in the transition state, resulting thus in the stereoinductive potential of the catalyst **PhPh₂OH** in KR of alcohols.

4.3.4 Introduction of a linker between the pyridine and urea moieties

One important drawback of all studied (4-aminopyridin-3-yl)-ureas is that the urea groups have only one free hydrogen, which is able to form hydrogen bonds. Using linkers as in model catalyst **cat81ur6** (Table 4.6) can solve this problem, providing thus two hydrogens in the urea moiety for hydrogen bond formation. Lower acetylation enthalpies for **cat81ur6** (−88.9 kJ/mol), as compared to parent derivative **5b** (−85.2 kJ/mol, Table 4.1), predicts that this derivative would potentially be very active in the acylation of alcohols.

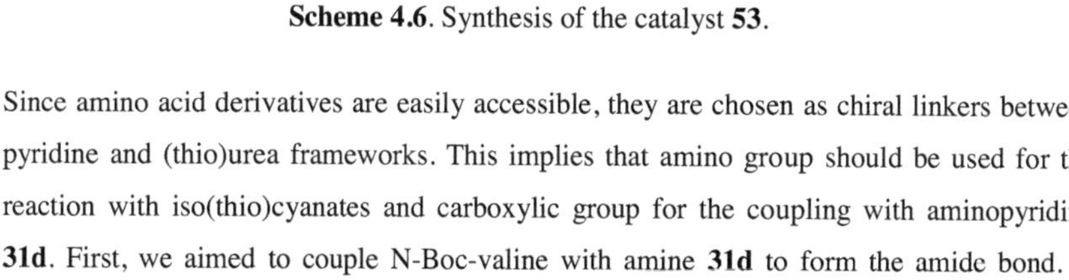

Scheme 4.6. Synthesis of the catalyst **53**.

Since amino acid derivatives are easily accessible, they are chosen as chiral linkers between pyridine and (thio)urea frameworks. This implies that amino group should be used for the reaction with iso(thio)cyanates and carboxylic group for the coupling with aminopyridine **31d**. First, we aimed to couple N-Boc-valine with amine **31d** to form the amide bond. A number of amide formation protocols were tried (chloroformate, EDAC/HOBt, TBTU/HOBt), but no product formation was observed. Then we turned our attention to acyl chloride. Since the N-Boc protecting group is labile under acidic conditions, which are normally used in synthesis of acyl chlorides, a Fmoc protecting group was introduced into the amino acid. Coupling of N-Fmoc-valine chloride with amine **31d**, performed in pyridine at 150 °C in MW or by deprotonation of the amine with BuLi or LDA, did not yield a desired product. Finally, a combination of phtalimide protecting group with deprotonation by NaHMDS (bis(trimethylsilyl)amide) was successful in coupling and the desired product **51** was obtained in 33% yield (Scheme 4.6). Deprotection by hydrazine proceeded smoothly, giving compound **52** in 83% yield. Reduction of **52** by $LiAlH_4/AlCl_3$ gave substantial amounts of amine **31d** instead of the desired product of amide group reduction. Therefore, compound **52** was reacted with 3,5-bis(trifluoromethyl)phenylisocyanate to give product **53**, which was then used for the KR experiments.

^{1}H NMR analysis of compound **53** reveals a very interesting feature of this substance: rotations around urea NH-CO bonds are relatively slow. Therefore the ^{1}H NMR spectrum of **53** measured at 23°C in $CDCl_3$ contained several sets of signals, which refer to different rotamers. Upon heating, these signals coalesce to one set of peaks. In order to estimate the free activation energy of these rotations, a coalescence experiment, *i.e.* measurement of NMR spectra at different temperatures, was carried out with a solution of compound **53** in C_6D_6 (Figure 4.9).

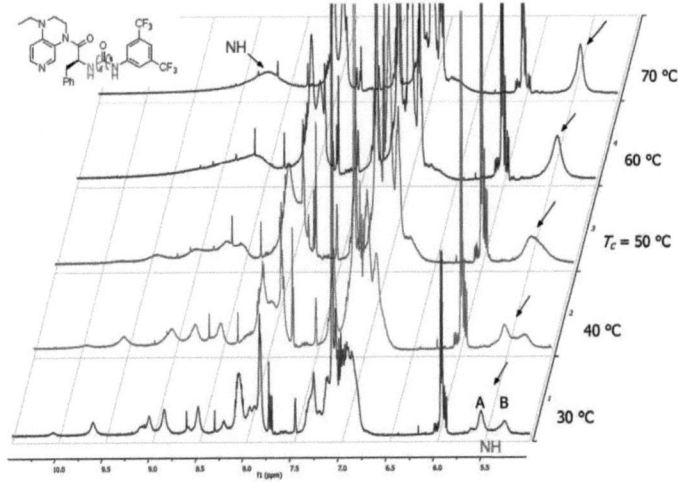

Figure 4.9. ^{1}H NMR spectra of catalyst **53** measured at different temperatures (in C_6D_6). Coalescence temperature $T_c = 50$ °C.

The only peak at 5.5 ppm, which is separated from other signals and therefore can be analyzed, is referred to the NH proton, marked in red on Figure 4.9, as was determined from the 2D NMR spectra (HMBC and HBQC). Coalescence of this peak was observed at 50 °C. Since this signal does not have an ideal form and is also relatively broad (the sample was not degassed before measurements), only an approximate estimate of the free activation energy was possible. The rate constant k at coalescence is given approximately by equation 1:[67a]

$$k_{T_c} = \frac{\pi}{2}|v_A - v_B| \qquad (1)$$

where v_A and v_B are the shifts of rotamers measured at low temperature, i.e. when the exchange is slow. Inserting equation 1 into the Eyring equation results in:

$$\frac{\pi}{2}|v_A - v_B| = \frac{RT_c}{hN_A}e^{\frac{-\Delta G^{\neq}}{RT_c}} \qquad (2)$$

or

$$\Delta G^{\neq} = RT_c \ln \frac{RT_c\sqrt{2}}{\pi hN_A|v_A - v_B|} \qquad (3)$$

If T_c is measured in K and the shifts v in Hz, then the free energy of activation is given (in kJ/mol) by equation 4 (using $N_A = 6.022 \cdot 10^{23}$ mol^{-1}, $R = 8.314$ J mol^{-1} K^{-1} and $h = 6.626 \cdot 10^{-34}$ J·s):

$$\Delta G^{\neq} = 19.1 \cdot 10^{-3} T_c (9.97 + \log T_c - \log|v_A - v_B|) \qquad (4)$$

For compound **53** T_c = 323 K and $v_A - v_B$ = 99.2 Hz (from the ^{1}H NMR spectrum at 30 °C), giving a free energy of activation of **$\Delta G^{\neq}$ = 65 kJ/mol**. This value is in a typical range for amide bond rotation (60–80 kJ/mol).[67b]

4.3.5 Potential of (4-aminopyridin-3-yl)-ureas in the kinetic resolution of alcohols

All studied chiral catalysts were then tested in the KR of several *sec*-alcohols (Table 4.7). Initial screening of catalysts was carried out with alcohol **alc1** in toluene. Results show that catalysts containing a diphenylcarbinol group are more selective than those with an ester group (compare entries 3 and 4 with 7 and 6, respectively). Furthermore, the phenylalanine-derived ureas are more selective than the valine and *tert*-leucine derivatives (entries 3–5). The best selectivities in the KR of alcohol **alc1** were achieved when catalysts **CyPhePh$_2$OH** and **PheNph$_2$OH** were used (selectivity values s up to 5, entries 10 and 11). When the cyclohexane bridged catalysts were used as diastereomeric mixtures with diastereomeric ratios 1.2:1, lower levels of selectivity were observed (entries 9 and 12).

Table 4.7. KR of alcohols **alc1-3** using chiral (4-aminopyridin-3-yl)-ureas.

Entry	Catalyst	Alcohol	Solvent	Time (h)	ee_A (%)[c]	ee_E (%)[d]	C (%)[a]	s [b]
1	PhEt	alc1	toluene	4	14.0	21.0	40	1.7[g]
2[e]	PhEt	alc1	toluene	3.5	11.3	24.5	32	1.8[g]
3	PheOMe	alc1	toluene	6	15.2	17.5	47	1.6
4	ValOMe	alc1	toluene	6	4.7	9.9	32	1.4
5	TleOMe	alc1	toluene	5	2.6	7.6	25	1.2[g]
6	ValPh$_2$OH	alc1	toluene	6.5	4.1	31.8	11	2.0
7	PhePh$_2$OH	alc1	toluene	10	34.8	39.5	47	3.2
8	PhePh$_2$OH	alc2	toluene	10	3.6	34.9	9	2.2
9[f]	CyPhePh$_2$OH	alc1	toluene	9.5	26.9	34.0	44	2.6
10	CyPhePh$_2$OH	alc1	toluene	7	26.1	53.2	33	4.2
11	PheNph$_2$OH	alc1	toluene	11	29.5	57.1	34	4.9
12[f]	CyPheNph$_2$OH	alc1	toluene	7	27.5	29.6	48	2.7
13	PheNph$_2$OH	alc3	toluene	5	7.3	12.5	37	1.4
14[h]	PheNph$_2$OH	alc3	DCM	5	35.1	32.8	52	2.7
15	CyPhePh$_2$OH	alc3	DCM	5	12.6	65.3	16	5.4
16	53	alc3	DCM	5	3.6	7.3	33	1.2
17	53	alc1	toluene	4	3.4	7.7	30	1.2

[a] Conversion $C = 100 \cdot ee_A/(ee_A+ee_E)$.
[b] Selectivity factor s was calculated as described in ref. 54.
[c] ee of recovered alcohol, established by CSP-HPLC.
[d] ee of ester, established by CSP-HPLC.
[e] The reaction was carried out with 0.5 % catalyst.
[f] The catalyst was used as a mixture of diastereomers with dr = 1.2:1.
[g] (S)-alcohol reacts faster.
[h] The reaction was carried out with 2 % catalyst.

Other alcohols **alc2** and **alc3** were also tested in KR experiments (Figure 4.10). Although alcohol **alc2** was found to be an ineffective substrate for these catalysts (entry 8), monoprotected diol **alc3** turned to be a suitable substrate for urea-pyridine catalysts. The KR of **alc3** proceeds more selectively in dichloromethane than in toluene (entries 13 and 14). The catalyst **CyPhePh$_2$OH** performed well in the KR of alcohol **alc3** in DCM, giving selectivity value s = 5.4 (entry 15). The catalyst **53** was found to be unselective in the KR of both alcohols **alc1** and **alc3** (entries 16 and 17).

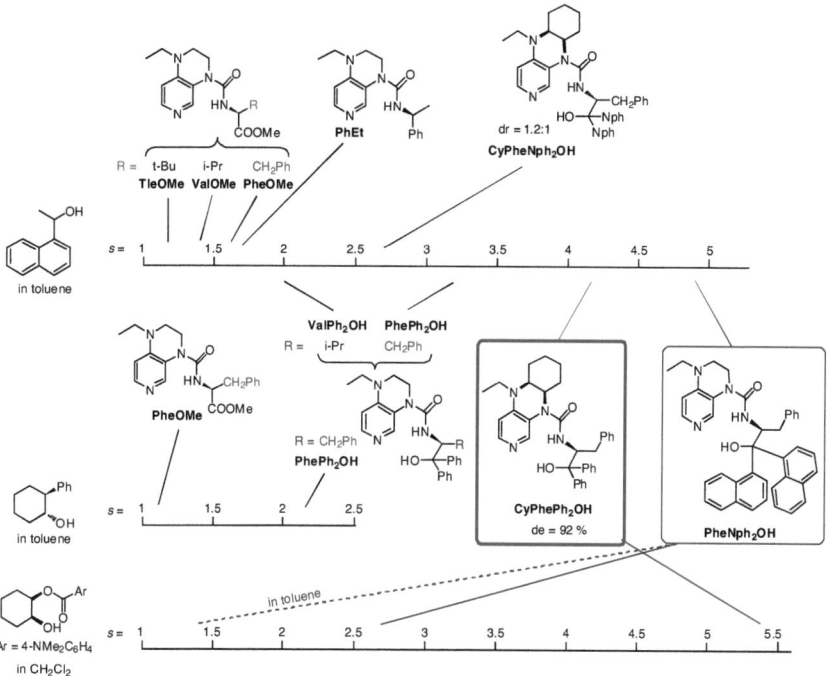

Figure 4.10. Selectivity scale of the (4-aminopyridin-3-yl)-urea derivatives in acylative KR of alcohols.

4.4 Conclusions

The new class of acylation catalysts, (4-aminopyridin-3-yl)-(thio)ureas, have been designed. Achiral (4-aminopyridin-3-yl)-(thio)ureas were shown to have a moderate activity in the alcohol acylation. The obtained data show that the derivatives with a cyclohexane bridge are generally more catalytically active than the catalysts with an ethylene bridge. Derivatives containing the 3,5-bis-(trifluoromethyl)phenyl group in the (thio)urea moiety (Scheme 4.7) are the most active catalysts among the new systems. The enhanced activity has been explained by the increased acidity of the NH hydrogen of the (thio)urea group due to the electron withdrawing character of the 3,5-bis-(trifluoromethyl)phenyl substituent. The X-ray analysis and concentration dependent NMR measurements indicate a hydrogen bonding interaction between the NH hydrogen and the pyridine nitrogen. However, catalyst aggregation does not influence the reaction rates of the catalyzed acetylation of alcohols.

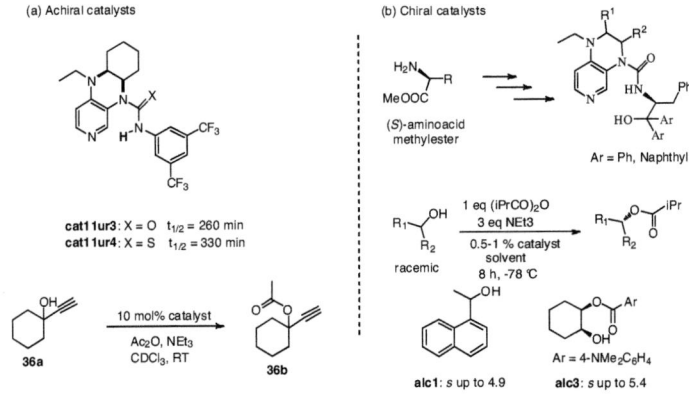

Scheme 4.7. (a) The most catalytically active achiral (4-aminopyridin-3-yl)-(thio)ureas. (b) Application of the chiral derivatives in the KR of *sec*-alcohols.

The chiral (4-aminopyridin-3-yl)-ureas have been prepared via a modular strategy from easily accessible amino acids (Scheme 4.7b). The potential of newly synthesized chiral derivatives was explored in the kinetic resolution (KR) of several secondary alcohols. The best selectivities were obtained with the phenylalanine-derived catalysts containing a diarylcarbinol group. Even though the selectivity values are moderate, the modular design allows variation of the urea substituents for further catalyst improvement.

5. Theoretical Prediction of Selectivity in KR of Secondary Alcohols

5.1 Introduction

Using chiral nucleophilic catalysts based on the DMAP motif major advances have recently been made in kinetic resolution experiments, in particular in those involving secondary alcohols as substrates.[20] Rationalizing the origin of selectivity in these resolution experiments represents an important step in the semi-rational catalyst optimization. Most of the resolution experiments involve acyl group transfer from the catalyst to the alcohol substrates in the rate- and selectivity-determining steps and some effort has thus been made to obtain a quantitative picture of the conformational properties of the acylpyridinium intermediates involved in this step. A limited number of experimental studies for the conformational preferences of acylated catalysts exist. Based on the ^{1}H NMR measurements of NOE effects and chemical shifts,[55a] Kawabata et al. have proposed an 'open' conformation for catalyst **45** in its neutral form and a 'closed' conformation for the acylpyridinium cation **45a** (Figure 5.1). The authors proposed that the 'closed' conformation of intermediate **45a** is necessary for controlling the π-facial reactivity of the N-acylpyridinium intermediate, which directs the enantioselectivity of the subsequent acylation of alcohols. Theoretical studies by Zipse et al. are in full support of this hypothesis.[56] It was also noted on this occasion[56] that DFT methods such as B3LYP are not able to describe stacking interactions induced through dispersion interactions properly.

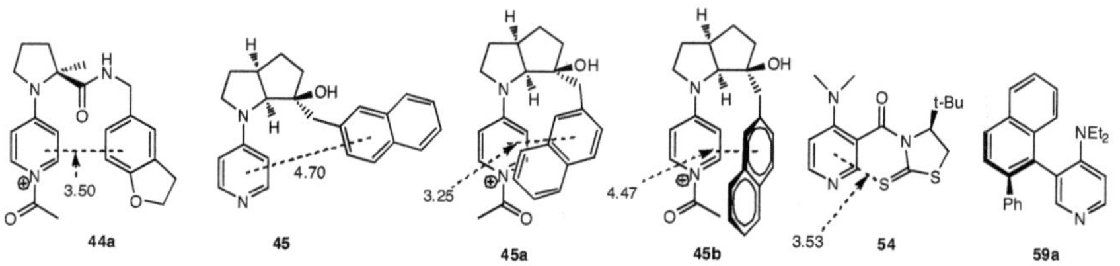

Figure 5.1. Structures of chiral catalysts used for the kinetic resolution of alcohols.[56] Distances between the center of the pyridine ring and selected substituents are given in Å.

The structures of a series of known chiral catalysts and their acyl derivatives have thus been optimized at the MP2(FC)/6-31G(d) level of theory (Figure 5.1) and it was established that the naphthalene and pyridine rings in the most favourable conformation of **45a** are π-stacked (distance 3.25 Å), but that a similarly stable second conformation **45b** exists with "side-on" stacking (distance 4.47 Å). Campbell's catalyst **44** also proved to have such a π-stacking conformation, both in the neutral form (distance 4.05 Å) and in the respective acetylpyridinium cation **44a** (distance 3.50 Å); however, mainly electrostatic effects dominate in this case. Yamada's catalyst **54** was shown to occupy a folded conformation in the neutral as well as the acylated form,[56] a result in full agreement with earlier studies at B3LYP/6-31G(d) level and with NOE experiments of the acyl intermediates.[68] No stacking interactions between the pyridine ring and the phenyl side chain were detected by calculations in the neutral or cationic form of catalyst **59a**.[56] The rotational barriers around the Ar-Ar bond were calculated for a series of systems based on catalyst **59a** at PM3 and RHF/STO-3G levels and then compared with the experimental values.[69]

Connon et al. have shown[57] that the ability of **43** to serve as an active and enantioselective acyl-transfer catalyst is due to a combination of aryl-pyridinium π-π interactions and substrate-catalyst H-bonding (Figure 5.2). The B3LYP/6-31G(d)-optimized methyl cation adduct of **43** was in this case found to resemble the corresponding benzyl cation adduct characterized by X-ray analysis. Conformational studies of acyl-transfer catalysts are, of course, not limited to pyridine derivatives. The conformational preferences of chiral phosphine **55**, for example, were studied by Vedejs et al. at the HF/6-31G(d) level of theory.[70b] The structures of the best conformations of the respective borane adduct were consistent with those found in X-ray analyses. Comparison of catalyst **55** and monocyclic phospholane structures suggested a possible explanation for the exceptional reactivity of the bicyclic phospholanes due to the better accessibility of the phosphorus atom in **55**.

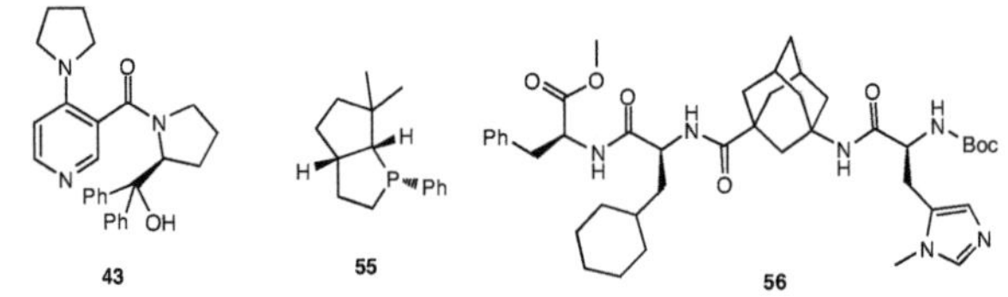

Figure 5.2. Structures of chiral catalysts used for the kinetic resolution of alcohols.[57,70,71]

Currently, only a small number of theoretical studies appears to exist in which not only conformational properties of the acylpyridinium intermediates have been studied, but a direct prediction of the outcome of kinetic resolution experiments with alcohols has been attempted. These studies deal, however, with chiral imidazole and amidine derivatives. Sunoj et al.[24a] have studied computationally the enantioselective acetylation of trans-cyclohexane-1,2-diol, catalyzed by N-methylimidazole-based peptide **56**, which was designed by Schreiner and co-workers (Figure 5.2).[71] N-methylimidazole itself was shown to be active as acylation catalyst,[72] but with lower activity than DMAP.[11c] Inclusion of the N-methylimidazole fragment into the chiral environment of small tripeptides still generates highly selective catalysts for the kinetic resolution of alcohols.[71b] The theoretical studies show that hydrogen bonding between the diol substrate and the peptide backbone plays an important role for enantioselectivity. The difference in reaction barriers for the two alcohol enantiomers, calculated at B3LYP/6-31G(d)//ONIOM2(B3LYP/6-31G(d):PM3) level, amounts to 19.3 kJ/mol and corresponds to 99% ee. This value is larger than experimentally reported (75%); however, the predominantly reacting alcohol enantiomer was predicted correctly. Houk *et al.* have optimized the transition states for the acylation of 1-phenylethanol by acetic and propionic anhydrides, catalyzed by compound **57**.[24b] Analysis of the transition state structures supports the π-interaction hypothesis of chiral recognition in the kinetic resolution of secondary benzylic alcohols (Figure 5.3). The differences in activation free energy between the alcohol enantiomers (at B3LYP/6-31G(d) level with CPCM single points for the solvent effects of $CHCl_3$) were shown to correlate well with the experimental values.

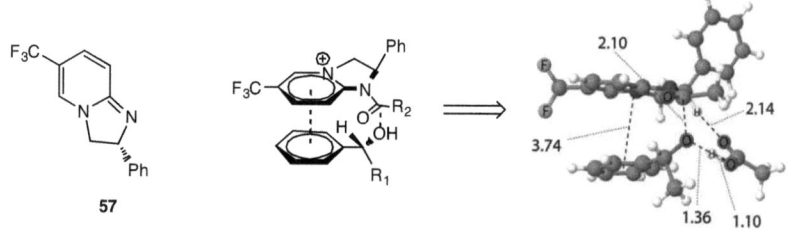

Figure 5.3. Birman's amidine-based catalyst **57** with the transition state model for the KR of alcohols.[24b] Distances are given in Å.

This chapter is organized as follows: first, experimental and theoretical methods will be used to gain mechanistic insight(s) into acylation reactions catalyzed by PPY. Then computations of the reaction profile will be applied for the rationalization and prediction of the selectivity for a series of derivatives of the Spivey's catalyst **59a**. In the last part several attempts to use less time-consuming computational models for the design of stereoselective pyridine derivatives are presented.

*The computations, described in sections 5.3.1 and 5.3.2, as well as calculations of transition states for catalysts **59b** and **59c**, described in section 5.3.3 (Table 5.7), were carried out by Dr. Y. Wei.*

5.2 Catalytic system with PPY

5.2.1 Determination of activation parameters for the PPY-catalyzed acylation reaction

Catalyst **59a** was shown to be very effective for the kinetic resolution of *sec*-alcohols.[73] However, the selectivity values seem to depend on at least two experimental parameters: $s = 24$ at 45% conversion with 2 eq. anhydride,[73] $s = 16.7$ at 12% and $s = 15.7$ at 27% conversion with 1 eq. anhydride[74] (for the reaction shown in Scheme 5.1). The anomalous increase in selectivity with increased conversion ($s = 32$ after 8 h vs $s = 25$ after 2 h) was observed in the kinetic resolution experiments with catalyst **59d** (Scheme 5.1).[74] There are also several communications of the conversion-dependent selectivity in the literature.[75] In the meantime it was noticed, that the reaction mixture at −78 °C is not homogeneous, while at room temperature no precipitation was observed.[73] This phenomenon can be explained by the formation of insoluble (in toluene) triethylammonium salt, which can in principle affect the reaction (rate or/and selectivity). Should this be the case, we must expect that the Arrhenius plot for the reaction rate or even selectivity is not linear anymore. In order to understand the role of the precipitation we have decided to study the reaction kinetics at different temperatures. As the model system the PPY-catalyzed isobutyrylation of 1-(1-naphthyl)ethanol **60** was chosen (Scheme 5.1).

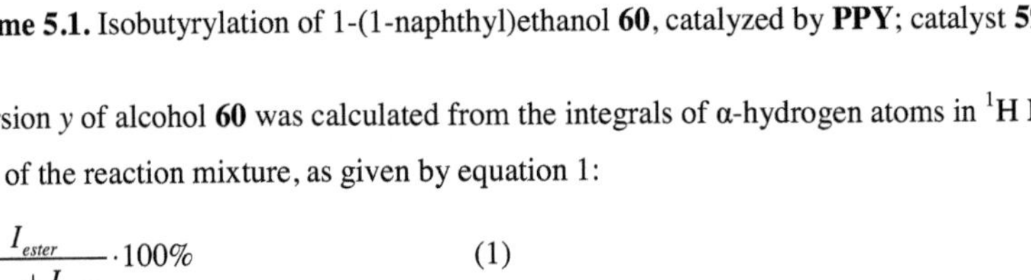

Scheme 5.1. Isobutyrylation of 1-(1-naphthyl)ethanol **60**, catalyzed by **PPY**; catalyst **59d**.

Conversion y of alcohol **60** was calculated from the integrals of α-hydrogen atoms in ^{1}H NMR spectra of the reaction mixture, as given by equation 1:

$$y = \frac{I_{ester}}{I_{ester} + I_{ROH}} \cdot 100\% \quad (1)$$

Dependence of conversion y vs time t was fitted by equation 2 for the second-order reaction kinetics:

$$y = y_0 \left(1 - \frac{1}{2e^{k(t-t_0)} - 1}\right) \quad (2)$$

$$k = k_2 [ROH]_0 \quad (3)$$

where k_2 is a rate-constant of the second-order reaction; t_0 has a meaning of time axis offset. With this parameter in the fitting process it is not necessary to measure the starting point of the reaction exactly. The variable y_0 allows rescaling of the conversion axis. The rate constant measurements were repeated at different temperatures (Table 5.1).

Table 5.1. Rate constants k_2 for the **PPY**-catalyzed acylation, measured at different temperatures.

T^{-1}, K^{-1}	0.0052	0.0051	0.0049	0.0047	0.0045	0.0043	0.0041	0.0040
T, K	193.15	198.15	203.15	213.15	223.15	233.15	243.15	248.15
T, °C	-80	-75	-70	-60	-50	-40	-30	-25
k_2, 10^{-4} M^{-1} s^{-1}	4.080	5.060	6.708	9.000	14.00	19.80	26.37	30.47

In order to get activation parameters, the obtained data were fitted with the Eyring equation (Figure 5.4).

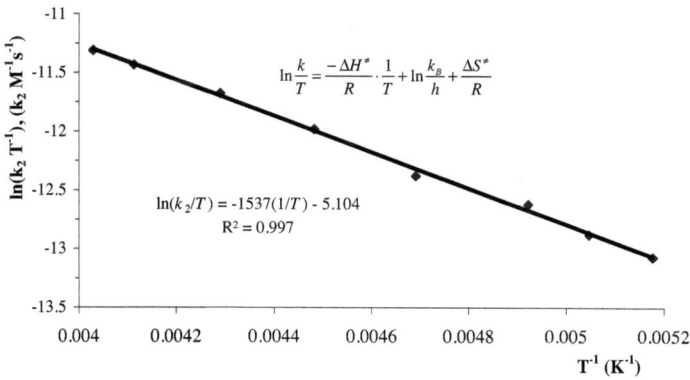

Figure 5.4. Eyring plot for the reaction shown in Scheme 5.1.

The fitted parameters are then transformed to activation parameters:
$\Delta H^{\neq} = \mathbf{12.8}$ kJ mol^{-1}; $\Delta S^{\neq} = \mathbf{-240}$ J mol^{-1} K^{-1}.
The free activation energy at 195 K can also be calculated from these data:
$\Delta G_{195}^{\neq} = \mathbf{59.6}$ kJ mol^{-1}

The obtained value of activation enthalpy $\Delta H^{\neq}$ is quite small for the reactions in solution, whereas the obtained negative value of the activation entropy is common for bimolecular reactions.[70b,82-84] As a comparison, the isobutyrylation of 1-(1-naphthyl)ethanol **60** (Scheme 5.1) catalyzed by chiral phosphane **55** (Figure 5.2) has been studied, and was found to have small activation enthalpies (in a range 5.8 – 12.5 kJ mol^{-1}), which depend on the reacting enantiomer of **60**.[70b] From the other side, activation entropies were shown to be *ca.* -308 J mol^{-1} K^{-1} for both enantiomers of alcohol **60**. This implies that the $\Delta G^{\neq}$ term that reflects enantiomer discrimination is dominated by differences in activation enthalpy, resulting in a significant temperature effect on enantioselectivity. Acyl transfer in acetonitrile has also been studied (catalyzed aminolysis of ethyl aryl carbonates), and was found to have very large, negative values for the entropy of activation (-272 to -280 J mol^{-1} K^{-1}) and small activation enthalpy (7.5 – 8.1 kJ mol^{-1}).[84] Moreover, DMAP-catalyzed phosphorylations have been investigated, and the kinetic parameters feature a similar combination of large, negative activation entropy, and minimal activation enthalpy.[83]

5.2.2 Theoretical study of the catalytic cycle with PPY

The DMAP-catalyzed acylation of alcohols by anhydrides and acyl chlorides is currently believed to proceed via the nucleophilic catalysis mechanism. This mechanism is also supported by the computational study of the DMAP-catalyzed acetylation of *tert*-butanol with acetic anhydride.[14] It was noticed, that in the case of primary alcohols the basic mechanism could become competitive with the nucleophilic.[19b] In order to investigate, which mechanism (basic or nucleophilic) is relevant for the PPY-catalyzed acylation of secondary alcohol **60**, we have studied the catalytic cycle computationally. The enthalpy profile as calculated at the B3LYP/6-311+G(d,p)//B3LYP/6-31G(d) level at the experimental temperature of 195.15 K is shown in Figure 5.5.

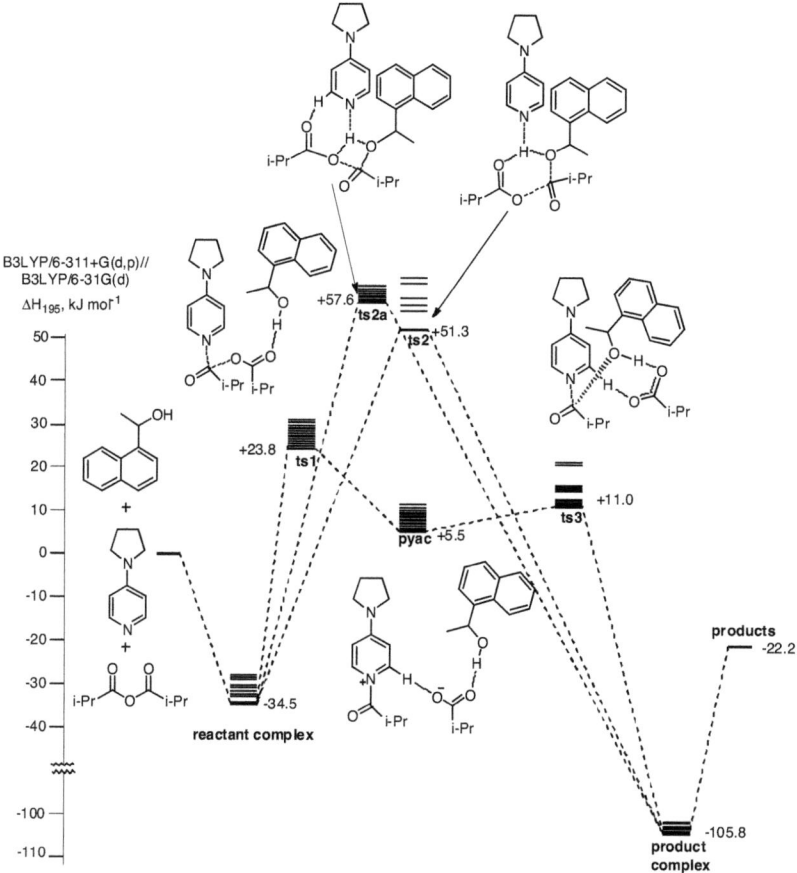

Figure 5.5. Enthalpy profile (ΔH_{195}) of the PPY-catalyzed acylation of 1-(1-naphthyl)ethanol **60** as calculated at B3LYP/6-311+G(d,p)//B3LYP/6-31G(d) level.

Both enthalpy and free energy values suggest the nucleophilic route through transition states **ts1** and **ts3** to be more favourable and the first step of this route to be rate-limiting. It sounds surprising, because the second step is commonly considered as the rate-determining step.[14] The six-membered transition state **ts2** of the basic route is slightly more favourable than the four-membered **ts2a**. We can then calculate activation parameters (relative to separate reactants and to reactant complex) for all transition states at B3LYP/6-311+G(d,p)//B3LYP/6-31G(d) level (Table 5.2). Data at the experimental temperature of 195.15 K as well as at 298.15 K are included in Table 5.2.

Table 5.2. Comparison of activation parameters at 195.15 K and at 298.15 K as calculated at B3LYP/6-311+G(d,p)//B3LYP/6-31G(d) level (in the gas phase). Activation parameters for the rate-limiting transition state **ts1** are marked in bold.

Activation parameter	Relative to separate reactants				Relative to reactant complex				Relative to PPY-alcohol complex and anhydride			
	ts1	ts2	ts2a	ts3	ts1	ts2	ts2a	ts3	ts1	ts2	ts2a	ts3
$\Delta H_{195}^{\neq}$, kJ mol^{-1}	**+23.8**	+51.3	+57.6	+11.0	**+58.3**	+85.8	+92.1	+45.5	**+53.6**	+81.1	+87.4	+40.8
$\Delta G_{195}^{\neq}$, kJ mol^{-1}	**+177.5**	+210.1	+216.5	+168.3	**+72.7**	+105.3	+111.7	+63.5	**+141.0**	+173.6	+180.0	+131.8
$\Delta S_{195}^{\neq}$, J mol^{-1} K^{-1}	**-788**	-814	-814	-806	**-74**	-100	-100	-92	**-448**	-474	-475	-466
$\Delta H_{298}^{\neq}$, kJ mol^{-1}	**+25.3**	+52.5	+58.9	+12.6	**+57.2**	+84.4	+90.8	+44.5	**+54.0**	+81.2	+87.6	+41.3
$\Delta G_{298}^{\neq}$, kJ mol^{-1}	**+124.7**	+161.8	+167.8	+118.8	**+80.3**	+117.4	+123.4	+74.4	**+114.7**	+151.8	+157.8	+108.8
$\Delta S_{298}^{\neq}$, J mol^{-1} K^{-1}	**-334**	-367	-365	-356	**-77**	-111	-109	-100	**-204**	-237	-235	-226

The experimental value of the activation enthalpy $\Delta H_{195}^{\neq}$ (+12.8 kJ mol^{-1}) is smaller than the theoretically predicted value for rate-limiting transition state **ts1** (+23.8 kJ mol^{-1}). Activation entropies, calculated for different transition states, have similar values (ca. -800 J mol^{-1} K^{-1} relative to separate reactants) and do not allow distinguishing different mechanisms. The large deviation from experiment can be explained by the fact that a trimolecular reaction is studied computationally, whereas the less negative experimental activation entropy value of -240 J mol^{-1} K^{-1} is more typical for bimolecular reactions. When activation parameters are calculated relative to the reactant complex, larger values of activation enthalpy and much less negative values of activation entropy (-74 J mol^{-1} K^{-1} at 195 K) are observed (Table 5.2). Noticeably, activation entropies calculated at 298 K are surprisingly less negative than calculated at 195 K. The reactant complex contains a relatively strong alcohol-PPY complex, whereas anhydride is weakly bonded. Activation entropy of **ts1** calculated thus relative to the alcohol-PPY complex and anhydride (-448 J mol^{-1} K^{-1} at 195 K) represents a bimolecular reaction and is therefore closer to the experimental value. Another reason for the less negative experimental activation entropy can be that the alcohol does not participate in **ts1** under experimental conditions. Not the least important factor is an inaccuracy of the present model for entropy calculation, which is based on the harmonic oscillator approximation.

In order to see whether the observation about the rate-limiting transition state also persists by other theoretical methods, we have carried out single point calculations at several other levels of theory again based on the B3LYP/6-31G(d) structures. The MP2 method, as well as DFT method with dispersion correction (DFT-D),[76] are chosen because we suppose that the dispersion interactions may exist and play some role since the studied transition states include several aromatic rings, although the used B3LYP functional cannot predict this type of interaction accurately. Single point calculations were done at different levels of theory for conformers whose populations are more than 1% at MP2(FC)/6-311+G(d,p)//B3LYP/6-31G(d) level. The obtained enthalpies and free energies for the best conformers of transition states and intermediates with respect to the separated reactants are shown in Table 5.3.

Table 5.3. Comparison of different levels for single-point calculations (enthalpies relative to separate reactants, kJ mol^{-1}).

Level of theory for single point with B3LYP/6-31G(d) geometry	ΔH_{195} (reactant complex)	ΔH_{195} (ts1)	ΔH_{195} (intermediate)	ΔH_{195} (ts3)	ΔH_{195} (ts2)	ΔH_{195} (ts2a)	ΔH_{195} (product complex)
B3LYP/6-31G(d)	-56.4 (+1.0)	+3.4 (+72.8)	-10.4 (+55.6)	-5.2 (+65.6)	+23.5 (+95.9)	+25.0 (+97.6)	-128.6 (-76.8)
B3LYP/6-311+G(d,p)	-34.5 (+104.8)	+23.8 (+177.5)	+5.5 (+157.3)	+11.0 (+168.3)	+51.3 (+210.1)	+57.6 (+216.5)	-105.8 (+29.8)
MP2(FC)/6-311+G(d,p)	-99.8 (-43.4)	-64.4 (+3.85)	-69.7 (-3.8)	-104.7 (-34.4)	-32.8 (+40.5)	-26.4 (+46.6)	-147.5 (-92.9)
MP2(FC)/6-31+G(2d,p)	-96.3 (-39.6)	-69.5 (-1.3)	-74.3 (-8.3)	-102.4 (-32.1)	-35.2 (+38.2)	-28.8 (+46.8)	-148.7 (-94.1)
B3LYP-D/6-311+G(d,p)	-88.2 (-54.1)	-55.4 (101.4)	-71.4 (+80.4)	-86.2 (+69.8)	-29.6 (+129.5)	-25.7 (+133.0)	-139.0 (-4.4)

[a] in brackets the corresponding ΔG_{195} values.

In order to avoid the basis set superposition error (BSSE), the relative enthalpies with respect to the reactant complex were also calculated (Table 5.4). The enthalpy differences between transition states **ts1** and **ts3** at different levels are also presented. The results show that using MP2 and B3LYP-D levels for single point calculations stabilizes all the transition states relative to the reactants and reactant complex. All these different theoretical methods predict the first step of the nucleophilic path to be rate-limiting and the basic route to be less favourable (by ca. 30–40 kJ mol^{-1}). Transition states of the basic route **ts2** and **ts2a** have similar energies at different levels of theory. Therefore, they are equally feasible for the basic mechanism. The enthalpy difference between transition states **ts1** and **ts3** increases from 12.8 kJ/mol at B3LYP/6-311+G(d,p)//B3LYP/6-31G(d) level up to 40.2 kJ mol^{-1} at MP2(FC)/6-311+G(d,p)//B3LYP/6-31G(d) level; the DFT method with dispersion correction B3LYP-D/6-311+G(d,p)//B3LYP/6-31G(d) level gives intermediate values of the enthalpy difference. The performance of the MP2 method for single point calculations does not depend significantly on the DFT level used for the geometry optimization (B3LYP or B98). Taking into account only the best conformers at B3LYP/6-311+G(d,p)//B3LYP/6-31G(d) level (*e.g.* of type I for **ts3**, see Figure 5.6) for the MP2 and B3LYP-D single point calculations, the enthalpy difference between transition states **ts1** and **ts3** remains in the range of the differences calculated at B3LYP levels.

Table 5.4. Enthalpies and free energies of transition states at different levels of theory (relative to reactant complex, kJ mol^{-1}).

Level of theory for single point with B3LYP/6-31G(d) geometry	ΔH_{195}(**ts1**)[a]	ΔH_{195}(**ts2**)[a]	ΔH_{195}(**ts2a**)[a]	ΔH_{195}(**ts3**)[a]	ΔH_{195} (**ts1-ts3**)[a]
B3LYP/6-31G(d)	+59.8 (+71.8)	+79.9 (+94.9)	+81.4 (+96.6)	+51.2 (+64.6)	8.6 (7.2)
B3LYP/6-311+G(d,p)	+58.3 (+72.7)	+85.8 (+105.3)	+92.1 (+111.7)	+45.5 (+63.5)	12.8 (9.2)
MP2(FC)/6-311+G(d,p)	+35.7 (+47.4)	+67.3 (+84.0)	+73.7 (+90.1)	-4.5 (+9.1)	40.2 (38.2) //9.2 (8.0)[b]
MP2(FC)/6-31+G(2d,p)	+26.8 (+38.3)	+61.1 (+77.8)	+67.5 (+86.4)	-6.1 (+7.5)	32.9 (30.7) //6.0 (4.8)[b]
B3LYP-D/6-311+G(d,p)	+32.9 (+47.3)	+58.7 (+75.4)	+62.5 (+78.9)	+2.1 (+15.7)	30.8 (31.6) //15.7 (14.2)[b]
MP2(FC)/6-31+G(2d,p)//B98/6-31G(d)	-	-	-	-	28.0 (28.9)

[a] In brackets the corresponding ΔG_{195} values.
[b] Taking the best conformers at B3LYP/6-311+G(d,p)//B3LYP/6-31G(d) level.

By more careful analysis of the conformational space of the studied transition states we found that relative energies of conformers at different levels of theory depend on the presence of stacking interactions in the system. The largest effect was found for the transition state **ts3** (Figure 5.6). The conformations of type **II** with stacking interactions between naphthyl and pyridine rings (the distance between the centers of the rings is 4.03 Å) have energies, which are comparable with type **I** conformational energies, when calculated by B3LYP methods. As mentioned above the B3LYP functional cannot predict dispersion interactions accurately. When using the MP2 method for single point calculations, type **II** conformations become much more stable than type **I** conformations by *ca.* 30 kJ mol^{-1}. However, one of the serious shortcomings of MP2 theory is a noticeable overestimation of the dispersion interaction energy.[65] Furthermore, the DFT functional with dispersion correction B3LYP-D can predict this type of interaction more accurately.[76] Indeed, the relative enthalpy difference between types **I** and **II** conformations becomes smaller when using B3LYP-D/6-311+G(d,p) level instead of MP2 level for the single point calculations (Figure 5.6). Thus we will use the economic DFT methods B3LYP/6-311+G(d,p)//B3LYP/6-31G(d) and B3LYP-D/6-311+G(d,p)//B3LYP/6-31G(d) to calculate the selectivity in section 5.3 in order to see which calculated results are in line with available experimental results.

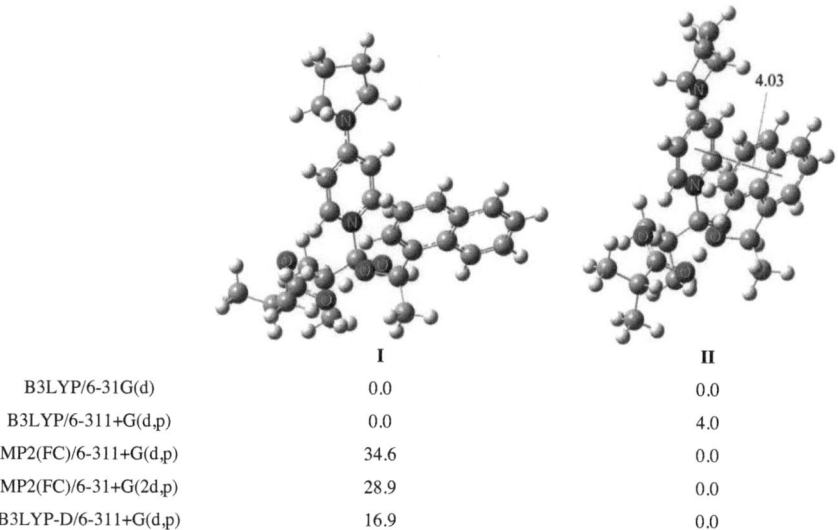

	I	II
B3LYP/6-31G(d)	0.0	0.0
B3LYP/6-311+G(d,p)	0.0	4.0
MP2(FC)/6-311+G(d,p)	34.6	0.0
MP2(FC)/6-31+G(2d,p)	28.9	0.0
B3LYP-D/6-311+G(d,p)	16.9	0.0

Figure 5.6. Relative enthalpies ΔH_{195} (kJ mol^{-1}) of two conformations of **ts3** as calculated at five levels of theory using geometries and thermal corrections at B3LYP/6-31G(d) level. Distances are given in Å.

Another possible way to distinguish experimentally different mechanisms is the measurement of the kinetic isotope effect. Since the transition states **ts2**, **ts2a** and **ts3** contain the hydrogen atom, which is involved in the broken and formed chemical bonds, measurements of the rate constant ratio k_H/k_D could help differentiate which step is rate-limiting. We carried out theoretical prediction of the isotope effect, since it can support the experimental data. The enthalpy and free energy differences upon substitution of the alcohol hydrogen H to deuterium D in the transition states with respect to the corresponding differences for the alcohol were calculated at B3LYP/6-31G(d) level at 195.15 K and 298.15 K (Table 5). The free energy differences were then transformed to the rate constant ratios, using equation 4:

$$\ln\frac{k_H}{k_D} = \frac{\Delta G(D-H)}{RT} \quad (4)$$

Table 5.5. Isotope effects for transition states **ts1, ts2, ts2a** and **ts3**, calculated at B3LYP/6-31G(d) level at 195.15 K and 298.15 K. ΔH(D-H) – difference between activation enthalpies with D and with H as the alcohol hydrogen.

T, K	ΔH(D-H), kJ mol^{-1}		ΔG(D-H), kJ mol^{-1}		k_H/k_D	
	298.15	195.15	298.15	195.15	298.15	195.15
ts3	3.23	3.26	3.83	3.62	4.68	9.30
ts1	-0.72	-0.76	-0.31	-0.46	0.88	0.75
ts2	0.61	0.63	1.23	1.01	1.64	1.86
ts2a	1.85	1.86	2.43	2.22	2.66	3.93

The obtained results show that the supposed rate-limiting transition state **ts1**, that does not contain the hydrogen atom, which is directly involved in the broken and formed chemical bonds, should give a secondary isotope effect that falls into the range between 0.75 at 195.15 K and 0.88 at 298.15 K. If the transition state **ts3** would be rate-limiting, the primary kinetic isotope effect should be much larger (up to 9.30 at 195.15 K). Transition states **ts2** and **ts2a** of the basic route give moderate isotope effects (up to 3.93 at 195.15 K). Comparison of these obtained data with experimental results can help to identify the rate-limiting transition state and thus the reaction mechanism.

5.3 Catalytic system with Spivey's catalyst.

In order to shed some light on the enantioselectivities of chiral DMAP-catalysts in acyl-transfer reactions, we have investigated theoretically the acylation of racemic secondary alcohols catalyzed by a series of Spivey's catalysts in detail. The most important question here is whether the enantioselectivities of chiral DMAP-catalyzed acyl-transfer reactions can be rationalized with the transition state in the rate-determining step that is also considered as the selectivity-determining step. The possible role of 4-dialkylamino substituents on the chiral transformation is discussed and a catalyst modification to improve the enantioselectivity is suggested.

5.3.1 The energy profile of the acylation catalyzed by catalyst 59a

In order to check whether the mechanism of the PPY catalyzed acylation of secondary alcohols also persists for the acylation catalyzed by the chiral catalyst **59a**, we have first investigated the nucleophilic and general base catalysis pathways for the reaction of racemic 1-(1-naphthyl)ethanol (**60**) with isobutyric anhydride (**61**) catalyzed by **59a** at the B3LYP/6-311+G(d,p)//B3LYP/6-31G(d) level of theory used in the previous theoretical studies of DMAP-catalyzed acetylation of alcohols.[14] All conformers of reactants **60** and **61**, and products **62** and **63** have been searched carefully and optimized at B3LYP/6-31G(d) level, and single point calculations were done at the B3LYP/6-311+G(d,p) level of theory in order to obtain the relative enthalpies at the B3LYP/6-311+G(d,p)//B3LYP/6-31G(d) level of theory. The systems investigated here are very flexible and have a large conformational space. A systematic conformational search of TSs **65** and **67** was first done using a modified OPLS-AA force field, and then the conformers identified by force field within the energy window of 40 kJ mol^{-1} were reoptimized at the B3LYP/6-31G(d) level of theory, and single point calculations were done at the B3LYP/6-311+G(d,p) level of theory. The IRC calculations have been run using the best conformers of TSs to obtain structures of the reactant complex, intermediate, and product complex. The TSs along the basic catalysis pathway were located based on the previously suggested "four-membered" and "six-membered" structures[14] and optimized at B3LYP/6-31G(d) level. Using these structures the relative enthalpies at the B3LYP/6-311+G(d,p)//B3LYP/6-31G(d) level of theory have been calculated. The nucleophilic and general base catalysis pathways are plotted in Figure 5.7 by using the lowest-energy conformer and the relative enthalpies for stationary points located on the potential energy surface are shown in Table 5.6. The diastereomeric transition states and intermediates are denoted as (R)-* and (S)-*, which represent the corresponding configuration of the involved alcohol.

The reaction is initiated through formation of a ternary complex **64** of reactants **60, 61** and catalyst **59a** for both the nucleophilic and general base catalysis pathways. Along the nucleophilic catalysis pathway, the reactant complex **64** passes through the first TS **65** to yield intermediates **66**, which then pass through the second TS **67** with concomitant proton transfer to product complex **68**. The alternative basic catalysis pathway proceeds through concerted TSs **69** to product complex **68** in one single step. The diastereomers including R-configuration alcohol are always a few kJ mol^{-1} lower than those including S-configuration alcohol. The most energetically favorable transition state **(R)-69** along the basis catalysis pathway is located 40 kJ mol^{-1} or so above the transition state **(R)-65** and 53 kJ mol^{-1} or so above the transition state **(R)-67** on the nucleophilic catalysis pathway. Single point calculations have also been done at the MP2/6-31G(d)//B3LYP/6-31G(d) level of theory for the best conformers of **(R)-65, (R)-67, (R)-69**. The energy of **(R)-69** is also higher than that of **(R)-65** and **(R)-67** by more than 30 kJ mol^{-1} at MP2/6-31G(d)//B3LYP/6-31G(d) level. This indicates that the nucleophilic catalysis pathway is more favorable than the general base catalysis pathway, which is in line with the results on the PPY catalyzed acylation discussed above.

Table 5.6. Relative enthalpies ΔH_{298} (in kJ mol^{-1}) for stationary points located on the potential energy surface at B3LYP/6-311+G(d, p)//B3LYP/6-31G(d) level in the gas phase.

Nucleophilic catalysis	(R)-	(S)-
59a+60+61	0.00	
64 (reactant complex)	-22.98	-22.54
65 (first TS)	26.80	34.29
66 (intermediate)	7.56	11.72
67 (second TS)	14.06	20.10
68 (product complex)	-87.35	-86.87
59a+(R)-62 +63	-21.61	
Basic catalysis (concerted)	(R)-	(S)-
59a+60+61	0.00	
64 (reactant complex)	-22.98	-22.54
69 (TS)	67.23	77.49
68 (product complex)	-87.35	-86.87
59a+(R)-62+63	-21.61	

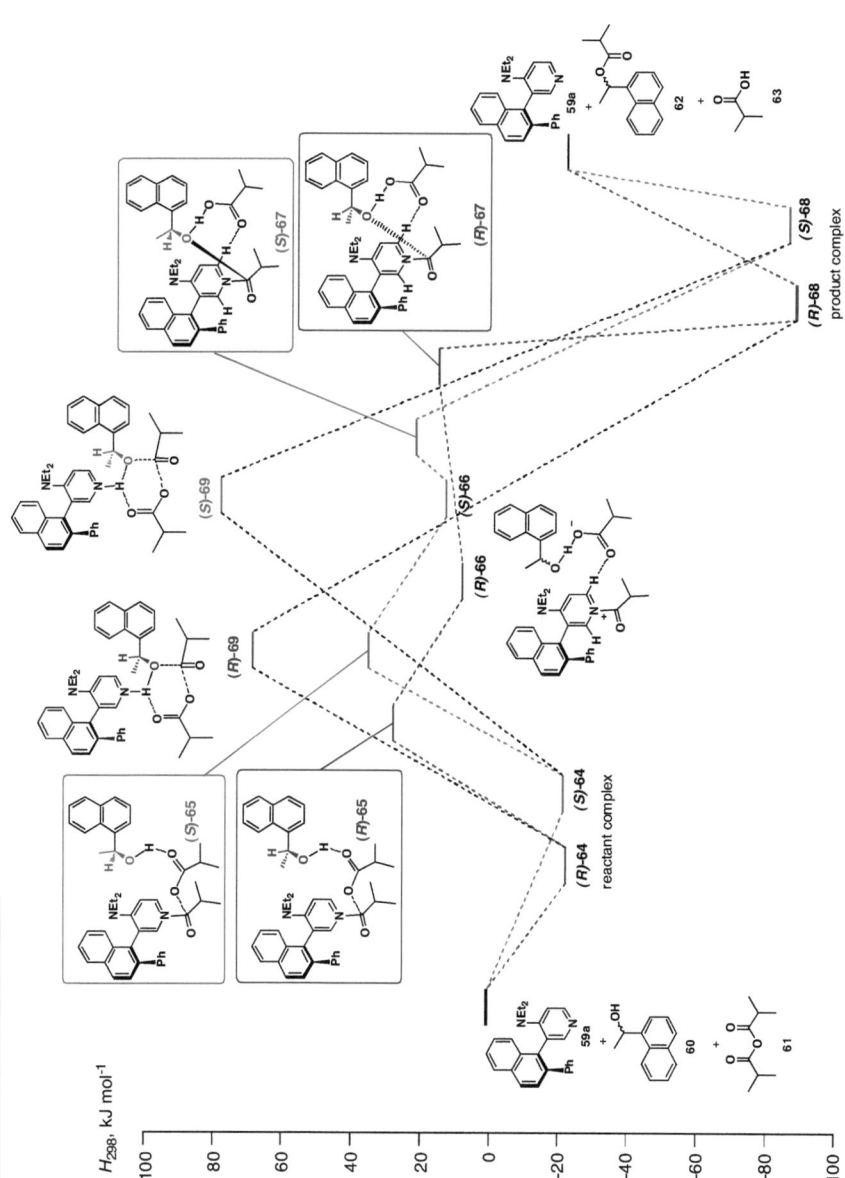

Figure 5.7. Gas phase enthalpy profile calculated at the B3LYP/6-311+G(d,p)//B3LYP/6-31G(d) level of theory.

5.3.2 Reaction barriers and conformational space of TSs

The energy difference between the diastereomeric TSs of the rate-determining step is the key point to predict the enantioselectivity. Surprisingly, the energy of first TS **65** in the formation of an acylpyridinium cation is higher than that of the second step commonly considered as the rate-determining step by *ca.* 13 kJ mol^{-1} at B3LYP/6-311+G(d,p)//B3LYP/6-31G(d) level. In order to see whether this observation also persists by other theoretical methods, we chose several other levels of theory to do single point calculations again based on the optimized B3LYP/6-31G(d) structures. The DFT methods with dispersion corrections (DFT-D)[76] and MP2 methods are chosen because we assume that the dispersion interactions may exist and play some role due to the system studied here including several aromatic rings, however, the popular B3LYP functional cannot predict this type of interaction accurately. The single point calculations were done at different levels of theory for conformers whose populations are more than 1% at B3LYP/6-311+G(d,p)//B3LYP/6-31G(d) level. The relative Boltzman-averaged enthalpies between **(R)-65** and **(R)-67** at different levels of theory are investigated carefully and compared (Figure 5.8). In order to avoid the basis set superposition error (BSSE), the relative enthalpies are calculated with respect to the reactant complex instead of the separated reactants.

The energy difference between **(R)-65** and **(R)-67** varies with theoretical methods, the variation is in the range of −14 kJ mol^{-1} and +14 kJ mol^{-1}. Thus, different theoretical methods predict different rate-determining steps using the same model system. At this point, it is hard to pin down which method is more reliable without higher level theoretical benchmark data that are too difficult to get for such a big system. MP2 results seem more basis sets dependent and they are much more computationally costly than DFT methods for the system studied here. We will use the economic DFT methods B3LYP/6-311+G(d,p)//B3LYP/6-31G(d) and B3LYP-D/6-311+G(d,p)//B3LYP/6-31G(d) to calculate the selectivity in the next section and to see which calculated results are in line with the experimental results.

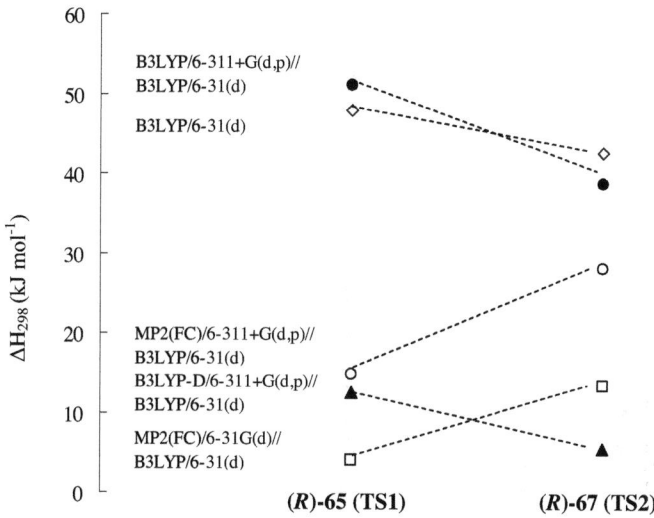

Figure 5.8. Relative energies of **(R)-65** and **(R)-67** with respect to the reactant complex at different levels of theory.

In principle, the rate-determining step is also considered to be the selectivity-determining step. Thus, it is difficult to predict which step is the selectivity-determining step due to the uncertainty of the rate-determining step described above. We have tried to calculate the free energy difference of diastereomers in these two steps to match the experimental value. It turns out that the free energy difference of the diastereomers of TS **67** is closer to experimental values. Moreover, the C–O bond formation between the alcohol and acetyl group proceeds in the second step, that supports TS **67** to be the selectivity-determining. The detailed theoretical prediction of catalytic selectivity is discussed in the next section. We focus our attention here on the structures and the energy difference of the diastereomers of TS **67** to investigate the possible factors influencing the stereoselectivity of catalyst **59a**.

Through analysis of the optimized geometries of transition state **67**, we found that all conformers can be classified into the four structural types as shown in Scheme 5.2. Figure 5.9 shows a pictorial representation of the relative energies of the conformers of **(R)-67** and **(S)-67**, respectively. Generally speaking, the carboxylate group is bonded to the left or right side of the pyridine ring by weak hydrogen bonding and the alcohol approaches the reaction center either from the front face or the back face of the pyridine ring. Type **67-I** shows that the carboxylate group is bonded to the right side of the pyridine ring and the alcohol approaches the reaction center from the back side. For this type the conformers with *R*-configuration alcohol are more stable than the conformer with *S*-configuration alcohol by more than 20 kJ mol^{-1}. In type **67-II** and **67-III**, the conformers with *S*-configuration alcohol are more stable than the conformers with *R*-configuration alcohol. Conformers in type **67-IV** have poor stabilities, no matter including either *R*-configuration or *S*-configuration alcohol. The most stable conformer with *R*-configuration alcohol belongs to the type **67-I**, which is more stable than the most stable conformer with *S*-configuration alcohol classified into the type **67-III** by 6.1 kJ mol^{-1}. Thus, calculations predict, that (*R*)-alcohol should react faster than (*S*)-alcohol, which is in line with experimental results.[73]

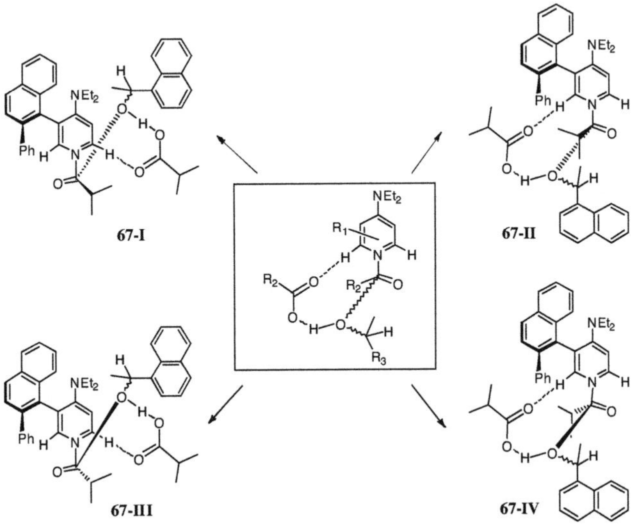

Scheme 5.2. The classified conformer types of TS **67**.

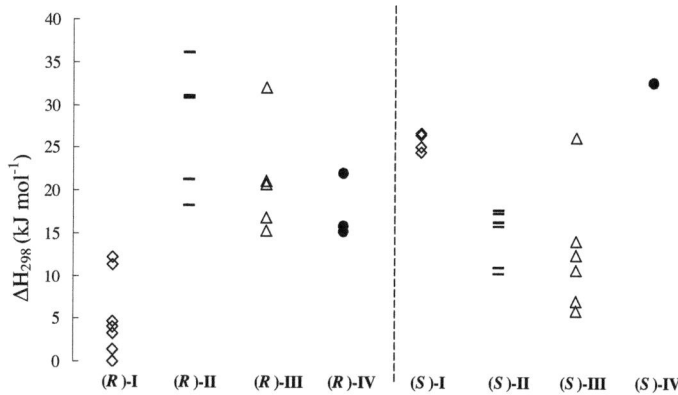

Figure 5.9. Relative enthalpies (kJ mol^{-1}) of conformers of TS **67a**, as calculated at B3LYP/6-311+G(d,p)//B3LYP/6-31G(d) level.

The B3LYP/6-31G(d) optimized structures of the most stable conformers of (**R**)-**67** and (**S**)-**67** are shown in Figure 5.10. Analysis of the structures reveals that the alcohol **60** (shown by light green color in Figure 5.10) approaches the reaction center from the back face of the pyridine ring in (**R**)-**67** and from the front face of the pyridine ring in (**S**)-**67**. There is no siginificant steric hindrance when alcohol approaches the reaction center from the back face of the pyridine in (**R**)-**67**. In contrast, alcohol approaching the reaction center from the front face of the pyridine in (**S**)-**67**, the steric repulsion between the tilted phenyl ring of the catalyst **59a** and the naphthyl ring of alcohol **60** may raise the energy of (**S**)-**67** relative to that of (**R**)-**67** by *ca.* 6 kJ mol^{-1}.

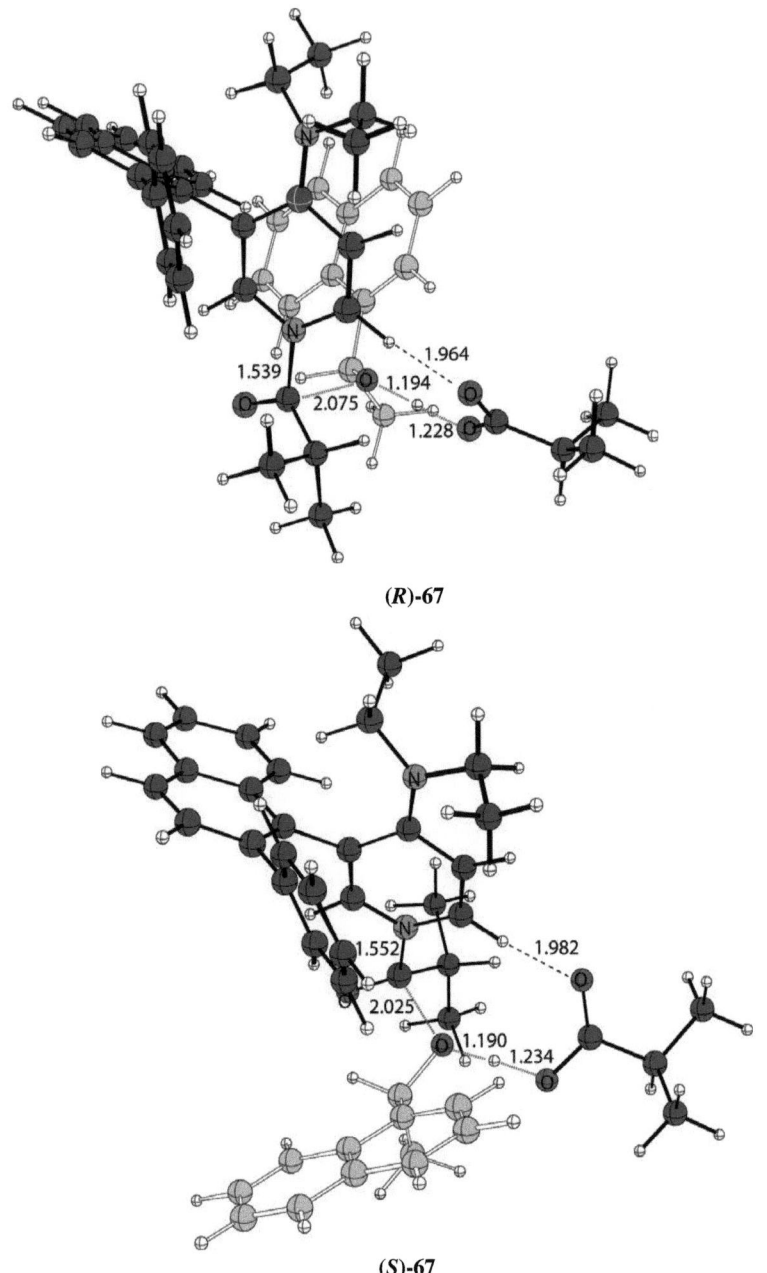

Figure 5.10. The most stable conformers of **(R)-67** and **(S)-67** at the B3LYP/6-311+G(d,p)// B3LYP/6-31G(d) level of theory. Distances are given in Å.

5.3.3 The selectivity rationalization: TS 67 for catalysts 59b, 59c and 59e

Spivey *et al.* have reported that varying the 4-dialkylamino substituents influences the selectivities of catalysts.[77a] Experimental results show that the selectivity decreases in the series **59e - 59a - 59b - 59c**, where the pyrrolidino-substituted catalyst **59c** is the least selective (Scheme 5.3). We choose a series of catalysts **59a – 59g** (shown in Scheme 5.3) and use the same substrate (1-(1-naphthyl)ethanol, **60**) to investigate their selectivity theoretically and compare it with experimental results.

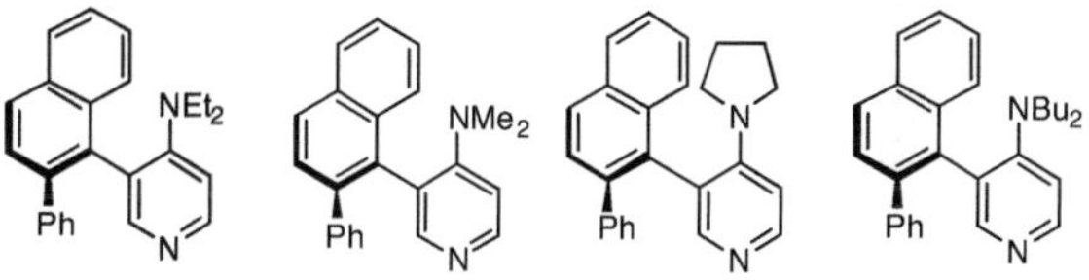

Scheme 5.3. Chiral pyridine catalysts used to model kinetic resolution of *sec*-alcohols: the known catalysts **59a-c, e** (the first row) and new derivatives **59d, f, g** (the second row). All catalysts have (*S*)-configuration.

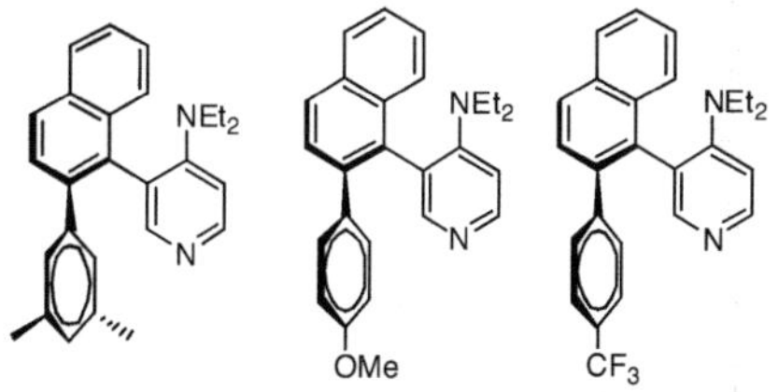

The enthalpy and free energy differences between the diastereomers of the TS **67** considered as the selectivity-determining TS were calculated for **59a-c, e** by DFT methods and listed in Table 5.7. The conformational space of TS **67b** and **67c** for **59b** and **59c**, respectively, were also searched in the similar way as for catalyst **59a** by modified OPLS-AA force field and then the identified conformers were reoptimized at B3LYP/6-31G(d) level. The conformational search for TS **67e** with catalyst **59e** was carried out as described below. The catalyst part in the most stable conformations of TS **67** with catalysts **59a** was modified to catalyst **59e** (with *trans* configuration of the added methylene and methyl groups) and the TSs were reoptimized at B3LYP/6-31G(d) level. The most stable conformations of TS **67e** were then used for the thermal corrections and single point calculations (the latter at B3LYP/6-311+G(d,p) and B3LYP-D/6-311+G(d,p) levels).

The calculated free energy differences ΔG_{298} for the TS **67** for catalysts **59a-c, e** calculated at B3LYP/6-31G(d) level cannot reproduce experimental values of ΔG_{195}, calculated using equation 5, and even predict the opposite result that the pyrrolidino-substituted catalyst **59c** should have higher selectivity than **59a** and **59b** (Table 5.7). From the other side, the calculated enthalpy differences ΔH_{298} do correlate with experimental selectivities (Figure 5.11). The thermal corrections recalculated at 195 K do not improve the correlation between experimental results and calculated enthalpies or free energies. In general, enthalpy and free energy differences between TSs **67** for catalysts **59a-c, e** calculated at B3LYP/6-31G(d) level predict (*R*)-alcohol to be more reactive than (*S*)-alcohol, which is in full agreement with experimental results.

$$\ln s = \ln \frac{k_R}{k_S} = -\frac{\Delta \Delta G^{\neq}_{195}}{RT} = \frac{\Delta G^{TS}_S - \Delta G^{TS}_R}{RT} = \frac{\Delta G^{TS}}{RT} \qquad (5)$$

Employing the combined DFT method B3LYP/6-311+G(d,p)//B3LYP/6-31G(d) does not yield better correlation of experimental selectivities with calculated free energies or enthalpies (Table 5.7). From the other side, inclusion of dispersion corrections (at B3LYP-D/6-311+G(d,p) level) significantly improves the correlation with calculated free energies. Moreover, after addition of the dispersion corrections, the enthalpy differences ΔH_{298} are significantly smaller. Noticeably, the enthalpy differences ΔH_{298} between TSs **67** for catalysts **59a-c, e** calculated at B3LYP-D/6-311+G(d,p)//B3LYP/6-31G(d) level can also be correlated with experimental enantioselectivities (correlation coefficient $R^2 = 0.7008$).

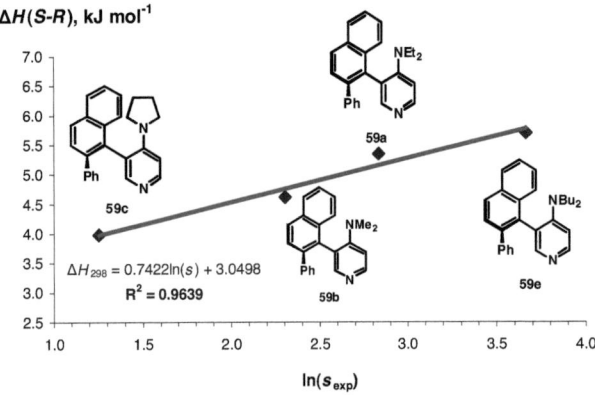

Figure 5.11. Correlation between experimental enantioselectivities and calculated enthalpy differences ΔH_{298} between TSs (*R*)- and (*S*)-**67**, as calculated at B3LYP/6-31G(d) level.

Table 5.7. Comparison of experimental and calculated energy differences $\Delta H(S-R)^f$ and $\Delta G(S-R)^f$ (in kJ mol^{-1}) of the diastereomers of TS **67** for catalysts **59a-g**. Positive numbers imply a preference for (*R*)-alcohol.

catalyst	s	ln s	$\Delta G_{195, \text{exp}}$	B3LYP/6-311+G(d,p)//B3LYP/6-31G(d) level				B3LYP-D/6-311+G(d,p)//B3LYP/6-31G(d) level				B3LYP/6-31G(d) level					
				ΔH_{298}	ΔG_{298}	ΔH_{195}	ΔG_{195}	ΔH_{298}	ΔG_{298}	ΔH_{195}	ΔG_{195}	ΔH_{298}	ΔH_{298}^d	ΔG_{298}^d	ΔH_{195}	ΔH_{195}^d	ΔG_{195}^d
59a	17[a]	2.83	4.50	6.13	5.65	5.92	8.40	1.42	6.09	1.79	4.31	5.35	5.59	4.19	5.48	5.55	6.77
59b	10[a]	2.30	3.66	6.12	5.60	6.50	6.83	1.45	1.93	4.48	1.99	4.62	4.35	3.67	4.80	4.64	5.33
59c	3.5[a]	1.25	1.99	5.82	9.90	6.12	9.48	0.17	4.18	0.00	3.30	3.99	4.40	8.23	4.31	4.48	8.32
59e	39[a]	3.66	5.82	6.01	11.99	6.12	9.55	9.49	15.30	9.70	13.72	5.70	5.96	10.27	5.75	6.00	8.29
Correlation coefficient R^2 [b]				0.3410	0.0407	0.0385	0.0058	0.7008	0.6095	0.7101	0.6044	0.9639	0.7700	0.0434	0.9496	0.8705	0.0005
59d	32	3.47	5.50	9.29	14.06	9.31	11.74	4.50	2.13	4.52	2.57	6.72	7.14	8.23	6.70	7.07	6.26
59f	11	2.40	3.81	6.14	7.01	6.25	6.68	4.22	5.85	4.29	5.28	4.41	4.07	4.78	4.51	4.16	4.76
59g	9	2.20	3.49	5.04	7.37	5.09	6.30	-0.98	0.91	-1.01	0.05	2.12	2.83	3.98	2.11	2.79	2.74
Correlation coefficient R^2 [c]				0.8906	0.5953	0.8390	0.9832	0.3666	0.0009	0.3800	0.0270	0.8599	0.9693	0.7604	0.8353	0.9601	0.6188
Correlation coefficient R^2 [e]				0.2759	0.2229	0.2138	0.1970	0.6010	0.2595	0.5594	0.3092	0.4933	0.5232	0.1360	0.4209	0.5030	0.0228

[a] Selectivities *s* are taken from the experimental results at 195 K.[75a]
[b] Correlation of the calculated energy differences $\Delta H(S-R)$ and $\Delta G(S-R)$ with experimental values ln*s* for catalysts **59a**, **59b**, **59c** and **59e**.
[c] Correlation of the calculated energy differences $\Delta H(S-R)$ and $\Delta G(S-R)$ with experimental values ln*s* for catalysts **59a**, **59d**, **59f** and **59g**
[d] Taking only the best conformers of (*R*) and (*S*) TS **67**.
[e] Correlation of the calculated energy differences $\Delta H(S-R)$ and $\Delta G(S-R)$ with experimental values ln*s* for all studied catalysts.
[f] Energy values are Boltzmann-averaged over the maximum available number of conformers (the actual numbers of conformers used for averaging are shown in Table A5.5).

In conclusion, enthalpy differences ΔH_{298} between TSs *(R)*- and *(S)*-**67** for catalysts **59a-c, e** calculated at B3LYP/6-31G(d) or combined B3LYP-D/6-311+G(d,p)//B3LYP/6-31G(d) levels can be used for the rationalization of experimentally measured enantioselectivities in KR experiments.

*5.3.4 The selectivity prediction for catalysts **59d**, **59f** and **59g***

Several derivatives of catalyst **59a** with different substituents in the phenyl ring have been suggested as potential catalysts for the KR of *sec*-alcohols (Scheme 5.3). In order to study whether the method described above would allow prediction of the selectivity for these new catalysts, we have calculated the corresponding enthalpy differences $\Delta H(S\text{-}R)$ between diastereomeric TSs **67**. In order to save computational time it would be desirable not to carry out the full conformational search for new TSs **67**, but use as a basis the conformations obtained for TS **67** with the parent catalyst **59a**. The variation of the catalyst part is assumed not to change dramatically the conformational space of TS **67**. First, the conformational search for TS **67** with catalyst **59d** was carried out. The catalyst part in the most stable conformations of TS **67** with catalysts **59a** of different types **I-IV** (Scheme 5.2) was modified to catalyst **59d** and the TSs were reoptimized at B3LYP/6-31G(d) level. The most stable conformations of TS **67d** were then used for the thermal corrections and single point calculations (the latter at B3LYP/6-311+G(d,p) and B3LYP-D/6-311+G(d,p) levels). Figure 5.12 shows a pictorial representation of the relative energies of the conformers of *(R)*-**67** and *(S)*-**67** with catalyst **59d**. The type *(R)*-**I** represents the most stable conformers of *(R)*-**67d**, as was observed for the catalyst **59a**. The most stable conformers of TS *(S)*-**67d** belong to types *(S)*-**II** and *(S)*-**III**, but in contrast to TS *(S)*-**67a** type *(S)*-**II** becomes now more stable due to the unfavorable interactions between naphthyl ring and 3,5-dimethylphenyl group in the type *(S)*-**III** transition states *(S)*-**67d** (compare Figures 5.9 and 5.12).

In order to save computational time it would be attractive to use only the best conformations to calculate enthalpy differences between diastereomeric TSs **67** for new catalysts instead of using Boltzmann-averaged values over all conformers. For this reason we studied in more detail the conformational space of TS **67** for catalyst **59d** with the goal to answer the question: how many conformations are necessary for the accurate prediction of enthalpy difference and thus the enantioselectivity?

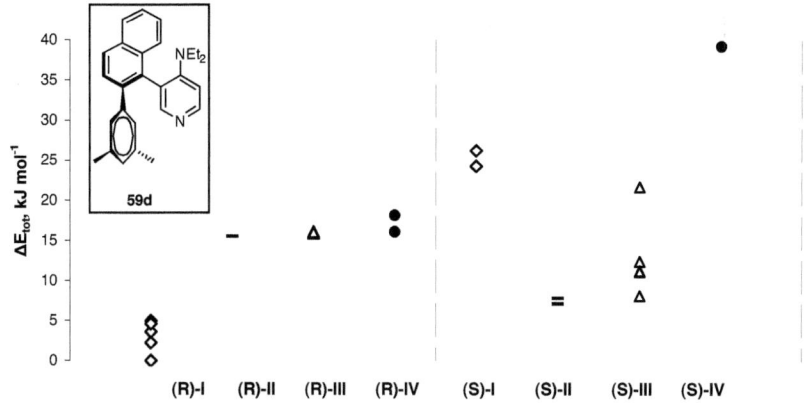

Figure 5.12. Relative energies (kJ mol^{-1}) of TS **67** conformers for catalyst **59d**, as calculated at B3LYP/6-31G(d) level.

The dependence of the calculated enthalpy difference on the number of conformers was studied. Figure 5.13 shows that at least five best conformers are required for Boltzmann-weighted averaging in order to get more accurate enthalpy values for TS *(R)-* and *(S)-*67d. Addition of the sixth conformer does not change the averaged enthalpy significantly. The enthalpy of the best conformer is by 1.4–1.7 kJ mol^{-1} lower than the averaged value, for that reason using only the best conformer is not accurate enough for the enthalpy difference calculation. The number of **(R)-67d** conformers used for the enthalpy averaging, when taking all conformers of **(S)-67d** for averaging, has a large effect on the enthalpy difference. Averaging over at least five conformers is necessary to get more accurate result. The same conclusion is also reliable for the averaging of TS **(S)-67d**. Since the error in the case of **(R)-67d** is positive and in the case of **(S)-67d** negative, these errors can cancel each other if the same numbers of conformers of TS **(R)-** and **(S)-67d** are used for the Boltzmann averaging. Indeed the analysis shows (Figure 5.13), that even averaging over the three most stable conformations gives the total error in the enthalpy difference of only 0.1 kJ mol^{-1}. This error is acceptable for the selectivity prediction. In the case of free energy differences the error of using only the three best conformations for averaging is larger (up to 0.8 kJ mol^{-1}).

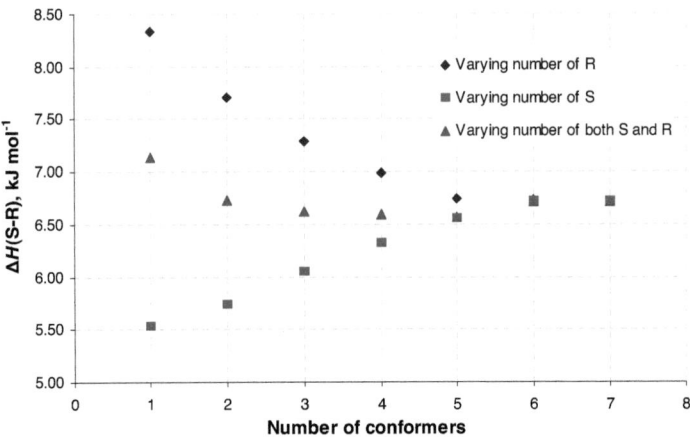

Figure 5.13. Dependence of the enthalpy difference $\Delta H_{298}(S\text{-}R)$ between **(R)**- and **(S)**-**67d** on the number of conformers for Boltzmann averaging as calculated at B3LYP/6-31G(d) level.

In conclusion, the enthalpy differences $\Delta H_{298}(S\text{-}R)$ between diastereomeric transition states **(R)**- and **(S)**-**67**, calculated at combined B3LYP-D/6-311+G(d,p)//B3LYP/6-31G(d) or B3LYP/6-31G(d) levels, can potentially be used for the prediction of the stereoselectivity for new derivatives of catalyst **59a**. Averaging over only the three most stable conformations gives accurate enough enthalpy differences. Taking the best conformations of TS **67** with catalyst **59a**, modifying the catalyst part and reoptimization of transition states at B3LYP/6-31G(d) level would save time for the conformational space search. Analysis of Figures 5.8 and 5.11 shows, that for TSs **(R)**-**67** taking into account type **(R)**-**I** conformations, and for **(S)**-**67** – both types **(S)**-**II** and **(S)**-**III** is necessary to find the most stable conformations of TSs **67**.

This method was subsequently used to calculate enthalpy differences $\Delta H_{298}(S\text{-}R)$ for other catalysts **59f** and **59g**, shown in Scheme 5.3 (Table 5.7). The obtained results show that the catalysts **59f** and **59g** are expected to show moderate selectivity, while the derivative **59d** should be more selective than the parent catalyst **59a**.

5.3.5 Synthesis and selectivity measurements for catalysts 59d, 59f and 59g

The new derivatives **59d**, **59f** and **59g** were synthesized in the Spivey group[74] by employing Suzuki-Miyaura cross-coupling of aryl boronic acids using Buchwald's S-Phos ligand **71**, which has recently been reported to efficiently cross couple sterically challenging coupling partners (Scheme 5.4).[78] The synthesis of precursor **70** has already been described by Spivey et al.[73]

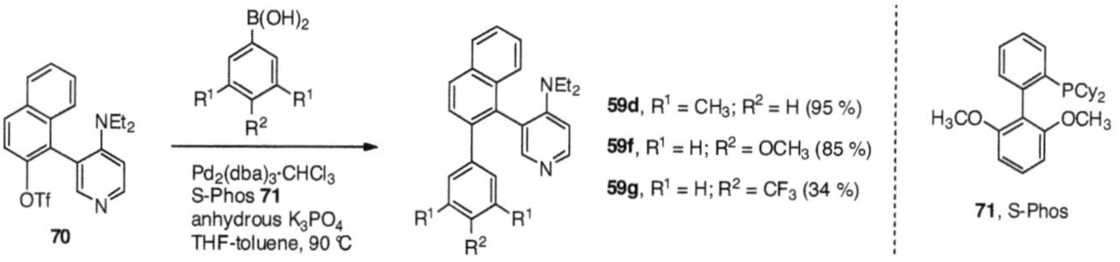

Scheme 5.4. Synthesis of **59a** analogues **59d**, **59f** and **59g** by Suzuki-Miyaura cross-coupling.

Compounds **59d-g** obtained as racemic mixtures were then resolved using a semi-preparative CSP-HPLC to obtain enantiomers of each catalyst with >99.9% ee purity. The new derivatives were then tested in the KR of alcohol **60** by isobutyric anhydride (Table 5.8).[74]

Table 5.8. KR of alcohol **60** by isobutyrylation catalyzed by **59a**, **59d**, **59f** and **59g**.[74]

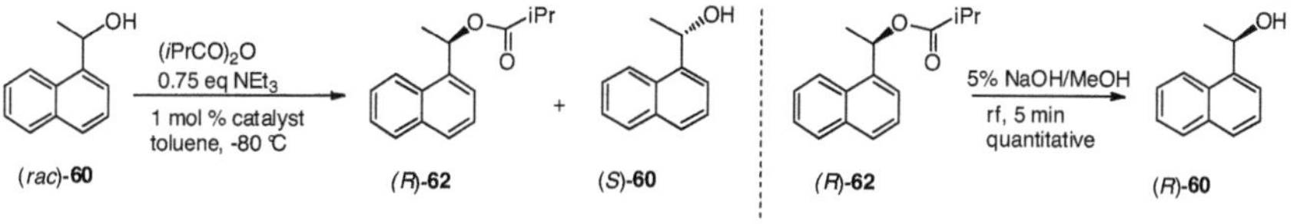

Entry	Catalyst (>99.9% ee)	Equiv. of anhydride	Time (h)	ee_A (%)[c]	ee_E (%)[d]	C (%)[a]	s[b]
1	(-)-**59a**	1.0	2	12.0	87.4	12.1	16.7
2	(-)-**59a**	1.0	8	30.5	84.3	26.6	15.7
3	(-)-**59d**	1.0	2	16.8	92.3	15.4	29.3
4	(-)-**59d**	1.0	8	55.2	87.8	38.6	26.7
5	(-)-**59d**	2.0	2	45.8	88.3	34.1	25.4
6	(-)-**59d**	2.0	8	98.2	75.1	56.7	32.0
7	(-)-**59f**	1.0	8	26.5	78.4	25.3	10.6
8	(-)-**59g**	1.0	8	7.0	77.9	8.3	8.6

[a] Conversion $C = 100*ee_A/(ee_A+ee_E)$.
[b] Selectivity factor s was calculated as described in ref. 54.
[c] ee of recovered alcohol **60**, established by CSP-HPLC.
[d] ee of ester **62**, established by CSP-HPLC on derived alcohol **60** following saponification by NaOH in MeOH.

Analysis of the obtained results shows that catalyst **59d** is indeed more selective than the parent derivative **59a**. This is in accordance with computational predictions. Catalyst **59d** is also more catalytically active than **59a**, giving 39 % conversion after 8 h (Entry 4). The *para*-

substituted derivatives **59f** and **59g** are less selective in accordance with theoretical predictions. The substantial decrease in the rate of reaction observed for catalyst **59g** may be attributed to the strongly electron-withdrawing 4-CF_3 group decreasing the electron-density on the pyridine nitrogen inductively thereby rendering it less nucleophilic for acyl-transfer. The anomalous increase in selectivity with increased conversion ($s = 32$ after 8 h cf $s = 25$ after 2 h) observed for catalyst **59d** cannot be explained at the moment, although there are isolated reports in the literature of the conversion-dependent selectivity.[75]

5.3.6 Comparison with theoretical predictions

With the experimentally measured selectivities s for selected derivatives of catalyst **59a** in hand, it is possible to quantify the predictive value of the theoretically calculated enthalpy differences $\Delta H_{298}(S\text{-}R)$ (Table 5.7). The enthalpy differences $\Delta H_{298}(S\text{-}R)$ calculated at B3LYP/6-31G(d) level can be correlated with experimental enantioselectivities (correlation coefficient: $R^2 = 0.8599$ for catalysts **59a, d, f, g**). Figure 5.14 shows that the derivatives of catalyst **59a**, substituted in the phenyl ring, have a larger slope of the correlation line comparing with the catalysts **59a-c, e**, which have different dialkylamino substituents. The thermal corrections recalculated at 195 K do not improve the correlation between experimental results and calculated enthalpies or free energies. Inclusion of dispersion corrections (at B3LYP-D/6-311+G(d,p) level) improves the overall correlation of experimental selectivities with calculated enthalpy differences for all studied catalysts **59a-g** (Table 5.7).

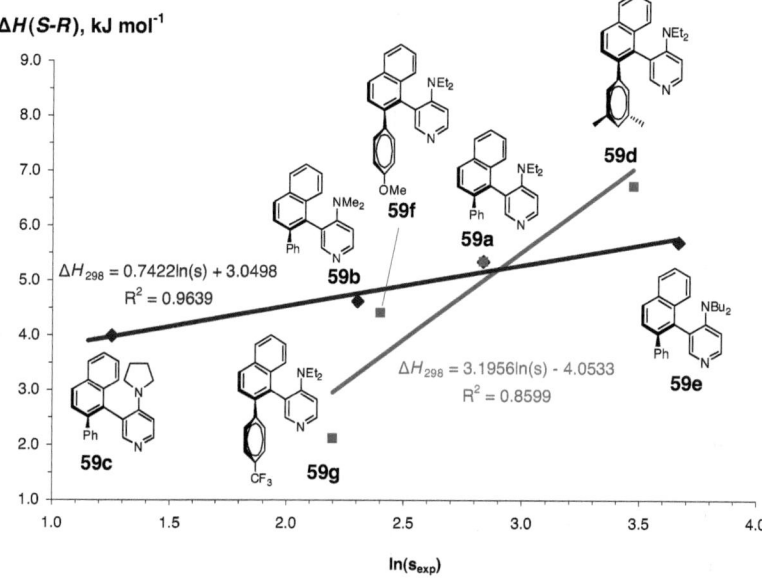

Figure 5.14. Correlation between experimental enantioselectivities and calculated enthalpy differences ΔH_{298} between TSs *(R)*- and *(S)*-**67**, as calculated at B3LYP/6-31G(d) level.

After independent optimization of the substitution pattern of the dialkylamino and the phenyl groups in terms of selectivity, even more selective catalysts can be designed by the combination of the best substitution patterns. Indeed, the combination of dibutylamino group (from **59d**) and 3,5-dimethylphenyl (from **59e**) in one molecule of **59h** leads to the most selective analogue of Spivey's catalyst **59a** (Figure 5.15).[74]

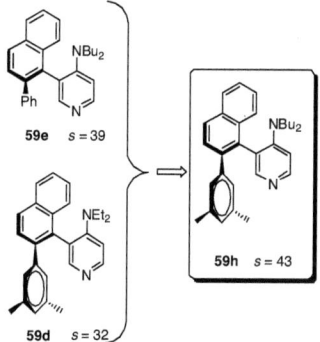

Figure 5.15. Design of the new highly selective catalyst **59h**.

5.4 Estimating the stereoinductive potential of the pyridines

In the previous section the computations of the reaction profile have been successfully applied for the rationalization and prediction of the selectivity for the series of chiral DMAP derivatives in the KR experiments. However, it would be more desirable to use less time-consuming computational models for the design of new chiral catalysts. Here we present two attempts of developing such models: a prochiral probe approach and the conformational analysis of transition states. First, several chiral 3,4-diaminopyridine derivatives were synthesized and tested in the KR of alcohol **60**.

5.4.1 Chiral 3,4-diaminopyridine derivatives

Catalyst (*rac*)-**5b** was resolved to enantiomers via cocrystallization with a chiral resolving agent. Several chiral acids such as *L*-tartaric, (–)-*O,O'*-dibenzoyl-*L*-tartaric and *D*-camphorsulfonic acids as well as different solvent mixtures were tested. The most effective resolution was achieved by using *L*-tartaric acid (1 eq.) in ethanol/ethyl acetate mixture (1:1) (Scheme 5.5). After two recrystallizations ee of catalyst **5b** was 98% (as determined by CSP-HPLC). On the basis of X-ray structure of **5b-salt**, the configuration of the enriched enantiomer relative to the configuration of *L*-tartaric acid was determined (Scheme 5.5).

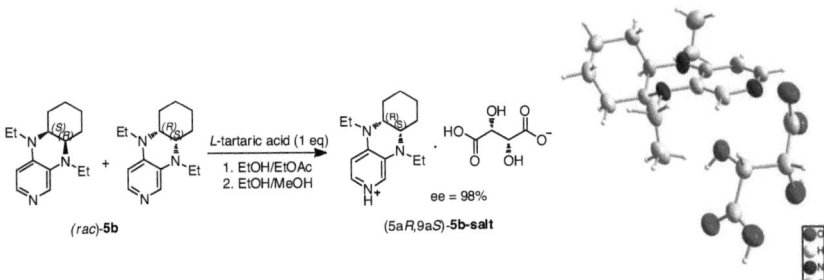

Scheme 5.5. Classical resolution of the catalyst (*rac*)-**5b** and X-ray structure of **5b-salt**.

The enantioenriched catalyst **5b**, as well as the camphor derivative **5l** (as a single diastereomer with >99% de, see Chapter 2), were then tested in the KR of alcohol **60** (Scheme 5.6). Since both 3,4-diaminopyridine derivatives are very catalytically active in acylation reaction (see Chapter 2), low loadings of catalysts (0.5 mol%) have been used for the KR of *sec*-alcohol **60**. However, very low levels of selectivity were measured for these catalysts (Scheme 5.6).

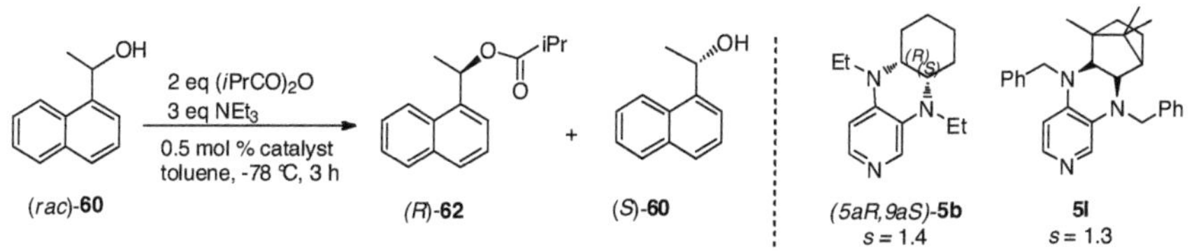

Chapter 5

Scheme 5.6. Kinetic resolution of alcohol **60** catalyzed by **5b** and **5l**.

5.4.2 Prochiral probe approach

As mentioned in Chapter 2, methyl cation affinities (MCA) can serve as the most simple model for a carbon basicity scale.[25b] The MCA values for a variety of N- and P- centered bases were shown to be correlated better with the experimentally observed catalytic efficiencies than proton affinities.[25c] Later on, this approach was extended to include affinity values towards a prochiral cation, formally derived from α-methoxy-α-trifluoromethyl-α-phenylacetic acid (MTPA, Mosher's acid, $Ph(OCH_3)(CF_3)C-CO_2H$) through decarboxylation.[80] The success of this latter acid as a derivatizing reagent for a wide range of chiral alcohols and amines suggests that the three substituents connected to C_2 (Ph, CF_3, OCH_3) provide a strongly differentiated environment in steric and electronic terms.[79] In order to emphasize the resemblance to Mosher's acid this cation is referred as "Mosher's cation" (or **MOSC**) and the corresponding reaction enthalpies at 298 K as "MOSCA" values.[80] As described in Scheme 5.7, reaction of **MOSC** with chiral nucleophiles can occur from the *re* or *si* face of the cation, leading to two diastereomeric adducts with two different affinity values MOSCA*re* and MOSCA*si*.

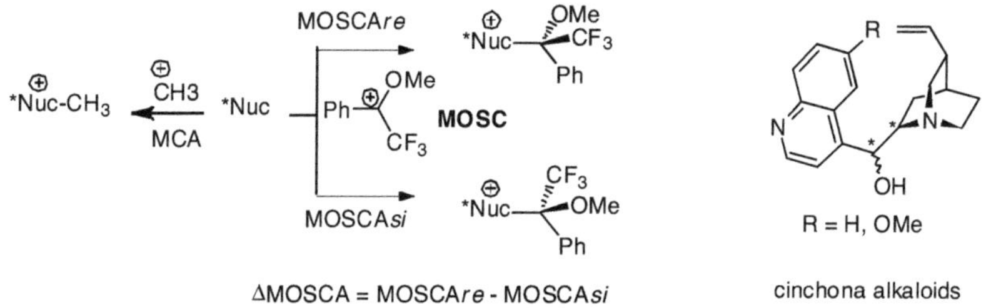

Scheme 5.7. MCA, MOSCA and ΔMOSCA values definition and general structure of cinchona alkaloids. Chiral centers are marked with star *.

This approach has successfully been applied for the rationalization of the activity (on the basis of MOSCA values) and stereoinductive potential (on the basis of ΔMOSCA values) for a number of cinchona alkaloids.[80] Here we would like to extend the prochiral probe approach to the estimation of the stereoinductive potential of chiral DMAP derivatives.

The MOSCA values for a series of substituted pyridines have been calculated at the MP2(FC)/6-31+G(2d,p)//B98/6-31G(d) level of theory (Table 5.9). The differences between *re* and *si* attack, ΔMOSCA values, calculated at the same level of theory in the gas phase and in chloroform are also listed in Table 5.9.

Table 5.9. MOSCA and ΔMOSCA values (in kJ mol^{-1}) for systems **5b, 5g, 5l** and **59a**.

Catalyst	Selectivity s^b	MOSCA *si* "MP2-5"[a]	MOSCA *re* "MP2-5"[a]	ΔMOSCA *re – si* "B98"[a]	ΔMOSCA *re – si* "MP2-5"[a]	ΔMOSCA *re – si* "MP2-5/solv"[a]
5b	1.3	312.21	314.08	1.30	1.87	1.83
5l	1.1	324.83	320.46	-1.06	-4.36	-6.62
59a	16.7	299.02	300.08	-0.25	1.07	1.94
5g		336.95	335.92	-2.88	-1.04	-0.14

[a] Levels of theory: "B98": B98/6-31G(d); "MP2-5": MP2(FC)/6-31+G(2d,p)//B98/6-31G(d); "MP2-5/solv": MP2(FC)/6-31+G(2d,p)//B98/6-31G(d) with PCM/UAHF/RHF/6-31G(d) solvation energies.
[b] Selectivity *s* in the acylative KR of alcohol **60** (see Scheme 5.6).

Analysis of the obtained data reveals that the MOSCA values for the studied pyridines are higher than for cinchona alkaloids (a typical range for the latter is 180–230 kJ mol^{-1}), indicating that these derivatives are more nucleophilic. The 3,4-diaminopyridine derivatives **5b**, **5l** and **5g** have higher MOSCA values than the Spivey's catalyst **59a**, in accordance with experimentally observed higher catalytic activity of these derivatives. The differences between *re* and *si* attack, ΔMOSCA values, are for studied pyridines generally lower than for cinchona alkaloids (2–9 kJ mol^{-1} for latter). It may be attributed to the larger distance between the chirality element and the nucleophilic center in DMAP derivatives as compared to cinchona alkaloids. The largest ΔMOSCA value, obtained for the camphor derivative **5l** (–4.4 kJ mol^{-1}), is inconsistent with the lower selectivity of this catalyst in KR of alcohols (s = 1.3). As can be seen from Table 5.9, generally ΔMOSCA values fail to correlate with experimental selectivities. Addition of the solvent effects at PCM/UAHF/RHF/6-31G(d) level does not improve the correlation. Notably, the quasienantiomeric derivatives **5b** and **5l** (in cyclohexane ring) have opposite signs of the MOSCA values.

5.4.3 Conformational analysis of transition states

Conformational preferences of the transition states have been shown to play an important role for the enantiodiscrimination in the KR of alcohols catalyzed by derivatives of the Spivey's catalyst **59a**. Classification of the selectivity-determining TSs into four types (see Scheme 5.2) was very helpful for the conformational search, since alcohols with different configuration prefer different directions to approach the reaction center: namely type **I** for (*R*)-alcohol and types **II** and **III** for (*S*)-alcohol are more relevant than other types (*cf* Figure 5.9). We envisioned that similar preferences could persist by using achiral alcohol, such as *tert*-butanol employed in Chapter 2 for the TS calculations. Conformational analysis of transition states presented here would then be a very practical approach since it would allow prediction both of catalytic activity and stereoselectivity of the chiral pyridine derivatives.

Through analysis of the optimized geometries of transition states of the acetylation catalyzed by chiral 3,4-diaminopyridines, we found that all conformers can be classified into the four structural types as shown in Scheme 5.8 (*cf* Scheme 5.2). Generally speaking, the carboxylate group is bonded to the left or right side of the pyridine ring by weak hydrogen bonding and the alcohol approaches the reaction centre either from the front face or the back face of the pyridine ring. Figure 5.16 shows a pictorial representation of the relative enthalpies of the conformers of transition states with the parent derivative **5b** (in "frozen TS" and "optimized TS" models).

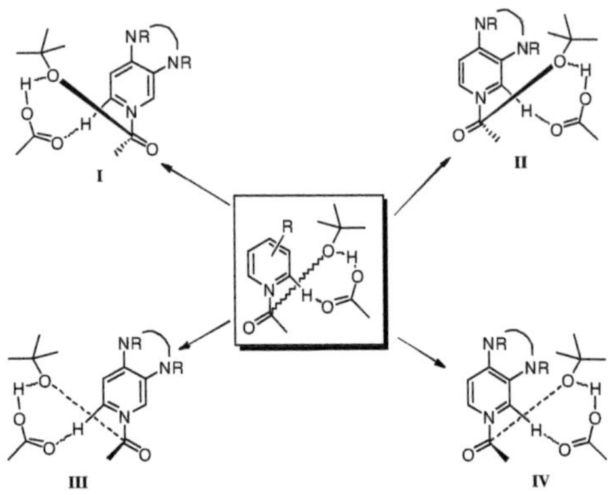

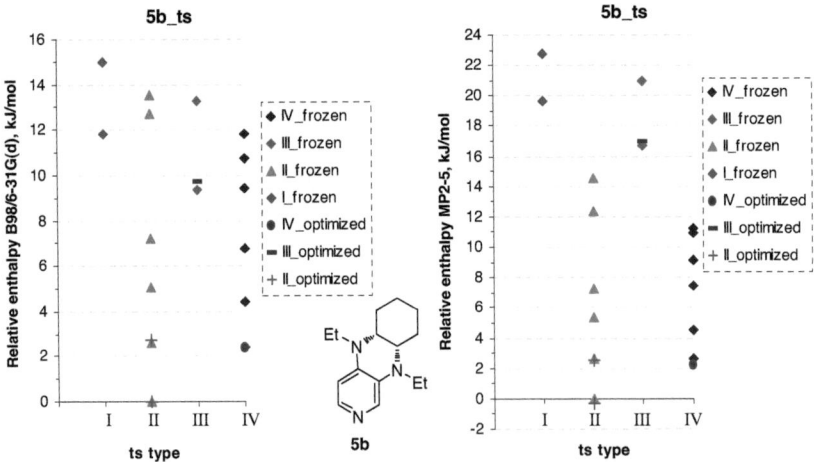

Scheme 5.8. The classified conformer types of transition states with 3,4-diaminopyridines.

As mentioned in Chapter 2, the transition states of types **II** and **IV** with acetate pointing towards the 3N-substituent are systematically more stable than TSs of types **I** and **III**. Analysis of Figures 5.16 and 5.17 reveals, that "frozen" and "optimized" TSs have very close relative enthalpies. Since the conformational space of "frozen" TSs has been studied more extensively than "optimized" TSs, only the "frozen" TSs enthalpies will be discussed in following. The plots obtained at B98 and MP2-5 levels are similar, except a larger spread of relative enthalpies calculated at MP2-5 level, probably due to the overestimation of the dispersion interactions by the MP2 method.[65]

Figure 5.16. Relative enthalpies (in kJ mol^{-1}) of conformers of TSs with catalyst **5b**, as calculated at B98/6-31G(d) (left) and at MP2(FC)/6-31+G(2d,p)//B98/6-31G(d) level (right).

The enthalpy difference between TSs of types **II** and **IV**, which is chosen as a guideline for the design of stereoselective catalysts, is relatively low for the studied pyridines: 2.5 kJ mol^{-1} for catalyst **5b** and 4.3 kJ mol^{-1} for dibenzyl substituted derivative **5k** (as calculated at MP2(FC)/6-31+G(2d,p)//B98/6-31G(d) level). The enthalpy differences calculated at B98/6-31G(d) level are even lower (Figure 5.17). It is in accordance with poor selectivity of the catalyst **5b** in the KR of alcohols.

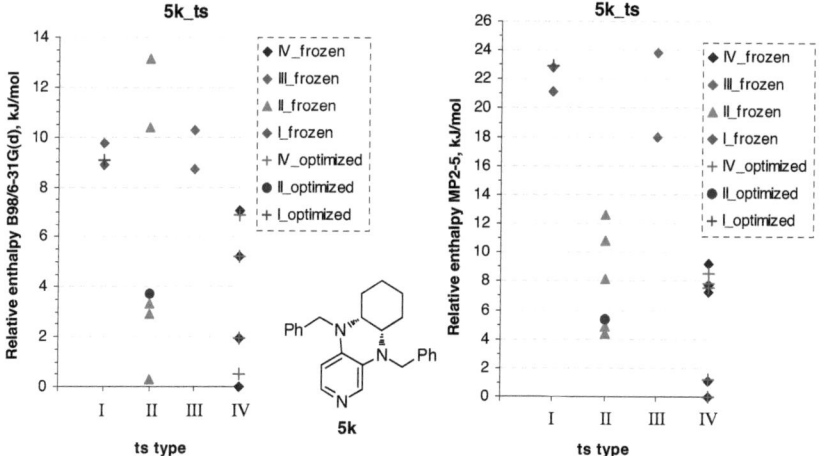

Figure 5.17. Relative enthalpies (in kJ mol^{-1}) of conformers of TSs with catalyst **5k**, as calculated at B98/6-31G(d) (left) and at MP2(FC)/6-31+G(2d,p)//B98/6-31G(d) level (right).

The enthalpy difference between TSs of types **II** and **IV** for the camphor derivative **5l** is also low (2.5 kJ mol^{-1}), when calculated at B98/6-31G(d) level, but becomes slightly higher (5.5 kJ mol^{-1}) and changes the sign, when calculated at MP2(FC)/6-31+G(2d,p)//B98/6-31G(d) level (Figure 5.18). This catalyst also displays poor selectivity in the KR of *sec*-alcohol **60**. The triazolyl derivative **5j** also has low enthalpy difference between **II** and **IV** (see Figure A5.3 in the Appendix).

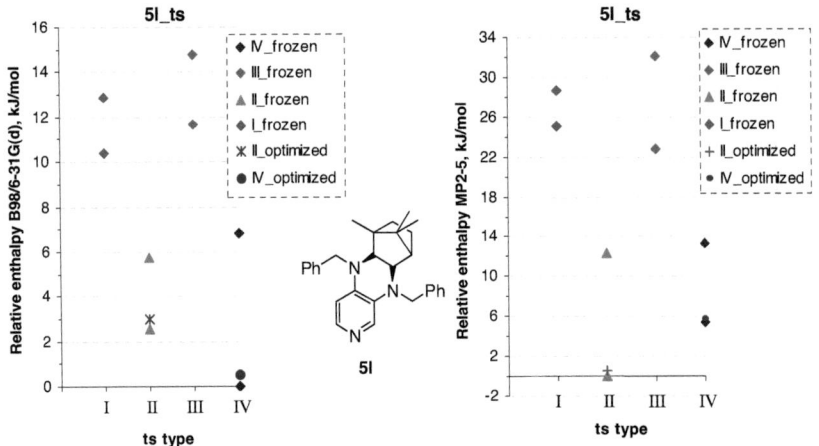

Figure 5.18. Relative enthalpies (in kJ mol^{-1}) of conformers of TSs with catalyst **5l**, as calculated at B98/6-31G(d) (left) and at MP2(FC)/6-31+G(2d,p)//B98/6-31G(d) level (right).

For the design of new chiral catalysts, the camphor derivative **5l** has been chosen as a general framework for several reasons: 1) it has a rigid structure, which narrows its conformational space; 2) the enthalpy difference between TSs of types **II** and **IV** is the largest among the studied 3,4-diaminopyridines (5.5 kJ mol^{-1}); 3) the camphor derivatives can be obtained as single diastereomers, whereas derivatives of the catalyst **5b** are obtained as racemates and further resolution to enantiomers is required.

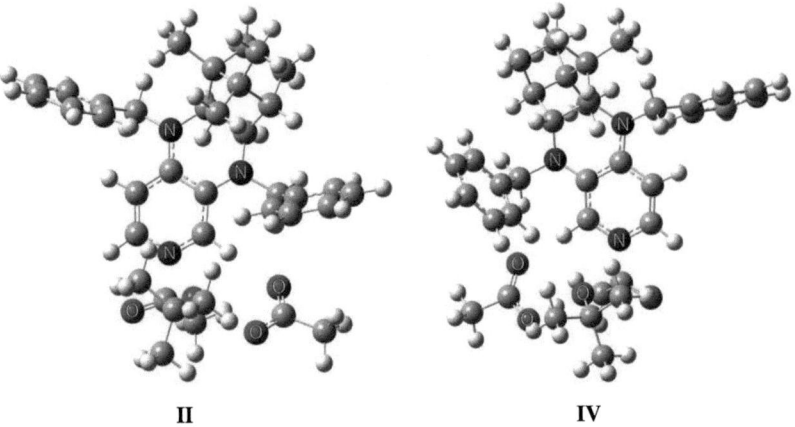

Figure 5.19. Structures of the most stable conformers of TSs with catalyst **5l**.

Through analysis of the best TSs conformers for catalyst **5l** (Figure 5.19), we envisioned that the introduction of the *meta*-substituents in the phenyl group in 3N-position would shield one of the pyridine ring sides, what can lead to the enhanced stereoselectivity in KR of alcohols. Since the synthetic route to **5l** allows introduction of the *meta*-substituents into both phenyl groups at the same time, two catalysts **5l-Me** and **5l-Ph** with *meta*-methyl and phenyl groups, respectively, were proposed. The relative acetylation and activation enthalpies for these derivatives, as well as for parent catalysts **5b** and **5l**, are collected in Table 5.10.

Table 5.10. Relative acetylation ΔH_{ac} and activation ΔH_{act} enthalpies (in kJ mol^{-1}) calculated in the gas phase and in chloroform for chiral 3,4-diaminopyridines.

Catalyst	$t_{1/2}$ [min] [b]	ΔH_{ac} [kJ mol^{-1}] [a]		ΔH_{act} [kJ mol^{-1}] [a]	
		gas	CHCl$_3$	gas	CHCl$_3$
5b	18	-119.6	-85.2	-59.17	-49.99
5l	138	-123.0	-75.4	-62.88	-41.41
5l-Me		-131.6	-77.8	-63.65	-38.31
5l-Ph		-114.3	-69.3	-36.94	-33.60

[a] Levels of theory: gas phase: MP2(FC)/6-31+G(2d,p)//B98/6-31G(d);
CHCl$_3$: MP2(FC)/6-31+G(2d,p)//B98/6-31G(d) with PCM/UAHF/RHF/6-31G(d) solvation energies.
[b] Reaction half-lives of the benchmark reaction (see Figure 2.7).

Inspection of the obtained data reveals that both proposed derivatives should still be quite catalytically active in acylation reactions. The phenyl substituted derivative **5l-Ph** is predicted to be less active than the parent **5l**. The methyl substituted **5l-Me** has less negative activation enthalpy than the catalyst **5l**, when solvation effects in chloroform are included, but lower acetylation and activation enthalpies in the gas phase. Therefore this derivative would have comparable catalytic activity in chloroform, and probably more effective than **5l** in less polar solvents like toluene. Analysis of the enthalpy differences between TSs of types **II** and **IV** for both derivatives (Figure 5.20) reveals that calculated at B98/6-31G(d) enthalpy differences are lower than for parent catalyst **5l** (*cf* Figure 5.18), whereas the values calculated at MP2(FC)/6-31+G(2d,p)//B98/6-31G(d) level are slightly higher (6 kJ mol^{-1} for **5l-Me** and 9 kJ mol^{-1} for **5l-Ph**).

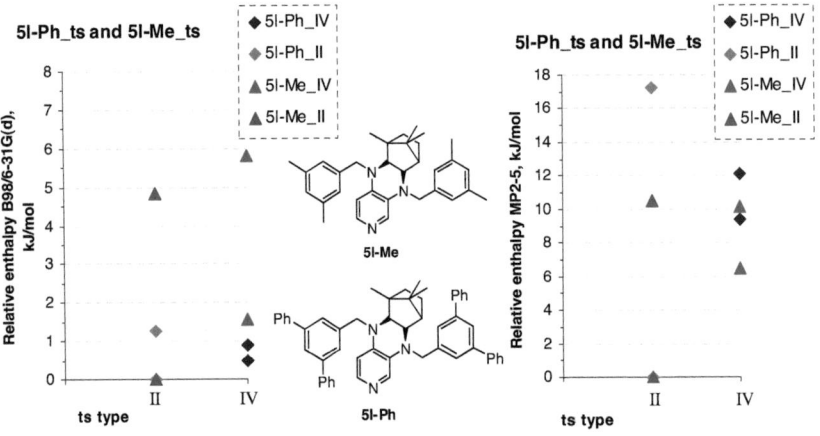

Figure 5.20. Relative enthalpies (in kJ mol^{-1}) of conformers of TSs with catalysts **5l-Me** and **5l-Ph**, as calculated at B98/6-31G(d) (left) and at MP2(FC)/6-31+G(2d,p)//B98/6-31G(d) level (right) (model "optimized transition states").

On the basis of computational results the methyl substituted derivative **5l-Me** was then chosen as potentially more active and selective catalyst. The synthesis of this derivative was straightforwardly carried out by employing 3,5-dimethylbenzoyl chloride as acylation reagent in the first step (Scheme 5.9), which was synthesized from the commercially available 3,5-dimethylbenzoic acid. Subsequent reduction of the amide groups by LiAlH$_4$/AlCl$_3$ gives the final product **5l-Me** in 40 % yield.

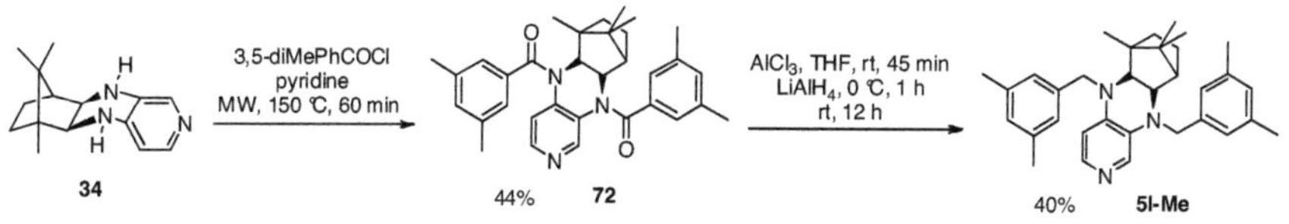

Scheme 5.9. Synthesis of the catalyst **5l-Me**.

The new derivative **5l-Me** was then tested in the benchmark acetylation reaction (Scheme 5.10), whereas the reaction proceeded to full conversion. The catalyst **5l-Me** was surprisingly two times more active than the parent camphor derivative **5l**. This is in contrast to less negative activation enthalpy for **5l-Me**, but in accordance with the slightly higher stability of the acetylpyridinium cation (Table 5.10). The KR of *sec*-alcohol **60**, carried out with 0.5 mol% of catalyst **5l-Me**, displays slightly higher level of selectivity ($s = 1.7$) than with the parent derivative **5l** (Scheme 5.10). This implies that the proposed approach based on TS conformational analysis can be used for the design of new chiral catalysts. However, further optimization of the substitution pattern of **5l** derivatives is necessary in order to get more selective catalyst.

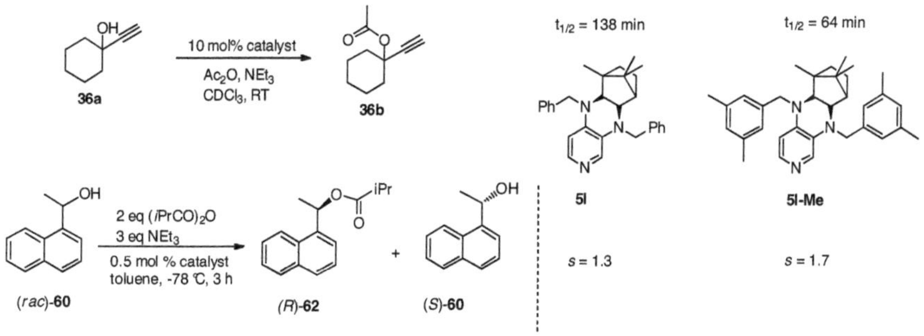

Scheme 5.10. Benchmark kinetics of alcohol **36a** acetylation and kinetic resolution of alcohol **60** catalyzed by **5l** and **5l-Me**.

5.5 Conclusions

Similar to acylation reactions with achiral catalysts (**PPY**), the commonly accepted nucleophilic mechanism is more favorable than the general base mechanism for the reaction of 1-(1-naphthyl)ethanol (**60**) with isobutyric anhydride, catalysed by chiral catalyst **59a**. The identified TS models in the selectivity-determining step can be classified into four types and reveal that alcohols with different configuration prefer different directions to approach the reaction center. The key TS model is applied to the selectivity rationalization and is helpful for the catalyst design, thus selectivities of several new catalysts are predicted theoretically and then examined experimentally, showing good agreement with computations. Less time-consuming models for selectivity prediction, based on prochiral probe approach and TSs conformational analysis, are also attempted.

Conclusions

Summary and General Conclusions

(1) A theoretical and experimental study of the structure-reactivity relationship of alcohol acetylation catalyzed by pyridine derivatives has been performed. Several 3,4-diaminopyridines were synthesized and their catalytic activity was elucidated in the benchmark acetylation reaction. The performance of the ground state and transition state models for the catalytic activity prediction of the large variety of pyridine catalysts has been analyzed, based on the correlation with relative acetylation rates. Using the combined MP2(FC)/6-31+G(2d,p)//B98/6-31G(d) level with inclusion of solvent effects at PCM/UAHF/RHF/6-31G(d) level for the calculation of relative acetylation enthalpies gives systematically better correlation with the catalytic activities than at the B3LYP level of theory (Scheme 6.1b). Even though the quality of the correlation is moderate, separate correlations of better fidelity exist for each of the catalyst families.

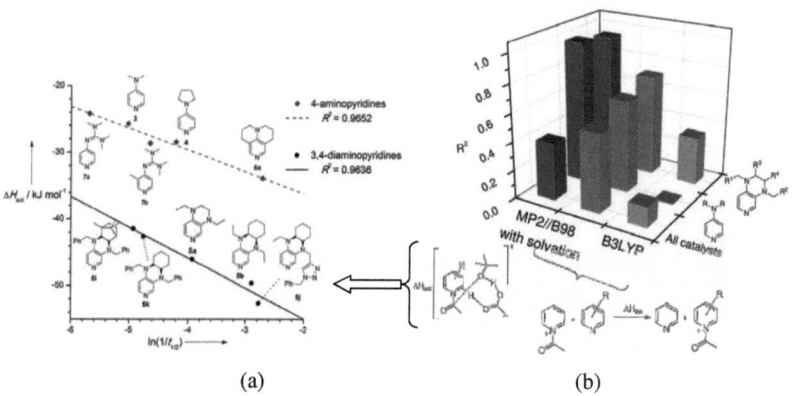

Scheme 6.1. (a) Correlation between relative activation enthalpies and relative reaction rates. (b) Comparison of different models for the prediction of catalytic activity.

Activation enthalpies (calculated at the same level of theory) give much better correlation with relative acylation rates ($R^2 = 0.96$ in each family) and can be used for the precise prediction of catalyst activity (Scheme 6.1a). The variation of the solvation model has a small influence on the overall correlation, but worsens the correlation in each catalysts family.

(2) The ground state model, *i.e.* acetylation enthalpy calculations, has been applied to the design of photoswitchable pyridines as well as planar-chiral aminopyridine derivatives, containing paracyclophane or ferrocenyl substituents. The model systems **diaza1** and **diaza2**

Conclusions

containing different *para*-substituents X were studied computationally (Scheme 6.2a). These systems were found to be potentially photoswitchable, since they have different acetylation enthalpies in the *cis* and *trans* states. The largest effects on the acetylation enthalpies were observed for the *p*-cyano substituted derivatives, which have electrostatic interactions between CN and COCH$_3$ groups in acetylated *cis*-derivatives (Scheme 6.2b). However, these effects are mainly electrostatic in nature. In order to further increase these effects, the azobenzene moiety should be directly introduced into the 5-position of the pyridine ring (Scheme 6.2c).

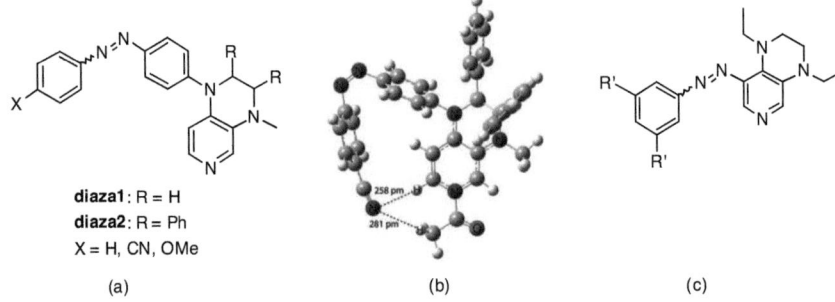

Scheme 6.2. The studied photoswitchable pyridines **diaza1** and **diaza2** (a) with the structure of the most stable conformer of the acetylated catalyst *cis*-*p*-cyano-**diaza2** (b). The suggested 3,4-diamino-5-azobenzene pyridine derivatives (c).

The relative acetylation enthalpies ΔH_{ac} for a series of 3-paracyclophane-4-aminopyridines show that the paracyclophane substituent decreases the relative stability of the acetylpyridinium cation (Scheme 6.3a), whereas the amide group in the pseudo-*ortho* position significantly lowers the acetylation enthalpy. Analysis of the relative acetylation enthalpies ΔH_{ac} for a series of ferrocenylpyridines shows, that the 2'-sulfoxidoferrocenyl group is a weak electron-donating substituent (Scheme 6.3b). More potent catalysts, which contain a 2'-sulfoxidoferrocenyl group in *meta* and dialkylamino group in *para* positions, have much lower acetylation enthalpies and are expected to be highly active catalysts in acylation reactions. In summary, the studied planar chiral catalysts are predicted to be active enough to catalyze the KR of alcohols. The additional electrostatic interactions between the sulfoxide oxygen and the neighbouring hydrogens observed for the acetylated species can be advantageous for the chiral recognition of alcohol enantiomers (Scheme 6.3c).

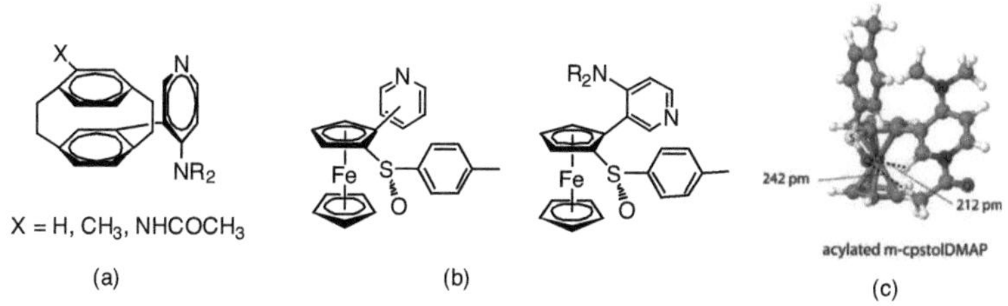

Conclusions

Scheme 6.3. Studied planar-chiral aminopyridine derivatives, containing paracyclophane (a) or ferrocenyl (b) substituents. Conformational analysis shows electrostatic interactions between the sulfoxide oxygen and the neighboring hydrogens (c).

(3) The design of a new class of acylation catalysts, (4-aminopyridin-3-yl)-(thio)ureas, has been described in Chapter 4. Achiral (4-aminopyridin-3-yl)-(thio)ureas were shown to have a moderate activity in the alcohol acylation. The obtained data show that the derivatives with a cyclohexane bridge are generally more catalytically active than the catalysts with an ethylene bridge. Derivatives containing the 3,5-bis-(trifluoromethyl)phenyl group in the (thio)urea moiety (Scheme 6.4a) are the most active catalysts among the new systems. The enhanced activity has been explained by the increased acidity of the NH hydrogen of the (thio)urea group due to the electron withdrawing character of the 3,5-bis-(trifluoromethyl)phenyl substituent. The X-ray analysis and concentration dependent NMR measurements indicate a hydrogen bonding interaction between the NH hydrogen and the pyridine nitrogen. However, catalyst aggregation does not influence the reaction rates of the catalyzed acetylation of alcohols.

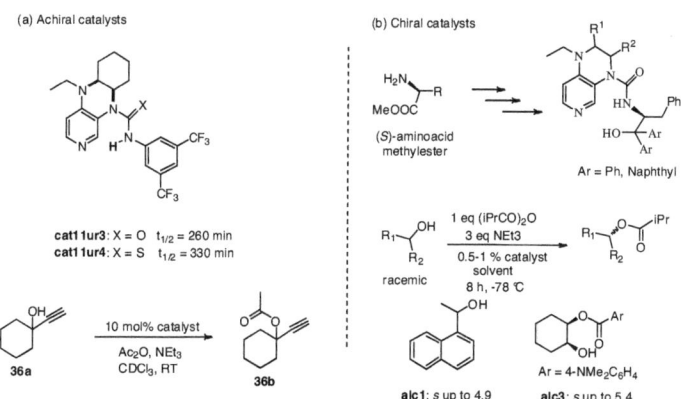

Scheme 6.4. (a) The most catalytically active achiral (4-aminopyridin-3-yl)-(thio)ureas. (b) Application of the chiral derivatives to the KR of *sec*-alcohols.

Conclusions

The chiral (4-aminopyridin-3-yl)-ureas have been prepared via a modular strategy from easily accessible amino acids (Scheme 6.4b). The potential of newly synthesized chiral derivatives was explored in the kinetic resolution (KR) of several secondary alcohols. The best selectivities were obtained with phenylalanine-derived catalysts containing a diarylcarbinol group. Even though the selectivity values are moderate, the modular design allows variation of the urea substituents for further catalyst improvement.

(4) Similar to acylation reactions with achiral catalysts (**PPY**), the commonly accepted nucleophilic mechanism is more favorable than the general base mechanism for the reaction of 1-(1-naphthyl)ethanol with isobutyric anhydride, catalyzed by the chiral catalyst **59a** (Scheme 6.5a). The identified TS models in the selectivity-determining step can be classified into four types and reveal that alcohols with different configuration prefer different directions to approach the reaction center (Scheme 6.5b). The key TS model is applied to the selectivity rationalization and is helpful for the catalyst design, thus selectivities of several new catalysts are predicted theoretically and then examined experimentally, showing good agreement with computations (Scheme 6.5c). Less time-consuming models for the selectivity prediction, based on the prochiral probe approach and TSs conformational analysis, were also attempted. However, these models did not give a good correlation with experimental selectivities.

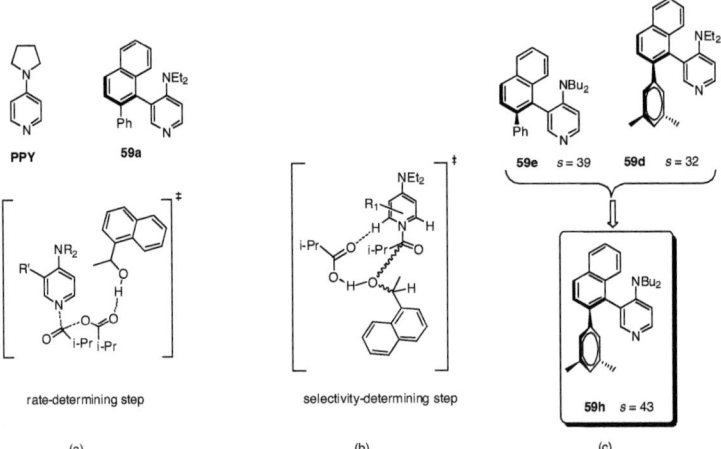

Scheme 6.5. The rate-limiting (a) and selectivity-determining (b) steps of the nucleophilic pathway for the isobutyrylation of 1-(1-naphthyl)ethanol; the key TS model was used to design the new highly selective catalysts for the KR of alcohols (c).

7. Experimental part

General information

All air and water sensitive manipulations were carried out under a nitrogen atmosphere using standard Schlenk techniques. Schlenk flasks were dried in the oven at 120 °C for at least 12 hours prior to use and then assembled quickly while still hot, cooled under nitrogen and sealed with a rubber septum. All commercial chemicals were of reagent grade and were used as received unless otherwise noted. CH_2Cl_2, THF and $CDCl_3$ were refluxed for at least one hour over CaH_2 and subsequently distilled. Acetic and isobutyric anhydride were stirred for one hour over K_2CO_3 and then distilled. ^1H and ^{13}C NMR spectra were recorded on Varian 300 or Varian INOVA 400 machines at ambient temperature. All ^1H chemical shifts are reported in ppm (δ) relative to $CDCl_3$ (7.26); ^{13}C chemical shifts are reported in ppm (δ) relative to $CDCl_3$ (77.16). ^1H NMR kinetic data were measured on a Varian Mercury 200 at 23 °C. HRMS spectra (ESI-MS) were carried out using a Thermo Finnigan LTQ FT instrument. IR spectra were measured on a Perkin-Elmer FT-IR BX spectrometer mounting ATR technology. Analytical TLC were carried out using aluminium sheets silica gel Si 60 F_{254}.

Chapter 2: Experimental details
E1.1 Synthesis of 3,4-diaminopyridine catalysts.

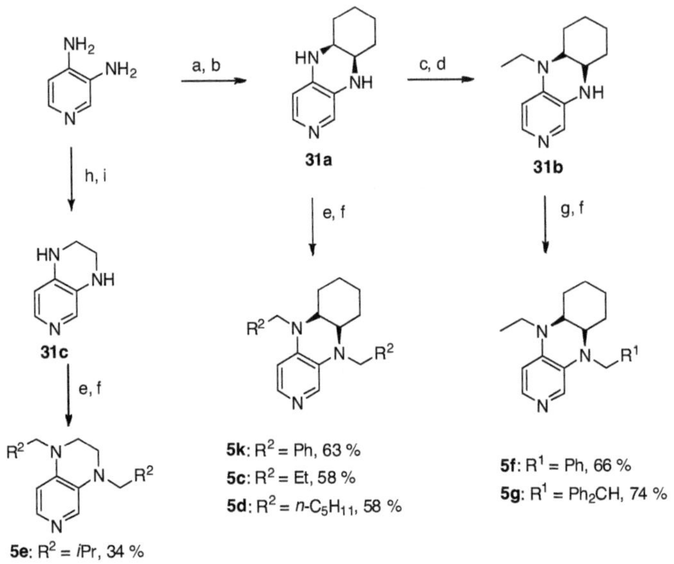

Scheme E1.1. Synthesis of 3,4-diaminopyridine catalysts: a) 1,2-cyclohexanedione, EtOH, 70 °C, 5 h, 90 %; b) LiAlH$_4$, THF, -40 °C, 30 min -> RT, 32 h, 79 %; c) Ac$_2$O, NEt$_3$, CH$_2$Cl$_2$, 5 mol% PPY, 96 %; d) AlCl$_3$, THF, RT, 45 min -> LiAlH$_4$, 0 °C, 1 h -> rf, 8h, 60 %; e) (R^{2}CO)$_2$O, pyridine, MW, 170 °C, 10 min, 60-86 %; f) AlCl$_3$, THF, RT, 45 min -> LiAlH$_4$, 0 °C, 1 h -> rf, 8h, 50-75 %; g) R^{1}COCl, pyridine, MW, 170 °C, 60 min, 98 %; h) glyoxal, EtOH, 70 °C, 2 h, 88 %; i) NaBH$_4$, EtOH, 40 °C, 15 h, 50 %.

General procedure I for the acylation of compounds 31a and 31b.

1.0 mmol **31a** or **31b**, 1.0 mL of dry pyridine and 4.0 mmol acyl anhydride (or acyl chloride) were placed into a 10 mL microwave vessel. The reaction vessel was sealed with a septum and placed into the microwave cavity. The corresponding program (250 W, 100% air-cooling, 170 °C, 20 min) was started. After cooling, the reaction mixture was quenched with MeOH. 2M NaOH solution (10 mL) was added. The aqueous layer was extracted with CH$_2$Cl$_2$ (3 × 20 mL) and dried over MgSO$_4$. The crude product was purified by flash-chromatography on SiO$_2$.

(*rac*)-(5-Benzoyl-5a,6,7,8,9,9a-hexahydro-5H-pyrido[3,4-b]quinoxalin-10-yl)-phenyl-methanone

Benzoyl chloride was used for the acylation of diamine **31a**. The crude product was purified by flash-chromatography on SiO_2 with $EtOAc/CHCl_3$ (1:3) as eluent to afford 58% of product.

^{1}H NMR (600 MHz, $CDCl_3$) δ 7.89 (d, ^3J = 5.4, 1H), 7.82 (s, 1H), 7.51 – 7.44 (m, 4H), 7.42 – 7.31 (m, 6H), 6.50 (d, ^3J = 5.4 Hz, 1H), 5.01 – 4.98 (m, 1H), 4.95 (m, 1H), 1.93 (s, 1H), 1.78 (m, 3H), 1.35 – 1.50 (m, 4H).
^{13}C NMR (151 MHz, $CDCl_3$) δ 169.77, 169.32, 147.11, 145.51, 140.79, 135.05, 134.87, 131.34, 130.77, 129.68, 128.62, 128.51, 128.31, 128.12, 119.03, 56.51, 56.48, 27.77, 27.56, 21.35, 21.26.
HRMS (EI): *m/z* calcd for $C_{25}H_{23}O_2N_3$ [M$^+$]: 397.1790; found: 397.1769.

(*rac*)-1,1'-(5a,6,7,8,9,9a-hexahydropyrido[3,4-b]quinoxaline-5,10-diyl)bis(hexan-1-one)

Hexanoic anhydride was used for the acylation of diamine **31a**. The crude product was purified by flash-chromatography on SiO_2 with EtOAc/hexane (1:1) as eluent to afford 86 % of product.

R_f = 0.2 (isohexane/EtOAc/NEt$_3$, 10:1:1).
^{1}H NMR (300 MHz, $CDCl_3$) δ 8.46 (s, 1H), 8.37 (d, J = 5.3 Hz, 1H), 7.22 (d, J = 5.3 Hz, 1H), 4.89 – 4.76 (m, 1H), 4.76 – 4.65 (m, 1H), 2.59 – 2.44 (m, 2H), 2.42 – 2.24 (m, 2H), 1.69 – 1.44 (m, 8H), 1.40 – 1.25 (m, 4H), 1.25 – 1.09 (m, 8H), 0.79 (q, J = 6.8 Hz, 6H).

^{13}C NMR (75 MHz, CDCl$_3$) δ 171.89, 171.83, 147.06, 146.91, 141.64, 130.81, 119.60, 55.71, 55.07, 34.63, 34.26, 31.25, 31.23, 28.37, 28.18, 25.17, 25.14, 22.32, 22.30, 21.65, 21.50, 13.81, 13.80.

IR (neat): 2954 (m), 2930 (s), 2859 (m), 1662 (vs), 1585 (m), 1564 (w), 1495 (m), 1452 (w), 1421 (m), 1383 (m), 1331 (m), 1291 (m), 1244 (m), 1194 (m), 1169 (m), 1138 (vw), 1113 (w), 1071 (w), 988 (vw), 920 (vw), 836 (w), 753 (w), 732 (w), 666 (vw), 645 (vw) cm^{-1}.

HRMS (EI): m/z calcd for C$_{23}$H$_{35}$O$_2$N$_3$ [M$^+$]: 385.2729; found: 385.2728.

(rac)-(5-ethyl-5a,6,7,8,9,9a-hexahydropyrido[3,4-b]quinoxalin-10(5H)-yl)(phenyl)methanone

Benzoyl chloride was used for the acylation of amine **31b**. The crude product was purified by flash-chromatography on SiO$_2$ with EtOAc/TEA (10:1) as eluent to afford 98% of brown oil.

^{1}H NMR (600 MHz, CDCl$_3$) δ 7.92 (d, J = 5.8 Hz, 1H), 7.54 (br s, 1H), 7.38 – 7.29 (m, 3H), 7.29 – 7.22 (m, 2H), 6.59 (d, J = 5.9 Hz, 1H), 4.68 (br s, 1H), 3.68 (br s, 1H), 3.59 (dq, ^{3}J = 7.1 Hz, ^{2}J = 14.5 Hz, 1H), 3.34 (dq, ^{3}J = 7.0 Hz, ^{2}J = 14.3 Hz, 1H), 2.21 (d, J = 15.1 Hz, 1H), 1.78 – 1.70 (m, 2H), 1.53 (s, 1H), 1.49 – 1.39 (m, 2H), 1.37 – 1.24 (m, 2H), 1.20 (t, ^{3}J = 7.1 Hz, 3H).

^{13}C NMR (151 MHz, CDCl$_3$) δ 146.09, 145.59, 143.80, 135.31, 130.37, 128.33, 106.46, 53.57, 39.65, 28.40, 25.59, 24.56, 18.76, 12.09.

HRMS (EI): m/z calcd for C$_{20}$H$_{24}$N$_3$O [M+H]$^+$: 322.1919; found: 322.1905.

(rac)-1-(5-ethyl-5a,6,7,8,9,9a-hexahydropyrido[3,4-b]quinoxalin-10(5H)-yl)-2,2-diphenylethanone

Experimental part

Diphenylacetyl chloride was used for the acylation of amine **31b**. The crude product was purified by flash-chromatography on SiO$_2$ with EtOAc/TEA (10:1) as eluent to afford 98% of product.

^{1}H NMR (600 MHz, CDCl$_3$) δ 8.14 – 8.03 (m, 2H), 7.37 (d, J = 7.5 Hz, 2H), 7.32 (t, J = 7.7 Hz, 2H), 7.29 – 7.21 (m, 2H), 7.18 – 7.08 (m, 2H), 6.94 (d, J = 6.9 Hz, 2H), 6.51 (d, J = 5.7 Hz, 1H), 5.57 (s, 1H), 4.89 (d, J = 12.7 Hz, 1H), 3.51 – 3.38 (m, 1H), 3.35 (s, 1H), 3.21 – 3.08 (m, 1H), 2.09 (d, J = 15.1 Hz, 1H), 1.69 (d, J = 12.7 Hz, 1H), 1.59 (d, J = 11.4 Hz, 1H), 1.49 (t, J = 14.4 Hz, 1H), 1.44 – 1.32 (m, 2H), 1.20 – 1.04 (m, 2H), 0.98 (t, J = 7.0 Hz, 3H).
^{13}C NMR (151 MHz, CDCl$_3$) δ 171.38, 147.39, 145.30, 145.05, 139.03, 138.94, 128.94, 128.88, 128.86, 128.77, 128.55, 128.39, 128.34, 127.02, 126.92, 120.81, 106.43, 53.97, 53.12, 49.30, 39.48, 28.02, 25.57, 24.59, 18.75, 11.92.
HRMS (EI): m/z calcd for C$_{27}$H$_{30}$N$_3$O [M+H]$^+$: 412.2389; found: 412.2375.

Experimental part

General procedure II for the reduction of amide intermediates.[26c]

AlCl$_3$ (1.3 eq per each amide bond) was suspended in 10 mL THF at r.t. After stirring for 30 min, the mixture was cooled to 0 °C and LiAlH$_4$ (2.2 eq per each amide bond) was added in small portions. The mixture was stirred for 15 min and then amide intermediate (1 eq, 1 mmol) was added. The reaction mixture was stirred for 1 h at 0 °C and then refluxed for 8 h. After cooling down to r.t. the reaction mixture was poured into ice water. The inorganic precipitate was filtered off and washed with CH$_2$Cl$_2$ (10 mL). The aqueous layer was made basic with NaOH solution to pH 12 and extracted with CH$_2$Cl$_2$ (3 × 10 mL). Organic layers were combined and dried over MgSO$_4$. The crude product was purified by flash-chromatography on SiO$_2$.

(*rac*)-5,10-Dibenzyl-5,5a,6,7,8,9,9a,10-octahydro-pyrido[3,4-b]quinoxaline, 5k

The crude product was purified by flash-chromatography on SiO$_2$ with EtOAc/MeOH (10:1) as eluent to afford 63% of product as oil.

^{1}H NMR (600 MHz, CDCl$_3$) δ 7.71 (d, ^3J = 5.5, 1H), 7.69 (s, 1H), 7.37 – 7.29 (m, 6H), 7.27 – 7.21 (m, 4H), 6.34 (d, ^3J = 5.5, 1H), 4.67 (d, J = 17.1, 1H), 4.65 (d, J = 16.5, 1H), 4.42 (d, J = 17.1, 1H), 4.31 (d, J = 16.5, 1H), 3.56 (br s, 1H), 3.45 (br s, 1H), 1.99 – 1.90 (m, 1H), 1.89 – 1.82 (m, 1H), 1.66 – 1.48 (m, 3H), 1.36 – 1.21 (m, 3H).

^{13}C NMR (151 MHz, CDCl$_3$) δ 140.91, 138.49, 137.65, 133.25, 131.25, 128.67, 128.60, 127.06, 126.89, 126.86, 126.60, 105.50, 60.37, 52.00, 50.99, 29.68, 27.63, 26.87, 21.03.

HRMS (EI): m/z calcd for C$_{25}$H$_{27}$N$_3$ [M$^+$]: 369.2205; found: 369.2203.

(*rac*)-5,10-dihexyl-5,5a,6,7,8,9,9a,10-octahydropyrido[3,4-b]quinoxaline, 5d

The crude product was purified by flash-chromatography on SiO$_2$ with EtOAc/NEt$_3$ (10:1) as eluent to afford 58% of product as off-white solid with m.p. 34.8-37.0 °C.

R$_f$ = 0.4 (basic aluminium oxide, i-hexane/NEt$_3$, 10:1).

^{1}H NMR (300 MHz, CDCl$_3$) δ 7.75 (d, J = 5.5, 1H), 7.70 (s, 1H), 6.32 (d, J = 5.5, 1H), 3.45 – 3.30 (m, 3H), 3.30 – 3.23 (m, 1H), 3.16 – 2.98 (m, 2H), 1.97 – 1.74 (m, 2H), 1.72 – 1.47 (m, 8H), 1.46 – 1.22 (m, 14H), 1.01 – 0.86 (m, J = 4.8, 6H).

^{13}C NMR (75 MHz, CDCl$_3$) δ 140.73, 140.12, 132.01, 130.83, 104.39, 56.49, 53.85, 47.31, 46.89, 31.68, 31.63, 27.57, 27.21, 26.89, 26.81, 26.34, 25.34, 22.76, 22.68, 22.64, 21.81, 14.03, 14.01.

IR (neat): 3029 (vw), 2927 (s), 2855 (s), 2178 (vw), 1618 (w), 1579 (vs), 1554 (m), 1512 (vs), 1486 (w), 1459 (m), 1445 (m), 1430 (w), 1407 (w), 1363 (m), 1319 (m), 1300 (m), 1265 (s), 1248 (s), 1217 (m), 1191 (m), 1157 (w), 1115 (w), 1095 (m), 1074 (m), 1058 (w), 1040 (w), 1021 (w), 964 (w), 927 (w), 875 (w), 838 (w), 797 (s), 749 (m), 727 (vs), 695 (w), 659 (w), 639 (w) cm^{-1}.

HRMS (EI): m/z calcd for C$_{23}$H$_{39}$N$_3$ [M$^+$]: 357.3144; found: 357.3141.

(rac)-10-benzyl-5-ethyl-5,5a,6,7,8,9,9a,10-octahydropyrido[3,4-b]quinoxaline, 5f

The crude product was purified by flash-chromatography on SiO$_2$ with EtOAc/NEt$_3$ (20:1) as eluent to afford 66% of product as oil.

^{1}H NMR (300 MHz, CDCl$_3$) δ 7.80 (d, ^3J = 5.5, 1H), 7.63 (s, 1H), 7.35 – 7.28 (m, 4H), 7.27 – 7.20 (m, 1H), 6.44 (d, ^3J = 5.5, 1H), 4.58 (d, ^2J = 16.4, 1H), 4.28 (d, ^2J = 16.4, 1H), 3.61 – 3.45 (m, 2H), 3.39 – 3.22 (m, 2H), 2.13 – 1.94 (m, 1H), 1.84 – 1.47 (m, 5H), 1.46 – 1.25 (m, 2H), 1.20 (t, ^3J = 7.1, 3H).

^{13}C NMR (75 MHz, CDCl$_3$) δ 140.87, 140.79, 138.49, 132.93, 131.21, 128.60, 126.91, 126.89, 104.69, 55.82, 54.72, 54.69, 52.26, 40.70, 28.17, 26.37, 22.81, 21.73, 11.44.

IR (neat): 2928 (m), 2855 (w), 1581 (m), 1553 (m), 1514 (s), 1494 (w), 1452 (m), 1353 (m), 1266 (m), 1246 (m), 1191 (m), 1095 (m), 1079 (m), 1028 (w), 972 (w), 801 (s), 729 (s) cm^{-1}.

HRMS (EI): m/z calcd for C$_{20}$H$_{26}$N$_3$ [M+H]$^+$: 308.2127; found: 308.2112.

Experimental part

(*rac*)-10-(2,2-diphenylethyl)-5-ethyl-5,5a,6,7,8,9,9a,10-octahydropyrido[3,4-b]quinoxaline, 5g

The crude product was purified by flash-chromatography on SiO_2 with EtOAc/TEA (20:1) as eluent to afford 92% of product (95% purity). Column chromatography on basic Al_2O_3 (Brockmann III) with EtOAc/hexanes (1:2) as eluent gave 74% yield of pure product.

IR (neat): 2931 (m), 1580 (m), 1554 (w), 1514 (m), 1450 (w), 1362 (w), 1269 (w), 1242 (w), 1098 (w), 955 (vw), 803 (vw), 699 (m).

^{1}H NMR (300 MHz, CDCl$_3$) δ 7.81 (d, *J* = 5.5, 1H), 7.71 (s, 1H), 7.37 – 7.30 (m, 4H), 7.30 – 7.16 (m, 6H), 6.46 (d, *J* = 5.6, 1H), 4.46 (dd, *J* = 4.7, 10.3, 1H), 4.04 (dd, *J* = 4.8, 14.2, 1H), 3.64 (dd, *J* = 10.3, 14.2, 1H), 3.50 (dq, *J* = 7.1, 14.2, 1H), 3.18 (dq, *J* = 7.0, 14.1, 1H), 3.00 – 2.91 (m, 1H), 2.73 (dt, *J* = 3.7, 10.6, 1H), 2.05 – 1.92 (m, 1H), 1.74 – 1.48 (m, 2H), 1.47 – 1.12 (m, 5H), 1.08 (t, *J* = 7.1, 3H).

^{13}C NMR (75 MHz, CDCl$_3$) δ 143.23, 142.19, 141.23, 140.04, 131.59, 130.11, 128.57, 128.55, 128.24, 128.03, 126.61, 126.48, 105.30, 58.33, 54.93, 48.02, 39.17, 28.55, 25.55, 24.15, 20.11, 10.60.

HRMS (EI): *m/z* calcd for $C_{27}H_{32}N_3$ $[M+H]^+$: 398.2596; found: 398.2583.

Experimental part

E1.2 Synthesis of camphor derivative 5l.

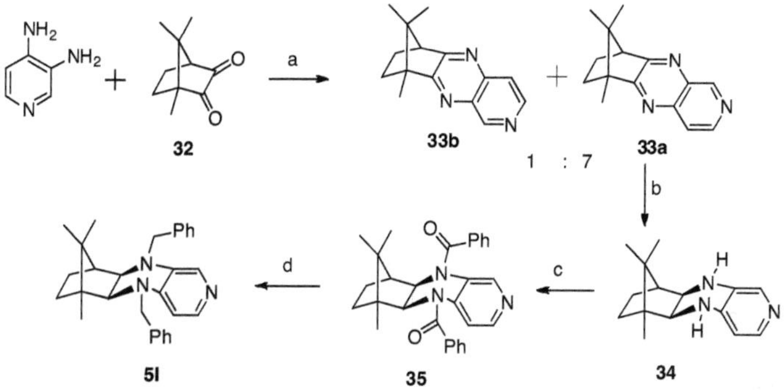

Scheme E1.2. Synthesis of camphor derivative **5l**: a) AcOH, rf, 6 h, 85 %; b) NaBH$_4$, BH$_3$, THF, 40 °C, 15 h, 55 %; c) PhCOCl, pyridine, MW, 150 °C, 60 min, 64 %; d) AlCl$_3$, THF, RT, 45 min -> LiAlH$_4$, 0 °C, 1 h -> rf, 8 h, 55 %.

(6R,9S)-6,11,11-trimethyl-6,7,8,9-tetrahydro-6,9-methanopyrido[3,4-b]quinoxaline, 33

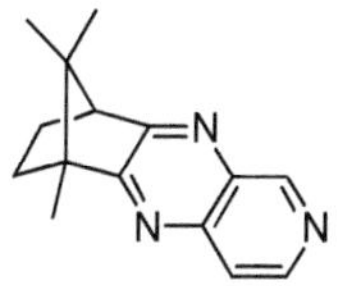

3,4-diaminopyridine (3.55 g, 33 mmol) and (S)-(+)-camphorquinone **32** (4.50 g, 27 mmol) were dissolved in 120 mL acetic acid. The reaction mixture was refluxed for 6 h. After cooling the reaction mixture to r.t. acetic acid was distilled off. The residue was neutralized with aq. Na$_2$CO$_3$ and extracted with CH$_2$Cl$_2$ (3 × 40 mL). The organic layers were combined and washed with aq. Na$_2$CO$_3$. After drying over MgSO$_4$ the solvents were removed in vacuum. The crude product was purified by flash-chromatography on SiO$_2$ with EtOAc/iso-hexane (2:1) as eluent to afford 5.5 g (85%) as a white solid (mixture of **33a** and **33b** in ratio 7:1). After recrystallization from cyclohexane/diethylether mixture (1:1) compound **33a** was obtained as a white solid with mp 136-138 °C.

$[\alpha]_D^{20}$ = +31.2° (c = 1, CHCl$_3$).

^{1}H NMR (400 MHz, CDCl$_3$) δ 9.31 (d, J = 0.7 Hz, 1H), 8.67 (d, J = 5.6 Hz, 1H), 7.81 (dd, J = 0.8, 5.6 Hz, 1H), 3.07 (d, J = 4.5 Hz, 1H, CH), 2.34 – 2.22 (m, 1H, CH$_2$), 2.11 – 1.99 (m, 1H, CH$_2$), 1.39 (s, 3H, CH_3), 1.43 – 1.34 (m, 2H, CH$_2$), 1.07 (s, 3H, CH$_3$CCH_3), 0.58 (s, 3H, CH_3CCH$_3$).

^{13}C NMR (101 MHz, CDCl$_3$) δ 170.05, 165.48, 152.83, 146.55, 144.86, 136.75, 121.60, 54.26, 54.11, 53.14, 31.53, 24.33, 20.26, 18.34, 9.83.

IR (neat): 3034 (w), 2956 (s), 2360 (w), 1589 (m), 1574 (m), 1477 (m), 1434(s), 1403 (s), 1392 (s), 1378 (m), 1355 (s), 1314 (m), 1262 (m), 1241 (m), 1175 (s), 1111 (s), 1076 (s), 1055 (s), 904 (s), 840 (vs), 828 (vs), 726 (s), 651 (s), 615 (vs), 600 (vs), 576 cm^{-1} (vs).

HRMS (EI): m/z calcd for C$_{15}$H$_{17}$N$_3$ [M$^+$]: 239.1423; found: 239.1419.

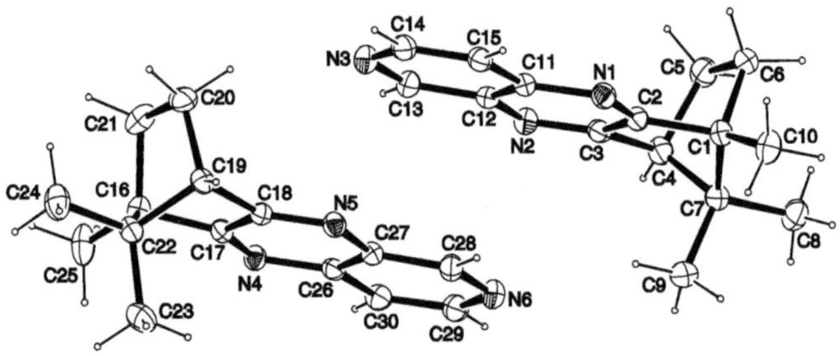

Figure E1.1 Representation of the X-ray structure of **33a**.

(5aS,6R,9S,9aR)-6,11,11-trimethyl-5,5a,6,7,8,9,9a,10-octahydro-6,9-methanopyrido[3,4-b]quinoxaline, 34

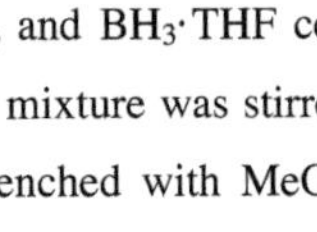

Compound **33a** (1.53 g, 6.4 mmol) and NaBH$_4$ (0.49 g, 12.8 mmol) were dissolved in THF (40 mL) in a 100 mL Schlenk flask, and BH$_3$·THF complex (25.6 mL, 25.6 mmol, 1M in THF) was slowly added at 0 °C. The mixture was stirred for 1 h at 0 °C and then 15 h at 40 °C. Afterwards, the mixture was quenched with MeOH (20 mL) and stirred for 1 h. 2M aqueous HCl was added and the mixture was stirred for another 1 h. After neutralization with 2M NaOH (aq) to pH 12, the mixture was extracted with CH$_2$Cl$_2$ (3 × 30 mL). After drying over MgSO$_4$ the solvents were distilled off. The crude product was purified by flash-chromatography on SiO$_2$ with EtOAc/MeOH/NEt$_3$ (20:10:1) as eluent to afford 0.86 g (55%) as a white solid with mp 195 °C (decomp). Determined by HPLC diastereomeric excess: de = 87 %. Conditions for HPLC: Daicel Chiralpak OD-H column analytical, hexane/i-propanol = 70/30, flow-rate 0.7 mL/min, detection at 254 nm, 25°C. Injection volumes: 5-10 μL. Retention times: 9.0 and 9.5 min (minor diastereomers), 13.3 min (major).

¹H NMR (400 MHz, CDCl₃) δ 7.69 (s, 1H), 7.69 (d, J = 5.2 Hz, 1H), 6.39 (d, J = 5.2, 1H), 4.32 (br s, 1H), 3.96 (br s, 1H), 3.47 (d, J = 7.2 Hz, 1H), 3.27 (dd, J = 3.6, 7.2, 1H), 1.80 – 1.69 (m, 2H), 1.55 (d, J = 8.6 Hz, 1H), 1.17 – 1.03 (m, J = 10.4 Hz, 2H), 1.00 (s, 3H), 0.99 (s, 3H), 0.82 (s, 3H).
¹³C NMR (75 MHz, CDCl₃) δ 139.94, 138.16, 133.29, 128.45, 107.76, 63.39, 59.48, 51.60, 48.59, 47.00, 34.79, 25.37, 22.32, 21.09, 20.52, 11.59.
IR (neat): 3235 (m), 2949 (s), 2869 (s), 1599 (s), 1527 (s), 1477 (m), 1351 (s), 1308 (vs), 1276 (vs), 1234 (m), 1175 (m), 1135 (m), 1105 (m), 1052 (m), 884 (w), 811 (s), 742 (w), 590 cm⁻¹ (w).
HRMS (EI): m/z calcd for $C_{15}H_{21}N_3$ [M⁺]: 243.1736; found: 243.1733.

((5aS,6R,9S,9aR)-6,11,11-trimethyl-5a,6,7,8,9,9a-hexahydro-6,9-methanopyrido[3,4-b]quinoxaline-5,10-diyl)bis(phenylmethanone), 35

Acylation of diamine **34** by benzoyl chloride was carried out according **general procedure I**. The crude product was purified by flash-chromatography on SiO₂ with EtOAc/hexanes (1:3) as eluent to afford 64% (400 mg) of **35** as a mixture of diastereomers (de = 70%). Conditions for HPLC: Daicel Chiralpak OD-H column analytical, hexane/i-propanol = 95/5, flow-rate 1.0 mL/min, detection at 254 nm, 25°C. Injection volumes: 5-10 μL. Retention times: 26.0 and 38.5 min (minor diastereomers), 34.1 min (major).
Recrystallization from ether/hexane mixture at 5 °C gives 190 mg of **35** with de = 97% as a white solid with mp 218-220 °C.

$[\alpha]_D^{20}$ = +92.0° (c = 1, CHCl₃).
¹H NMR (600 MHz, CDCl₃) δ 7.73 (d, J = 5.5 Hz, 1H), 7.71 (s, 1H), 7.51 – 7.41 (m, 6H), 7.40 – 7.33 (m, 4H), 6.28 (d, J = 5.5 Hz, 1H), 5.01 (d, J = 10.8 Hz, 1H), 4.74 (d, J = 10.8 Hz, 1H), 2.36 (d, J = 4.8 Hz, 1H), 1.99 – 1.90 (m, 1H), 1.71 – 1.61 (m, 1H), 1.57 – 1.47 (m, 1H), 1.44 – 1.36 (m, 1H), 1.00 (s, 3H), 0.83 (s, 3H), 0.15 (s, 3H).

Experimental part

^{13}C NMR (151 MHz, CDCl$_3$) δ 170.73, 170.48, 147.46, 144.20, 138.91, 135.21, 135.04, 131.16, 131.14, 130.39, 128.69, 128.64, 128.60, 128.52, 128.46, 128.44, 120.09, 68.16, 66.90, 52.49, 51.60, 44.94, 36.19, 27.57, 23.97, 20.94, 11.39.

IR (neat): 2957 (w), 1662 (m), 1652 (m), 1576 (w), 1509 (w), 1446 (w), 1342 (m), 1318 (m), 1256 (w), 1134 (w), 836 (w), 779 (w), 695 (m) cm^{-1}.

HRMS (EI): m/z calcd for C$_{29}$H$_{30}$N$_3$O$_2$ [M+H]$^+$: 452.2338; found: 452.2316.

(5aS,6R,9S,9aR)-5,10-dibenzyl-6,11,11-trimethyl-5,5a,6,7,8,9,9a,10-octahydro-6,9-methanopyrido[3,4-b]quinoxaline, 5l

Reduction of compound **35** (190 mg) was carried out according **general procedure II**. The crude product was purified by flash-chromatography on SiO$_2$ with EtOAc/TEA (40:1) as eluent to afford 42 % (74 mg) of **5l** as off-white oil. Diastereomeric excess determined by HPLC: de > 99%.

Conditions for HPLC: Daicel Chiralpak OD-H column analytical, hexane/*i*-propanol = 70/30, flow-rate 0.7 mL/min, detection at 254 nm, 25°C. Injection volumes: 5-10 μL. Retention times: 5.8 and 7.4 min (minor diastereomers), 8.9 min (major).

$[α]_D^{20}$ = +33.8° (c = 1, CHCl$_3$).

^{1}H NMR (300 MHz, CDCl$_3$) δ 7.69 (s, 1H), 7.66 (d, *J* = 5.5 Hz, 1H), 7.39 – 7.24 (m, 8H), 7.23 – 7.16 (m, 2H), 6.31 (d, *J* = 5.6 Hz, 1H), 4.79 (dd, *J* = 11.0 Hz, *J* = 17.5 Hz, 2H), 4.58 (dd, *J* = 8.0 Hz, *J* = 17.4 Hz, 2H), 3.42 (dd, *J* = 8.0 Hz, *J* = 18.7 Hz, 2H), 2.13 – 2.05 (m, 1H), 1.85 – 1.71 (m, 1H), 1.62 – 1.49 (m, 1H), 1.28 (t, *J* = 7.1 Hz, 1H), 1.23 – 1.11 (m, 2H), 1.07 (s, 3H), 1.05 (s, 3H), 0.82 (s, 3H).

^{13}C NMR (75 MHz, CDCl$_3$) δ 139.42, 138.70, 138.09, 137.64, 131.89, 129.06, 128.63, 126.88, 126.81, 126.53, 126.32, 107.02, 69.41, 64.20, 56.91, 54.43, 51.31, 48.78, 46.41, 35.96, 25.66, 22.19, 20.62, 14.15.

HRMS (EI): m/z calcd for C$_{29}$H$_{34}$N$_3$ [M+H]$^+$: 424.2753; found: 424.2734.

Experimental part

E1.3 General procedure for kinetic measurements.

Three stock solutions were prepared in dry calibrated 5 mL flasks. Stock solution A: 1.2 M in acetic anhydride (612.5 mg) and 0.6 M in 1,4-dioxane. Stock solution B: 1.8 M in NEt$_3$ (910.7 mg) and 0.6 M in ethynylcyclohexanol (372.5 mg). Stock solution C: 0.06 M in catalyst (10 mol% relative to alcohol). A nitrogen-flushed dry NMR tube was filled with 0.2 mL each of stock solution A, B and C. The NMR tube was fused with a gas burner and immediately injected into a Varian Mercury 200 MHz NMR. FIDs were collected in defined time intervals and the obtained multiple FID ^{1}H NMRs analyzed by integration of the product peaks with a self-written subroutine using the VNMR software package (Figure E1.2).

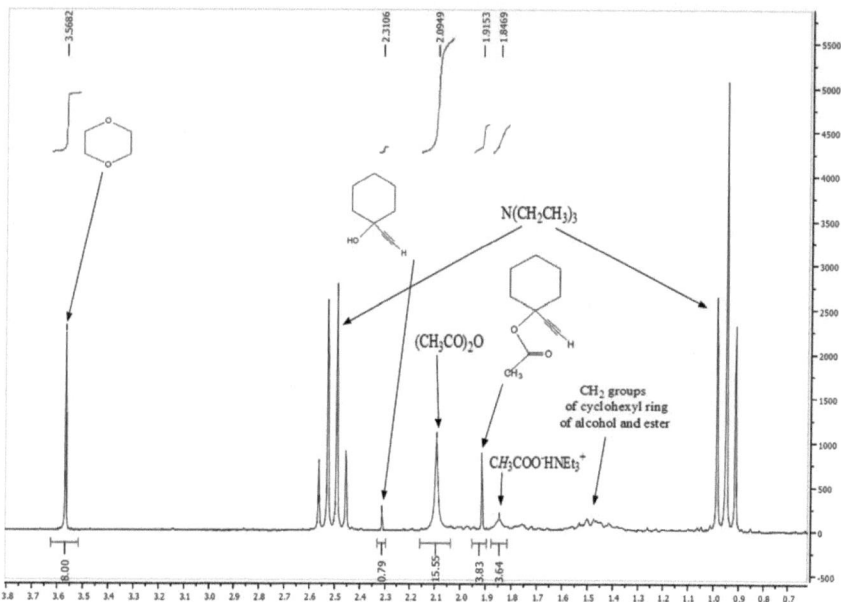

Figure E1.2. ^{1}H NMR (200 MHz) spectrum of reaction mixture during the kinetic measurement.

Experimental part

The conversion can either be calculated directly from the intensities of reactant and product peaks, or by using dioxane as an internal standard. The following two equations for quantifying the conversion then apply:

with standard:

$$\text{conv.} = \frac{(I_{Ester}/3)}{(I_{Dioxane}/8)} \cdot 100\% \qquad (1)$$

without standard:

$$\text{conv.} = \frac{I_{Ester}}{0.25 \cdot (I_{Ac_2O} + I_{HNEt_3OAc} + I_{Ester})} \cdot 100\% \qquad (2)$$

Dependence of the conversion vs. time t was fitted by equation (3) for the second-order reaction kinetics:

$$\text{conv.} = y_0 \left(1 - \frac{1}{2e^{k(t-t_0)} - 1}\right) \qquad (3)$$

$$k = k_2 [ROH]_0 \qquad (4)$$

where k_2 is a rate-constant of the second-order reaction; t_0 has a meaning of time axis offset. With this parameter in the fitting process it's not necessary to measure the starting point of the reaction exactly. The variable y_0 allows for rescaling of the conversion axis.

A half-life can be calculated with equation (5):

$$\tau_{1/2} = \frac{\ln 1.5}{k_2 [ROH]_0} \qquad (5)$$

Experimental part

Table E1.1. Catalytic activity of the 3,4-diamino and 4-aminopyridines.[a]

catalyst	run	k_2, l mol^{-1} s^{-1}	Δk_2	c_1	t_0, min	$t_{1/2}$, min	$\Delta t_{1/2}$, min
DMAP[b]	1	2.18E-04	1.39E-06	1.010	-28	**155.0**	**1.7**
PPY	1	5.15E-04	5.52E-06	1.082	-28	**66.6**	**1.0**
5b	1	1.90E-03	3.03E-05	1.088	-8	**17.8**	**0.3**
5f	1	6.35E-04	9.04E-06	1.011	-8	53.2	0.8
5f	2	7.20E-04	1.77E-05	1.070	-8	46.9	1.2
5f	3	7.07E-04	2.04E-06	0.999	-9	47.8	0.1
5f	averaged					**49.3**	**1.0**
5g	1	7.37E-04	4.17E-06	0.953	-22	45.9	0.3
5g	2	7.96E-04	1.04E-05	0.997	-15	42.5	0.6
5g	averaged					**44.2**	**1.0**
5h						**35.0**	**1.0**
5l	1	2.35E-04	1.12E-06	0.983	-18	143.9	0.7
5l	2	2.51E-04	8.71E-07	1.055	-20	134.7	0.5
5l	3	2.52E-04	7.10E-07	1.037	-19	133.9	0.4
5l	averaged					**137.5**	**2.0**
5k	1	2.99E-04	1.44E-06	0.911	-60	112.9	0.5
5k	2	2.78E-04	2.17E-06	1.043	-17	121.5	0.9
5k	3	2.96E-04	8.84E-07	1.052	-21	114.0	0.3
5k	averaged					**116.1**	**2.0**
5d	1	1.90E-03	2.41E-05	1.015	-8	17.8	0.2
5d	2	1.70E-03	3.70E-05	1.087	-9	19.9	0.4
5d	averaged					**18.8**	**0.5**

[a] Conditions: 0.2 M alcohol, 2.0 equiv of Ac$_2$O, 3.0 equiv of NEt$_3$, 0.1 equiv catalyst, CDCl$_3$, 23.0±1.0 °C.
[b] Data from ref. 26c.

Experimental part

Fitted conversion plots.

Fitting parameters:
$ka = k_2[ROH]_0$, where k_2 is the rate-constant of the second-order reaction;
t0 has a meaning of time axis offset (t_0 in equation 3);
y0 allows rescaling of the conversion axis (y_0 in equation 3).

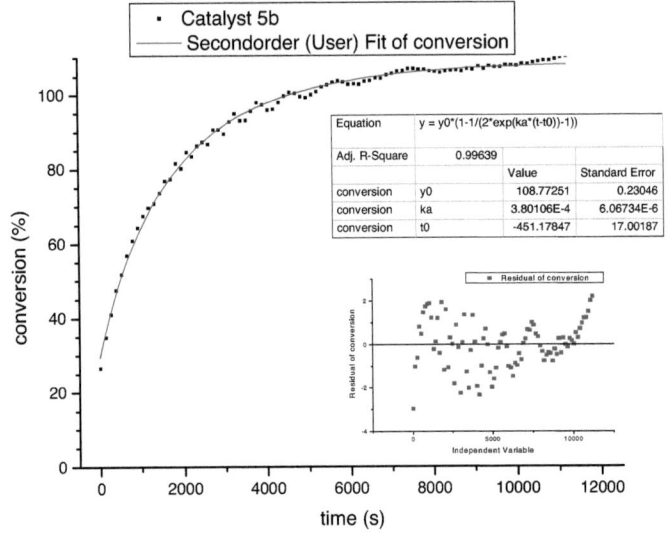

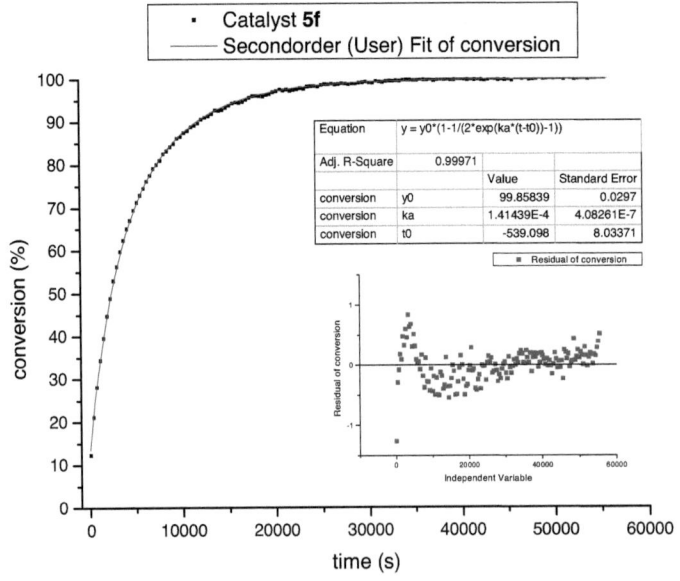

Experimental part

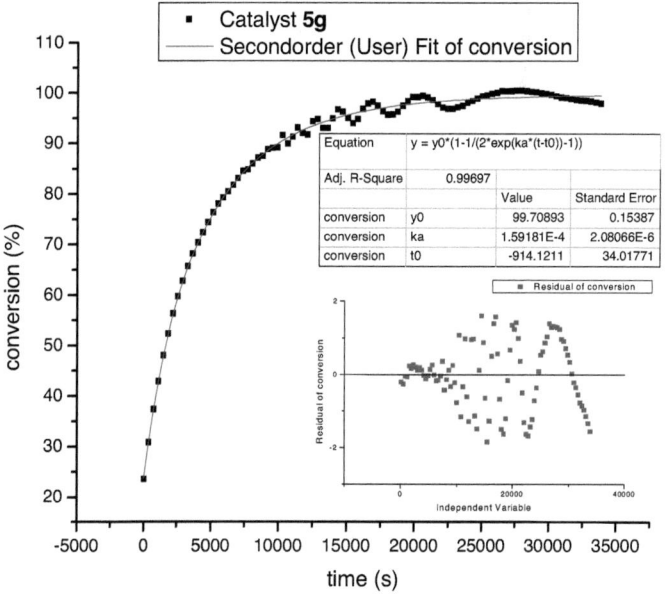

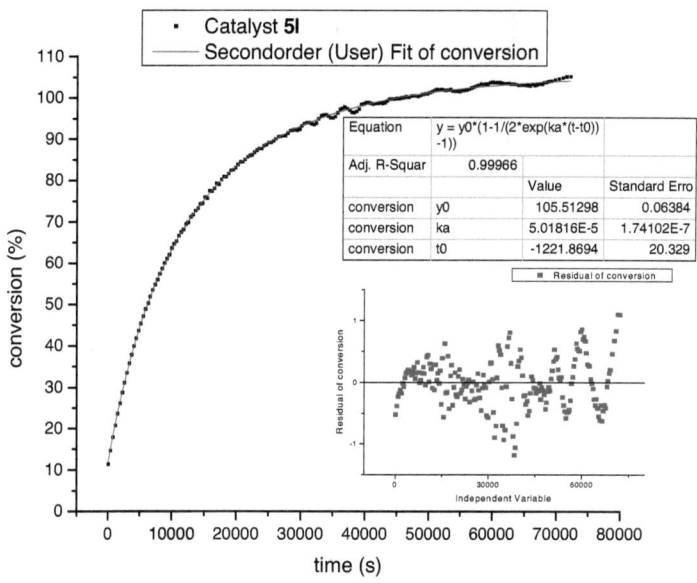

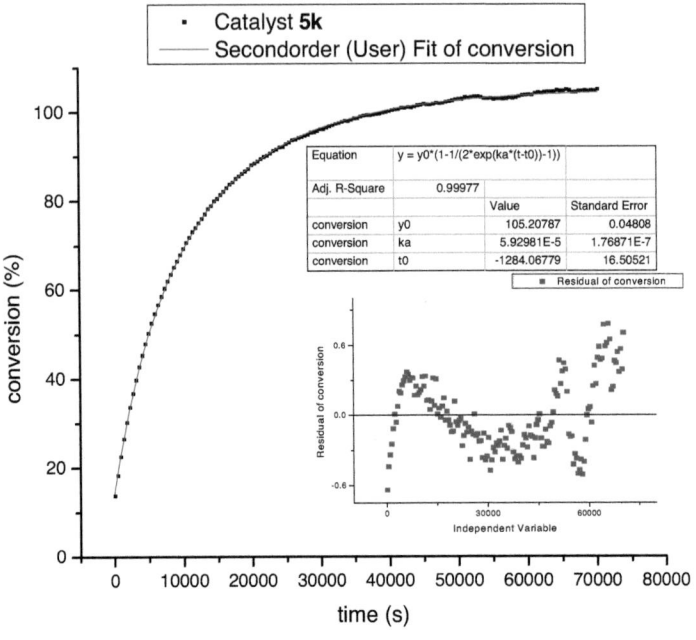

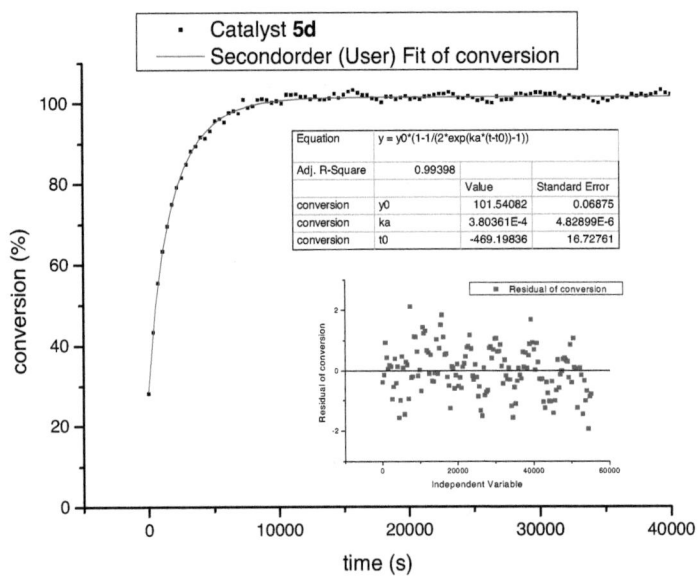

Experimental part

Table E1.2. Kinetic measurements with 5 mol% of the catalyst **5d**.[a]

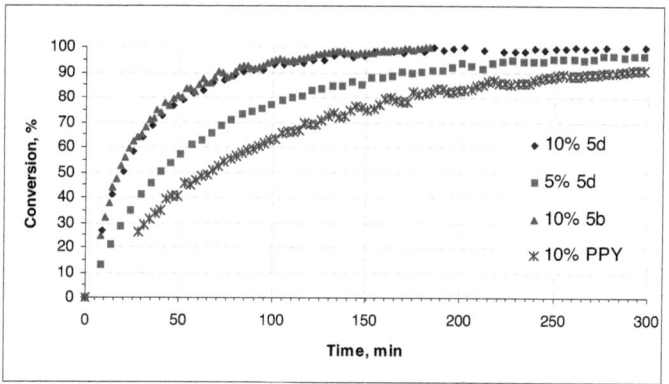

catalyst	run	k_2, l mol^{-1} s^{-1}	Δk_2	c_1	t_0, min	$t_{1/2}$, min	$\Delta t_{1/2}$, min
5d	1	8.23E-04	2.89E-06	1.018	-8	41.1	0.5
5d	2	8.90E-04	3.04E-06	1.024	-8	38.0	0.5
5d	averaged					**40.0**	**0.7**

[a] Conditions: 0.2 M alcohol, 2.0 equiv of Ac_2O, 3.0 equiv of NEt_3, 0.05 equiv catalyst, $CDCl_3$, 23.0±1.0 °C.

Figure E1.5. Conversion-time plots for catalysts **PPY**, **5b** and **5d** (5 and 10%).

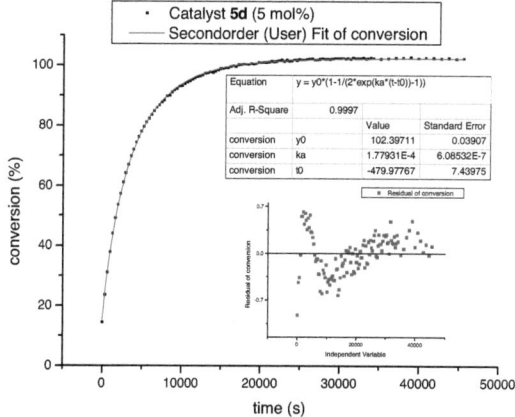

153

Chapter 4: Experimental details
E4.1. Achiral catalysts.

General procedure III. Synthesis of urea-catalysts.

Compound (*rac*)-**31b** or **31d** (1 mmol) was dissolved in freshly distilled CH_2Cl_2 (20 mL). To this solution the corresponding iso(thio)cyanate (1 mmol) was added. After stirring 12 h at RT the reaction mixture was quenched with methanol (1 mL). The solvent was removed under reduced pressure and the residue was purified by flash column chromatography.

Physical data of catalysts.

1-ethyl-N-phenyl-2,3-dihydropyrido[3,4-b]pyrazine-4(1H)-carboxamide, cat81ur1

The crude product was purified by flash column chromatography on SiO_2 with EtOAc as eluent to afford 99% of **cat81ur1** as pale yellow foam.

^{1}H NMR (300 MHz, CDCl$_3$): δ = 8.25 (s, 1H, CH), 8.01 (d, *J* = 5.8 Hz, 1H, CH), 7.39 (m, 3H, NH and Ph), 7.26 (m, 2H, Ph), 7.02 (m, 1H, Ph), 6.55 (d, *J* = 5.9 Hz, 1H, CH), 3.84 (m, 2H, CH$_2$), 3.40 (m, 4H, CH$_2$), 1.19 (t, *J* = 7.2 Hz, 3H, CH$_3$).

^{13}C NMR (75 MHz, CDCl$_3$): δ = 152.98, 147.05, 143.82, 143.56, 138.39, 128.85, 123.37, 121.43, 119.72, 105.45, 77.58, 77.36, 77.16, 76.73, 47.49, 44.57, 38.91, 11.09.

HRMS (ESI): *m/z* calcd for $C_{16}H_{19}N_4O$ [M+H]$^+$: 283.1559; found: 283.1551.

1-ethyl-N-phenyl-2,3-dihydropyrido[3,4-b]pyrazine-4(1H)-carbothioamide, cat81ur2

The crude product was purified by flash column chromatography on SiO_2 with EtOAc as eluent to afford 13% of **cat81ur2**.

^{1}H NMR (600 MHz, CDCl$_3$): δ = 8.25 (br s, 2H, CH and NH), 8.00 (d, *J* = 5.9 Hz, 1H, CH), 7.36 (m, 2H, Ph), 7.31 (m, 2H, Ph), 7.18 (m, 1H, Ph), 6.57 (d, *J* = 5.9 Hz, 1H, CH), 4.46 (br s, 2H, CH$_2$), 3.49 (t, *J* = 5.3 Hz, 2H, CH$_2$), 3.41 (m, 2H, CH$_2$), 1.22 (t, *J* = 7.2 Hz, 3H, CH$_3$).

^{13}C NMR (151 MHz, CDCl$_3$): δ = 180.96 (C=O), 147.68, 144.40, 143.59, 138.97, 128.69, 126.12, 125.18, 120.81, 105.96, 47.27, 44.44, 44.33, 11.15.

HRMS (ESI): *m/z* calcd for $C_{16}H_{19}N_4S$ [M+H]$^+$: 299.1330; found: 299.1324.

N-benzyl-1-ethyl-2,3-dihydropyrido[3,4-b]pyrazine-4(1H)-carboxamide, cat81ur3

The crude product was purified by flash column chromatography on SiO_2 with EtOAc as eluent to afford 73% of **cat81ur3** as pale yellow foam.

R_f = 0.28 (SiO_2; EtOAc/NEt$_3$=10:1).

^{1}H NMR (300 MHz, CDCl$_3$): δ = 8.19 (s, 1 H, H-5), 7.98 (d, ^3J = 5.8 Hz, 1 H, H-7), 7.27 (m, 5 H, Phenyl), 6.53 (d, ^3J = 5.9 Hz, 2 H, H-8), 5.75 (m, 1 H, NH), 4.44 (d, ^3J = 5.8 Hz, 2 H, H-15), 3.82 (t, ^3J = 4.9 Hz, 2 H, H-3), 3.40 (m, 4 H, H-2 and H-11), 1.19 (t, ^3J = 7.1 Hz, 3 H, H-12).

^{13}C NMR (75 MHz, CDCl$_3$): δ = 155.7 (C=O), 146.6 (7), 143. 6 (C$_q$), 143.6 (5), 138.9 (C$_q$), 128.6 (Phenyl), 127.5 (Phenyl), 127.3 (Phenyl), 121.7 (C$_q$), 105.3 (8), 47.4 (2), 44.9 (15), 44.6 (11), 39.1 (3), 11.0 (12).

IR (neat): 3183 (w), 2925 (w), 1652 (s), 1598 (m), 1518 (s), 1431 (m), 1351 (s), 1323 (s), 1286 (m), 1270 (m), 1200 (m), 1178 (m), 1120 (w), 1179 (m), 1044 (m), 960 (w), 924 (w), 878 (m), 815 (m), 730 (m), 712 (s) cm^{-1}.

HRMS (ESI): *m/z* calcd for $C_{17}H_{21}N_4O$ [M+H]$^+$: 297.1710; found: 297.1709; *m/z* calcd for $C_{17}H_{19}N_4O$ [M-H]$^+$: 295.1564; found: 295.1560.

N-benzyl-1-ethyl-2,3-dihydropyrido[3,4-b]pyrazine-4(1H)-carbothioamide, cat81ur4

The crude product was purified by flash column chromatography on SiO_2 with EtOAc as eluent to afford 92% of **cat81ur4**.

^{1}H NMR (300 MHz, CDCl$_3$): δ = 8.08 (s, 1H, CH), 7.91 (d, J =5.9 Hz, 1H, CH), 7.28 (m, 5H, Ph), 7.12 (br s, 1H, NH), 6.53 (d, J = 5.9 Hz, 1H, CH), 4.88 (d, J = 5.4 Hz, 2H, CH_2Ph), 4.44 (br s, 2H, CH$_2$), 3.42 (m, 4H, CH$_2$), 1.19 (t, J = 7.2 Hz, 3H, CH$_3$).

^{13}C NMR (75 MHz, CDCl$_3$): δ = 182.11 (C=O), 147.46, 144.40, 143.74, 137.47, 128.73, 127.74, 127.61, 120.60, 105.94, 50.10, 47.27, 44.76, 44.40, 11.13.

HRMS (ESI): *m/z* calcd for $C_{17}H_{21}N_4S$ [M+H]$^+$: 313.1487; found: 313.1480.

N-(3,5-bis(trifluoromethyl)phenyl)-1-ethyl-2,3-dihydropyrido[3,4-b]pyrazine-4(1H)-carboxamide, cat81ur1f

The crude product was purified by flash column chromatography on SiO$_2$ with EtOAc as eluent to afford 59% of **cat81ur1f**.

^{1}H NMR (300 MHz, CDCl$_3$): δ = 8.34 (s, 1H, NH), 8.20 (s, 1H, CH), 8.02 (s, 2H, 2CH), 7.93 (d, *J* = 5.8 Hz, 1H, CH), 7.53 (s, 1H, CH), 6.54 (d, *J* = 5.9 Hz, 1H, CH), 3.90 (t, *J* = 5.1 Hz, 2H, CH$_2$), 3.58 – 3.37 (m, 4H, CH$_2$), 1.25 (t, *J* = 7.1 Hz, 3H, CH$_3$).

^{13}C NMR (75 MHz, CDCl$_3$): δ = 152.64 (C=O), 147.19, 143.70, 140.42, 132.04 (q, $^2J_{C\text{-}F}$ = 33 Hz, CCF$_3$), 123.2 (q, $^1J_{C\text{-}F}$ = 273 Hz, CF$_3$), 120.73, 119.64, 119.55, 116.29, 105.57, 47.57, 44.62, 39.00, 11.08.

HRMS (ESI): *m/z* calcd for $C_{18}H_{17}F_6N_4O$ [M+H]$^+$: 419.1307; found: 419.1296.

(*rac*)-5-ethyl-N-phenyl-5a,6,7,8,9,9a-hexahydropyrido[3,4-b]quinoxaline-10(5H)-carboxamide, cat11ur1

The crude product was purified by flash column chromatography on SiO$_2$ with EtOAc as eluent to afford 70% of **cat11ur1** as white foam.

R_f = 0.24 (SiO$_2$; EtAc/NEt$_3$=10:1).

^{1}H NMR (600 MHz, CDCl$_3$): δ = 8.24 (s, 1 H, H-1), 8.09 (d, ^3J = 5.8 Hz, 1 H, H-3), 7.36 (m, 2 H, H-20), 7.25 (m, 2 H, H-21), 7.06 (s, 1 H, N-H), 7.01 (tt, ^3J = 8.4 Hz, ^4J = 1.2 Hz, 1 H, H-22), 6.66 (d, ^3J = 5.9 Hz, 1 H, H-11), 4.71 (dt, ^3J = 12.7 Hz, ^3J = 4.2 Hz, 1 H, H-13), 3.58 (m, 1 H, H-15), 3.56 (m, 1 H, H-12), 3.34 (m, 1 H, H-15), 2.20 (dt, ^3J = 15.1 Hz, ^3J = 2.7 Hz, 1 H, H-Cy), 1.74 (d, ^3J = 11.3 Hz, 1 H, H-Cy), 1.67 (d, ^3J = 12.7 Hz, 1 H, H-Cy), 1.59 (tt, ^3J = 12.1 Hz, ^4J = 3.2 Hz, 1 H, H-Cy), 1.46 (m, 2 H, H-Cy), 1.20 (m, 5 H, H-16 and 2x H-Cy).

^{13}C NMR (150 MHz, CDCl$_3$): δ = 152.6 (17), 147.0 (3), 145.0 (11), 144.4 (1), 138.3 (19), 128.9 (21), 123.3 (22), 120.5 (14), 119.4 (20), 106.7 (4), 53.0 (12), 49.4 (13), 39.7 (15), 28.3 (Cy), 26.1 (Cy), 24.8 (Cy), 18.9 (Cy), 12.2 (16).

IR (neat): 3183 (w), 2930 (w), 2861 (w), 1667 (s), 1596 (m), 1512 (s), 1460 (w), 1440 (m), 1399 (w), 1378 (m), 1355 (m), 1318 (m), 1282 (m), 1230 (s), 1197 (w), 1169 (m), 1121 (w), 1097 (w), 1070 (w), 1024 (m), 973 (w), 898 (w), 815 (w), 763 (m), 747 (m), 734 (m), 689 (m) cm^{-1}.

HRMS (ESI): m/z calcd for C$_{20}$H$_{25}$N$_4$O [M+H]$^+$: 337.2023; found: 337.2021; m/z calcd for C$_{20}$H$_{23}$N$_4$O [M-H]$^+$: 335.1877; found: 335.1872.

(rac)-N-benzyl-5-ethyl-5a,6,7,8,9,9a-hexahydropyrido[3,4-b]quinoxaline-10(5H)-carboxamide, cat11ur2

The crude product was purified by flash column chromatography on SiO$_2$ with EtOAc as eluent to afford 47% of **cat11ur2** as white foam.

R_f = 0.48 (SiO$_2$; EtAc/NEt$_3$=10:1).

^{1}H NMR (400 MHz, CDCl$_3$): δ = 8.14 (s, 1 H, H-1), 7.98 (d, ^3J = 5.8 Hz, 1 H, H-3), 7.24 (m, 5 H, Phenyl), 6.58 (d, ^3J = 5.9 Hz, 1 H, H-4), 5.56 (m, 1 H, N-H), 4.66 (dt, ^3J = 12.7 Hz, ^3J = 4.2 Hz, 1 H, H-13), 4.47 (dd, ^2J = 14.9 Hz, ^3J = 5.9 Hz, 1 H, H-19), 4.32 (dd, ^2J = 14.9 Hz, ^3J = 5.6 Hz, 1 H, H-19), 3.54 (m, 1 H, H-15), 3.48 (m, 1 H, H-12), 3.31 (m, 1 H, H-15), 2.18 (m, 1 H, H-Cy), 1.63 (m, 2 H, H-Cy), 1.44 (m, 2 H, H-Cy), 1.15 (m, 5 H, H-16 and 2x H-Cy).

^{13}C NMR (100 MHz, CDCl$_3$): δ = 155.6 (17), 146.6 (3), 144.9 (11), 144.6 (1), 138.9 (20), 128.6 (Phenyl), 127.5 (Phenyl), 127.2 (Phenyl), 120.8 (14), 106.6 (4), 52.7 (12), 49.6 (13), 44.8 (19), 39.5 (15), 28.3 (Cy), 26.1 (Cy), 24.8 (Cy), 18.9 (Cy), 12.0 (16).

IR (neat): 2938 (w), 2861 (w), 1656 (m), 1595 (m), 1510 (m), 1474 (m), 1434 (m), 1379 (m), 1357 (m), 1332 (m), 1280 (s), 1236 (s), 1168 (s), 1124 (s), 1028 (m), 1000 (w), 970 (m), 942 (w), 882 (m), 807 (m), 767 (m), 723 (m), 702 (m), 682 (s) cm^{-1}.

HRMS (ESI): m/z calcd for C$_{21}$H$_{27}$N$_4$O [M+H]$^+$: 351.2179; found: 351.2179.

(rac)-N-(3,5-bis(trifluoromethyl)phenyl)-5-ethyl-5a,6,7,8,9,9a-hexahydropyrido[3,4-b]quinoxaline-10(5H)-carboxamide, cat11ur3

The crude product was purified by flash column chromatography on SiO_2 with EtOAc as eluent to afford 66% of **cat11ur3** as off-white solid with mp 112-114°C.

R_f = 0.51 (SiO_2; EtAc/NEt_3=10:1).

^{1}H NMR (300 MHz, $CDCl_3$): δ = 8.24 (s, 1H, N-H), 8.17 (s, 1H, H-1), 8.01 (s, 2H, H-20), 7.95 (d, ^3J = 5.9 Hz, 1H, H-3), 7.51 (s, 1H, H-22), 6.64 (d, ^3J = 5.9 Hz, 1H, H-4), 4.76 (dt, ^3J = 12.6 Hz, ^3J = 4.2 Hz, 1H, H-13), 3.62 (m, 2H, H-12 and H-15), 3.41 (m, 1H, H-15), 2.26 (d, ^2J = 15.1 Hz, 1H, H-Cy), 1.68 (m, 2H, H-Cy), 1.53 (m, 3H, H-Cy), 1.25 (m, 5H, H-16 and 2H-Cy).

^{13}C NMR (75 MHz, $CDCl_3$): δ = 152.3 (17), 146.8 (3), 145.0 (11), 144.0 (1), 140.4 (19), 132.0 (q, $^2J_{C-F}$ = 33 Hz, 21), 123.0 (q, $^1J_{C-F}$ = 273 Hz, 23), 119.8 (20), 119.4 (14), 116.2 (22), 106.8 (4), 53.3 (12), 49.7 (13), 39.6 (15), 28.3 (Cy), 25.9 (Cy), 24.7 (Cy), 18.9 (Cy), 12.1 (16).

IR (neat): 3205 (w), 2964 (w), 2924 (w), 2859 (w), 1653 (s), 1594 (m), 1510 (s), 1452 (w), 1432 (w), 1366 (m), 1314 (m), 1283 (s), 1265 (m), 1234 (s), 1215 (m), 1172 (m), 1145 (w), 1125 (m), 1088 (w), 1036 (m), 968 (m), 895 (w), 842 (w), 810 (m), 766 (m), 722 (s), 708 (m), 660 (m) cm^{-1}.

HRMS (ESI): m/z calcd for $C_{22}H_{23}F_6N_4O$ [M+H]$^+$: 473.1771; found: 473.1767; m/z calcd for $C_{22}H_{21}F_6N_4O$ [M-H]$^+$: 471.1625; found: 471.1614.

Experimental part

NOESY:

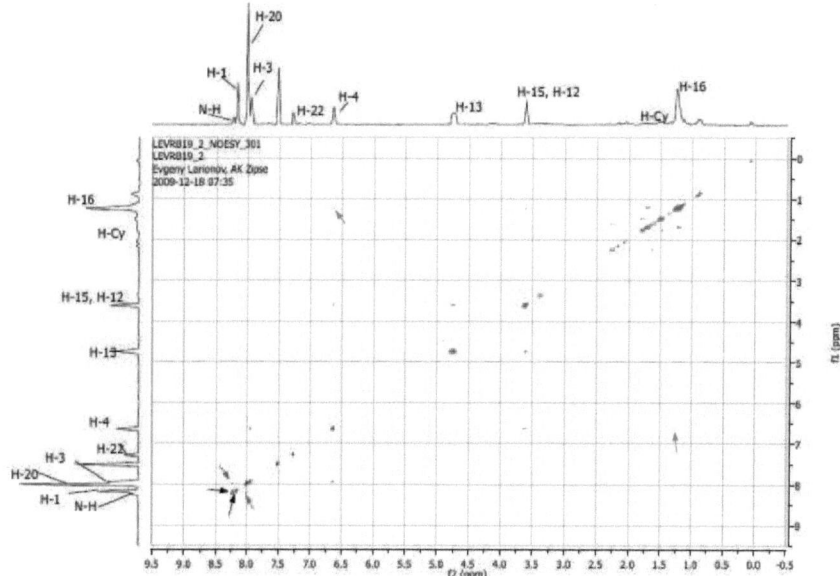

(rac)-N-(3,5-bis(trifluoromethyl)phenyl)-5-ethyl-5a,6,7,8,9,9a-hexahydropyrido[3,4-b]quinoxaline-10(5H)-carbothioamide, cat11ur4

The crude product was purified by flash column chromatography on SiO_2 with EtOAc as eluent to afford 40% of **cat11ur4** as off-white solid with m.p. 128-130 °C.

^{1}H NMR (400 MHz, $CDCl_3$): δ = 8.98 (br s, 1H, NH), 8.05 (s, 1H, CH), 7.89 (s, 2H, CH), 7.85 (d, J = 5.9 Hz, 1H, CH), 7.58 (s, 1H, CH), 6.60 (d, J = 6.0 Hz, 1H, CH), 5.65 (dt, J = 4.2 Hz, J = 12.7 Hz, 1H, CH_{Cy}), 3.66 (m, J =3.4, 1H, CH), 3.58 (m, 1H, CH_2), 3.37 (m, 1H, CH_2), 2.23 (d, J = 15.1 Hz, 1H, CH_2), 1.75 (m, 2H, CH_2), 1.63 (m, 1H, CH_2), 1.50 (m, 2H, CH_2), 1.23 (t, J = 7.1 Hz, 3H, CH_3).

^{13}C NMR (101 MHz, CDCl$_3$): δ = 180.14 (C=O), 147.03, 144.93, 144.60, 141.12, 131.43, 124.58, 121.68, 119.74, 118.54, 107.16, 104.04, 55.30, 53.32, 39.94, 28.29, 25.11, 24.36, 18.81, 12.38.

IR (neat): 2934 (w), 2360 (w), 1597 (m), 1512 (s), 1470 (w), 1443 (w), 1378 (m), 1359 (s), 1285 (s), 1240 (s), 1166 (s), 1138 (m), 1124 (s), 1067 (w), 1000 (m), 973 (m), 884 (w), 846 (w), 804 (m), 771 (m), 680 (s) cm^{-1}.

HRMS (ESI): *m/z* calcd for C$_{22}$H$_{23}$F$_6$N$_4$S [M+H]$^+$: 473.1548; found: 489.1537.

E4.2 Chiral catalysts.

E4.2.1 The catalysts **PhEt** and **CyPhEt** were prepared from (*S*)-(1-isocyanatoethyl)benzene and amines **31d** and **31b**, respectively, according to general procedure III.

(*S*)-1-Ethyl-*N*-(1-phenylethyl)-2,3-dihydropyrido[3,4-b]pyrazine-4(1H)-carboxamide, PhEt

The crude product was purified by flash-chromatography on SiO_2 with EtOAc as eluent to afford 69% of **PhEt** as white foam.

$[\alpha]_D^{22} = +37.4°$ (c = 1, $CHCl_3$).

$R_f = 0.26$ (SiO_2; EtAc/NEt_3=10:1).

^{1}H NMR (300 MHz, $CDCl_3$): δ = 8.23 (s, 1 H, H-5), 8.06 (d, ^3J = 5.8 Hz, 1 H, H-7), 7.30 (m, 5 H, Phenyl), 6.56 (d, 7.30, ^3J = 5.8 Hz, 1 H, H-8), 5.49 (d, ^3J = 7.4 Hz, 1 H, N-H), 5.04 (m, 1 H, H-15), 3.78 (m, 2 H, H-3), 3.40 (q, ^3J = 7.3 Hz, 2 H, H-11), 3.36 (m, 2 H, H-2), 1.47 (d, ^3J = 6.9 Hz, 3 H, H-20), 1.20 (t, ^3J = 7.2 Hz, 3 H, H-12).

^{13}C NMR (75 MHz, $CDCl_3$): δ = 154.8 (13), 146.8 (7), 143.9 (C_q), 143.7 (C_q), 143.5 (5), 128.7 (19 or 18), 127.2 (19 or 18), 125.9 (17), 121.8 (8), 105.4 (C_q), 50.5 (15), 47.4 (2), 44.6 (11), 38.9 (3), 22.5 (20), 11.1 (12).

IR (neat): 3205 (w), 2971 (w), 2930.7 (w), 1675 (m), 1594 (s), 1517 (s), 1449 (m), 1337 (s), 1449 (m), 1235 (s), 1220 (m), 1181 (s), 1120 (m), 1092 (m), 1043 (m), 941 (w), 911 (w), 883 (m), 800 (m), 749 (s), 680 (s), 663 (m) cm^{-1}.

HRMS (ESI): m/z calcd for $C_{18}H_{23}N_4O$ [M+H]$^+$: 311.1866; found: 311.1866; m/z calcd for $C_{18}H_{21}N_4O$ [M-H]$^+$: 309.1721; found: 309.1718.

(rac)-5-Ethyl-N-((S)-1-phenylethyl)-5a,6,7,8,9,9a-hexahydropyrido[3,4-b]quinoxaline-10(5H)-carboxamide, CyPhEt

The crude product was purified by flash-chromatography on SiO_2 with EtOAc as eluent to afford 80% of **CyPhEt** as off-white foam. The product is obtained as a diastereomeric mixture with de = 1:1 (determined by NMR).

R_f = 0.54 (SiO_2; EtAc/NEt$_3$=10:1).

^{1}H NMR (300 MHz, CDCl$_3$): δ = 8.23 (s, 1 H, H-1′), 8.13 (s, 1 H, H-1), 8.05 (d, ^3J = 5.8 Hz, 1 H, H-3′), 8.04 (d, ^3J = 5.8 Hz, 1 H, H-3), 7.34 (m, 4 H, Phenyl+Phenyl′), 7.26 (m, 4 H, Phenyl+Phenyl′), 7.19 (m, 2 H, Phenyl+Phenyl′), 6.62 (d, ^3J = 5.8 Hz, 1 H, H-4′), 6.61 (d, ^3J = 5.9 Hz, 1 H, H-4), 5.35 (m, 2 H, 2xN-H), 5.00 (m, 2 H, H-19+H-19′), 4.62 (m, 2 H, H-13+H-13′), 3.56 (m, 2 H, H-15+H-15′), 3.49 (m, 1 H, H-12′), 3.39 (m, 1 H, H-12), 3.32 (m, 2 H, H-15+H-15′), 2.16 (m, 2 H, H-Cy+H-Cy′), 1.62 (m, 8 H, H-Cy+H-Cy′), 1.49 (d, ^3J = 6.9 Hz, 3 H, H-24′), 1.41 (m, 6 H, H-Cy+H-Cy′), 1.36 (d, ^3J = 7.0 Hz, 3 H, H-24), 1.19 (t, ^3J = 7.1 Hz, 3 H, H-16′), 1.15 (t, ^3J = 7.2 Hz, 3 H, H-16).

^{13}C NMR (75 MHz, CDCl$_3$): δ = 154.8 (17′), 154.6 (17), 146.6 (3′), 146.5 (3), 144.9 (11+11′), 144.4 (1′), 144.3 (1), 144.0 (20+20′), 128.7 (22′), 128.6 (22), 127.2 (23′), 127.1 (23), 126.0 (21′), 125.9 (21), 121.0 (14′), 120.9 (14), 106.8 (4′), 106.7 (4), 52.9 (12′), 52.8 (12), 50.5 (19′), 50.4 (19), 49.6 (13′), 49.4 (13), 39.7 (15′), 39.5 (15), 28.4 (Cy′), 28.2 (Cy), 26.1 (Cy′), 26.0 (Cy), 24.8 (Cy+Cy′), 22.6 (24′), 22.4 (24), 19.0 (Cy′), 18.9 (Cy), 12.1 (16′), 12.0 (16).

IR (neat): 3208 (w), 2971 (w), 2933 (w), 2861 (w), 1737 (w), 1649 (m), 1592 (m), 1506 (s), 1448 (m), 1353 (m), 1317 (m), 1281 (s), 1234 (s), 1212 (m), 1173 (m), 1144 (w), 1096 (m), 1066 (m), 1029 (m), 970 (m), 910 (w), 802 (m), 760 (m), 730 (s), 690 (s) cm^{-1}.

HRMS (ESI): m/z calcd for $C_{22}H_{29}N_4O$ [M+H]$^+$: 365.2336; found: 365.2333; m/z calcd for $C_{22}H_{27}N_4O$ [M-H]$^+$: 363.2190; found: 363.2186.

Experimental part

E4.2.2 General procedure IV. Synthesis of the isocyanato carboxylic acid esters[66] and subsequent reaction with amine 31b or 31d.

A solution of (S)-aminoacid methyl ester (1.43 mmol) in CH_2Cl_2 (2 mL) was added to a solution of di-*tert*-butyl dicarbonate (436 mg, 2.0 mmol) and DMAP (87 mg, 0.71 mmol) in CH_2Cl_2 (2 mL) and the reaction mixture was stirred for 10 min at RT. Then a solution of amine **31b** or **31d** (1.43 mmol) in CH_2Cl_2 (2 mL) was added. The reaction mixture was stirred overnight at RT and then quenched with 1 mL methanol. The solvents were removed under reduced pressure and the crude mixture was purified by flash column chromatography on SiO_2.

(S)-Methyl 2-(1-ethyl-1,2,3,4-tetrahydropyrido[3,4-b]pyrazine-4-carboxamido)-3-phenylpropanoate, PheOMe

The crude product was purified by flash column chromatography on SiO_2 with EtOAc/MeOH (10:1) as eluent to afford 50% of product as off-white oil.

$[\alpha]_D^{22} = +9.6°$ (c = 1, $CHCl_3$).

^{1}H NMR (300 MHz, $CDCl_3$): δ = 8.05 (d, ^3J = 5.8 Hz, 1H, H-2), 8.01 (s, 1H, H-1), 7.28 (m, 3H, H-12), 7.09 (m, 2H, H-12), 6.52 (d, ^3J = 5.8 Hz, 1H, H-3), 5.55 (d, ^3J = 7.9 Hz, 1H, NH-8), 4.76 (td, ^3J = 5.6 Hz, ^3J = 7.6 Hz, 1H, H-9), 3.95 (m, 1H, H-7), 3.72 (s, 3H, H-11), 3.52 (m, 1H, H-7), 3.37 (q, ^3J = 7.1 Hz, 2H, H-5), 3.29 (m, 2H, H-6), 3.18 (dd, ^3J = 5.5 Hz, ^2J = 13.9 Hz, 1H, H-10), 3.00 (dd, ^3J = 7.4 Hz, ^2J = 13.8 Hz, 1H, H-10), 1.18 (t, ^3J = 7.1 Hz, 3H, H-4).

^{13}C NMR (75 MHz, $CDCl_3$): δ = 172.5 (C=O), 155.1, 147.1, 144.0, 143.4, 136.0, 129.0, 128.7, 127.1, 121.3, 105.3, 54.8, 52.3, 47.3, 44.5, 39.1, 37.9, 11.1.

IR (neat): 2973 (w), 2933 (w), 1741 (m), 1654 (m), 1597 (m), 1512 (s), 1453 (w), 1434 (w), 1346 (m), 1286 (m), 1237 (m), 1198 (s), 1176 (s), 1090 (w), 1043 (m), 931 (w), 884 (w), 804 (w), 700 (s), 664 (m) cm^{-1}.

HRMS (ESI): m/z calcd for $C_{20}H_{25}N_4O_3$ [M+H]$^+$: 369.1927; found: 369.1919.

(S)-Methyl 2-(1-ethyl-1,2,3,4-tetrahydropyrido[3,4-b]pyrazine-4-carboxamido)-3-methylbutanoate, ValOMe

The crude product was purified by flash column chromatography on SiO_2 with EtOAc/MeOH (10:1) as eluent to afford 68% of product as off-white oil.

$[\alpha]_D^{22}$ = +18.4° (c = 1, $CHCl_3$).

^{1}H NMR (300 MHz, $CDCl_3$): δ = 8.28 (s, 1H, CH), 8.04 (d, J = 5.8 Hz, 1H, CH), 6.54 (d, J = 5.8 Hz, 1H, CH), 5.56 (d, J = 8.6 Hz, 1H, NH), 4.41 (dd, J = 5.1 Hz, J = 8.6 Hz, 1H, CHNH), 4.02 (d, J = 13.0 Hz, 1H, CH₂), 3.69 (s, 3H, COOCH₃), 3.49 (m, 1H, CH₂), 3.36 (dd, J = 5.1 Hz, J = 12.0 Hz, 4H, 2CH₂), 2.12 (m, 1H, CH(CH₃)₂), 1.18 (t, J = 7.1 Hz, 3H, CH₂CH₃), 0.95 (d, J = 6.8 Hz, 3H, CH(CH₃)₂), 0.85 (d, J = 6.9 Hz, 3H, CH(CH₃)₂).

^{13}C NMR (75 MHz, $CDCl_3$): δ = 172.9 (C=O), 155.4, 146.9, 143.7, 143.4, 121.5, 105.4, 58.9, 52.04, 47.4, 44.6, 39.1, 31.0, 19.2, 18.0, 11.0.

IR (neat): 2965 (w), 2874 (w), 2362 (w), 1738 (m), 1657 (s), 1595 (s), 1502 (s), 1433 (m), 1347 (s), 1287 (m), 1235 (m), 1179 (s), 1118 (m), 1078 (w), 1042 (m), 996 (w), 926 (w), 883 (w), 804 (m), 743 (m), 730 (s) cm^{-1}.

HRMS (ESI): m/z calcd for $C_{16}H_{25}N_4O_3$ [M+H]$^+$: 321.1927; found: 321.1920.

(S)-Methyl 2-(1-ethyl-1,2,3,4-tetrahydropyrido[3,4-b]pyrazine-4-carboxamido)-3,3-dimethylbutanoate, TleOMe

The crude product was purified by flash column chromatography on SiO_2 with EtOAc/MeOH (10:1) as eluent to afford 34% of product as yellow oil.

$[\alpha]_D^{22}$ = +21.4° (c = 1, $CHCl_3$).

^{1}H NMR (300 MHz, $CDCl_3$): δ = 8.21 (s, 1H, H-1), 7.98 (d, J = 5.8 Hz, 1H, H-2), 6.49 (d, J = 5.8 Hz, 1H, H-3), 5.59 (d, J = 9.0 Hz, 1H, NH-8), 4.25 (d, J = 9.1 Hz, 1H, H-9), 3.99 (m, 1H, H-7), 3.62 (s, 3H, H-11), 3.41 (m, 1H, H-7), 3.31 (m, 4H, H-5 and H-6), 1.12 (t, J = 7.1 Hz, 3H, H-4), 0.89 (s, 9H, H-10).

¹³C NMR (75 MHz, CDCl₃) δ = 172.4 (C=O), 155.3, 147.0, 143.7, 143.4, 121.3, 105.4, 61.9, 51.6, 47.3, 44.5, 39.1, 34.3, 26.7, 11.0.

IR (neat): 2964 (w), 2872 (w), 1736 (m), 1667 (s), 1594 (m), 1502 (s), 1476 (w), 1432 (w), 1340 (s), 1286 (m), 1237 (m), 1208 (s), 1180 (s), 1164 (s), 1089 (w), 1041 (m), 990 (w), 880 (w), 801 (w), 754 (m), 732 (m) cm⁻¹.

HRMS (ESI): *m/z* calcd for C₁₇H₂₇N₄O₃ [M+H]⁺: 335.2083; found: 335.2076.

(*S*)-Methyl 2-((*cis*)-5-ethyl-5,5a,6,7,8,9,9a,10-octahydropyrido[3,4-b]quinoxaline-10-carboxamido)-3-phenylpropanoate, CyPheOMe

The crude product was purified by flash column chromatography on SiO₂ with EtOAc/MeOH (20:1) as eluent to afford 59% of **CyPheOMe** as yellow oil. Diastereomeric ratio determined by ¹H NMR: de = 1.2:1.

¹H NMR (300 MHz, CDCl₃): δ = 8.14 (s, 1H, Py), 8.06 (d, *J* = 6.1 Hz, 1H, Py), 8.04 (d, *J* = 5.9 Hz, 1H, Py'), 7.31 (m, 4H, Ph+Ph'), 7.16 (m, 4H, Ph+Ph'), 6.96 (m, 2H, Ph+Ph'), 6.63 (d, *J* = 6.0 Hz, 1H, Py), 6.60 (d, *J* = 5.7 Hz, 1H, Py'), 5.55 (d, *J* = 7.6 Hz, 1H, NH), 5.40 (d, *J* = 7.9 Hz, 1H, NH'), 4.73 (dt, *J* = 6.4 Hz, *J* = 12.9 Hz, 2H, CH+CH'), 4.57 (dq, *J* =4.0 Hz, *J* = 12.1 Hz, 2H, CH+CH'), 3.73 (s, 3H, COOCH₃), 3.68 (s, 3H, COOCH₃'), 3.56 (m, 2H), 3.45 (m, 1H), 3.32 (m, 3H), 3.19 (dd, *J* = 5.5 Hz, ²*J* = 13.9 Hz, 1H, CH₂'), 3.10 (dd, *J* = 5.7 Hz, ²*J* = 13.8 Hz, 1H, CH₂), 2.98 (m, 2H), 2.16 (m, 2H), 1.68 (m, 2H), 1.55 (m, 4H), 1.41 (m, 4H), 1.17 (m, 10H).

¹³C NMR (75 MHz, CDCl₃): δ = 172.7 (C=O), 172.2 (C=O), 155.1, 154.8, 146.9, 146.7, 144.9, 144.86, 144.7, 144.6, 136.1, 136.0, 129.1, 129.0, 128.8, 128.4, 127.2, 126.9, 120.6, 120.3, 106.6, 106.6, 55.0, 54.7, 52.9, 52.6, 52.3, 52.2, 49.7, 39.8, 39.5, 38.0, 37.8, 28.3, 28.2, 26.0, 24.8, 18.9, 18.9, 12.2, 12.0.

HRMS (ESI): *m/z* calcd for C₂₄H₃₁N₄O₃ [M+H]⁺: 423.2396; found: 423.2389.

E4.2.3. General procedure V. Grignard reaction.

A solution of Grignard reagent PhMgBr or NphMgBr (25 mmol) in THF (25 mL) was added to a solution of the corresponding methyl ester (1.4 mmol) in THF (10 mL) at 0 °C. After stirring for 48 h at RT the reaction mixture was quenched by adding 20 ml of saturated aq. NH_4Cl, and the precipitate was filtered off. The filtrate was extracted with CH_2Cl_2 (3 × 20 ml) and the organic layer was dried over $MgSO_4$. The crude product was purified by the column chromatography on SiO_2.

(S)-N-(1-Benzyl-2-hydroxy-2,2-diphenylethyl)-1-ethyl-2,3-dihydropyrido[3,4-b]pyrazine-4(1H)-carboxamide, PhePh₂OH

The crude product was purified by flash-chromatography on SiO_2 with EtOAc/MeOH (10:1) as eluent to afford 27% of **PhePh₂OH** as off-white foam.

$[\alpha]_D^{22}$ = -19.6° (c = 1, CH_2Cl_2).

¹H NMR (300 MHz, $CDCl_3$): δ = 7.71 (m, 2H, 2CH), 7.63 (d, J = 7.2 Hz, 2H, 2CH), 7.52 (s, 1H, CH_{Py}), 7.24 (m, 12H, CH), 6.25 (d, J = 6.0 Hz, 1H, CH_{Py}), 6.09 (d, J = 8.9 Hz, 1NH), 5.19 (m, 1H, CHNH), 3.67 (m, 2H, CH_2), 3.27 (m, 3H, CH_2), 2.99 (m, 3H, CH_2), 1.29 (s, 1H, OH), 1.09 (t, J = 7.1 Hz, 3H, CH_3).

¹³C NMR (75 MHz, $CDCl_3$): δ = 171.1 (C=O), 156.0, 146.8, 145.6, 145.0, 143.6, 142.8, 139.7, 129.4, 128.3, 128.2, 127.8, 126.6, 126.4, 126.2, 126.0, 125.6, 121.6, 105.1, 80.8, 60.3, 47.2, 44.6, 39.4, 29.7, 11.1.

IR (neat): 2917 (w), 2874 (w), 2361 (w), 1734 (w), 1654 (m), 1600 (s), 1514 (s), 14483 (m), 1343 (s), 1288 (m), 1236 (s), 1177 (s), 1118 (m), 1062 (m), 1044 (s), 963 (w), 886 (w), 803 (m), 748 (s), 739 (s) cm^{-1}.

HRMS (ESI): m/z calcd for $C_{31}H_{33}N_4O_2$ [M+H]⁺: 493.2603; found: 493.2594.

Experimental part

(S)-N-(2-Hydroxy-1-isopropyl-2,2-diphenylethyl)-1-ethyl-2,3-dihydropyrido[3,4-b]pyrazine-4(1H)-carboxamide, ValPh₂OH

The crude product was purified by flash-chromatography on SiO_2 with EtOAc/MeOH (10:1) as eluent to afford 63% of **ValPh₂OH** as off-white solid with m.p. 195-198 °C.

$[\alpha]_D^{22}$ = -36.8° (c = 1, CHCl₃).

^{1}H NMR (300 MHz, CDCl₃): δ = 7.97 (s, 1H, CH$_{Py}$), 7.59 (t, J = 6.5 Hz, 4H, m-CH$_{Ph}$), 7.22 (m, 6H, o- and p-CH$_{Ph}$), 6.37 (m, 2H, CH$_{Py}$ and NH), 5.96 (d, J = 5.7 Hz, 1H, CH$_{Py}$), 5.18 (d, J = 9.4 Hz, 1H, CHNH), 4.70 (d, J = 12.9 Hz, 1H, CH$_2$NCO), 3.37 (m, 1H, CH$_2$), 3.13 (m, 3H, CH$_2$), 2.85 (m, 1H, CH$_2$), 2.00 (m, 1H, CH(CH$_3$)$_2$), 1.17 (d, J = 6.5 Hz, 3H, CH(CH_3)$_2$), 1.04 (t, J = 6.9 Hz, 3H, CH$_2$CH_3), 0.96 (d, J = 6.8 Hz, 3H, CH(CH_3)$_2$).

^{13}C NMR (75 MHz, CDCl₃): δ = 156.4, 148.0, 146.2, 145.2, 143.4, 143.3, 128.0, 127.6, 126.2, 126.1, 126.0, 125.4, 121.7, 105.4, 81.8, 60.7, 48.0, 44.5, 39.5, 29.1, 23.4, 19.4, 11.2.

HRMS (EI): m/z calcd for C$_{27}$H$_{32}$N$_4$O$_2$ [M$^+$]: 444.2525; found: 444.2513.

(cis)-N-((S)-1-Benzyl-2-hydroxy-2,2-diphenylethyl)-5-ethyl-5a,6,7,8,9,9a-hexahydropyrido[3,4-b]quinoxaline-10(5H)-carboxamide, CyPhePh₂OH

The crude product was purified by column chromatography on SiO_2 with EtOAc/NEt₃ (20:1) as eluent to afford 83% of product as 1.2:1 diastereomeric mixture. Diastereomeric ratio can be determined by NMR, as well as by chiral HPLC (OD-H column), giving the same results. This mixture can be resolved by the recrystallization from EtOH/ethylacetate mixture, giving a single diastereomer with de = 92% (determined by chiral HPLC) as white solid with m.p. 200-205 °C (decomp).

Conditions for HPLC: Daicel Chiralpak OD-H column analytical, hexane/i-propanol = 95/5, flow-rate 1.0 mL/min, detection at 254 nm, 25°C. Injection volumes: 5-10 μL. Retention times: 12.3 min (minor diastereomer), 13.8 min (major).

Experimental part

$[\alpha]_D^{22}$ = -22.2° (c = 1, CH_2Cl_2).

^{1}H NMR (400 MHz, $CDCl_3$): δ = 7.68 (d, J = 7.5 Hz, 2H), 7.61 (m, 3H), 7.39 – 7.06 (m, 11H), 6.62 (br s, 1H), 6.17 (d, J = 8.9 Hz, 1H, NH), 6.09 (d, J = 5.9 Hz, 1H), 5.27 (br s, 1H, CHNH), 4.56 – 4.44 (m, 1H, CH), 3.48 – 3.33 (m, 1H, CH_2), 3.25 – 3.06 (m, 3H, CH_2), 2.95 (d, J = 11.5 Hz, 1H, CH_2), 2.12 (d, J = 15.0 Hz, 1H, CH_2), 1.74 – 1.57 (m, 1H), 1.57 – 1.32 (m, 4H), 1.12 – 0.95 (m, 4H), 0.96 – 0.84 (m, 1H).

^{13}C NMR (75 MHz, $CDCl_3$): δ = 171.1, 155.7, 147.1, 145.6, 145.0, 144.6, 144.1, 139.7, 129.4, 128.2, 127.7, 126.5, 126.3, 126.2, 126.1, 125.6, 120.7, 106.4, 80.8, 52.6, 49.6, 38.7, 28.2, 25.8, 24.7, 18.8, 11.9.

IR (neat): 3201 (w), 2936 (w), 2918 (w), 2850 (w), 2362 (w), 1725 (w), 1644 (m), 1626 (m), 1598 (m), 1509 (s), 1448 (m), 1343 (s), 1316 (m), 1282 (s), 1235 (s), 1172 (m), 1062 (m), 1036 (m), 968 (w), 897 (w), 800 (m), 745 (s), 730 (s) cm^{-1}.

HRMS (ESI): m/z calcd for $C_{35}H_{39}N_4O_2$ [M+H]$^+$: 547.3073; found: 547.3063.

(S)-N-(1-Benzyl-2-hydroxy-2,2-di(1-naphthyl)-ethyl)-1-ethyl-2,3-dihydropyrido[3,4-b]pyrazine-4(1H)-carboxamide, PheNph$_2$OH

The crude product was purified by column chromatography on SiO_2 with EtOAc/MeOH (20:1) as eluent to afford 34% of **PheNph$_2$OH** as off-white foam.

$[\alpha]_D^{22}$ = -8.4° (c = 1, CH_2Cl_2).

^{1}H NMR (300 MHz, $CDCl_3$): δ = 8.31 (d, J = 20.5 Hz, 2H, 2CH), 7.78 (m, 11H, CH), 7.47 (m, 6H, CH), 7.21 (m, 1H, CH), 7.06 (br s, 1H, CH), 6.21 (d, J = 9.0 Hz, 1H, OH), 6.02 (d, J = 6.0 Hz, 1H, CH_{Py}), 5.44 (br s, 1H, NH), 3.78 (m, 2H, CH_2), 3.24 (m, 2H, CH_2), 3.01 (m, 2H, CH_2), 2.81 (m, 3H), 0.74 (t, J = 7.1 Hz, 3H, CH_3).

^{13}C NMR (75 MHz, $CDCl_3$): δ = 156.1 (C=O), 145.6, 144.1, 143.7, 143.3, 142.7, 139.5, 133.3, 133.1, 132.3, 132.2, 132.0, 129.5, 128.4, 128.3, 128.3, 128.1, 127.4, 127.4, 127.3, 126.2, 126.0, 125.8, 125.8, 125.6, 125.3, 124.3, 124.2, 121.5, 105.1, 81.1, 46.8, 44.1, 39.4, 10.6.

Experimental part

IR (neat): 3055 (w), 2925 (w), 2360 (w), 1657 (m), 1600 (s), 1506 (s), 1452 (m), 1340 (s), 1270 (m), 1237 (m), 1178 (m), 1121 (m), 1081 (m), 1044 (s), 886 (w), 817 (m), 745 (s), 695 (s) cm^{-1}.

HRMS (ESI): m/z calcd for $C_{39}H_{37}N_4O_2$ [M+H]$^+$: 593.2917; found: 593.2906.

(*cis*)-N-((*S*)-1-Benzyl-2-hydroxy-2,2-di(1-naphthyl)-ethyl)-5-ethyl-5a,6,7,8,9,9a-hexahydropyrido[3,4-b]quinoxaline-10(5H)-carboxamide, CyPheNph$_2$OH

The crude product was purified by column chromatography on SiO$_2$ with EtOAc/NEt$_3$ (20:1) as eluent to afford 60% of **CyPheNph$_2$OH** as 1.2:1 diastereomeric mixture. Diastereomeric ratio can be determined by NMR, as well as by chiral HPLC (OD-H column), giving the same results.

Conditions for HPLC: Daicel Chiralpak OD-H column analytical, hexane/*i*-propanol = 75/25, flow-rate 0.3 mL/min, detection at 254 nm, 25°C. Injection volumes: 5-10 μL. Retention times: 18.6 min and 20.6 min.

^{1}H NMR (300 MHz, CDCl$_3$): δ = 8.31 (m, 8H), 7.79 (m, 33H), 7.42 (m, 16H), 7.26 (m, 18H), 6.45 (s, 1H, CH), 6.26 (m, 1H, NH), 6.07 (s, 1H, CH), 5.68 (m, 1H, NH), 4.39 (m, 3H), 2.99 (m, 12H), 2.65 (m, 2H), 2.12 (m, 2H), 1.91 (m, 4H), 1.62 (m, 4H), 1.30 (m, 19H), 0.89 (m, 11H).

^{13}C NMR (75 MHz, CDCl$_3$): δ = 155.8, 144.9, 144.1, 143.9, 143.2, 142.8, 142.6, 139.5, 139.5, 133.3, 133.2, 133.1, 133.0, 132.2, 132.2, 132.1, 132.0, 129.4, 129.4, 128.6, 128.4, 128.3, 128.3, 128.3, 128.2, 128.2, 128.1, 128.0, 127.9, 127.6, 127.5, 127.4, 127.3, 126.2, 126.0, 125.8, 125.8, 125.6, 125.5, 125.4, 124.9, 124.3, 124.2, 124.1, 124.1, 124.1, 124.0, 120.7, 106.3, 106.1, 81.3, 80.9, 60.4, 52.5, 52.2, 49.8, 49.8, 49.6, 39.7, 38.5, 38.4, 28.0, 27.9, 25.9, 25.7, 24.6, 18.9, 18.9, 18.8, 11.8, 11.1, 8.7.

HRMS (ESI): m/z calcd for $C_{43}H_{43}N_4O_2$ [M+H]$^+$: 647.3386; found: 647.3375.

E4.3. Synthesis of the catalyst 53.

(S)-2-(1-Benzyl-2-(1-ethyl-2,3-dihydropyrido[3,4-b]pyrazin-4(1H)-yl)-2-oxo-ethyl)isoindoline-1,3-dione, 51

1.04 g of **31d** (6.35 mmol, 1.0 equiv.) were dissolved in 15 mL of dry THF. 4.1 mL (8.25 mmol, 1.3 equiv.) of a 2.0 M solution of sodium bis(trimethylsilyl)amide in THF were added dropwise to this solution through a rubber septum at -78° C. After 30 minutes (S)-2-(1,3-dioxoisoindolin-2-yl)-3-phenylpropanoyl chloride **PhtPheCl** (8.25 mmol, 1.3 equiv.) in 10 mL THF was added dropwise and the reaction was allowed to warm to room temperature overnight. The reaction was quenched by adding 5 mL of saturated aq. NH$_4$Cl. The aqueous layer was extracted with EtOAc (3 × 20 ml) and CH$_2$Cl$_2$ (3 × 20 ml). The combined organic layers were dried over Na$_2$SO$_4$, filtered and evaporated under reduced pressure to yield an oil that was purified through column chromatography on silica gel with EtOAc/NEt$_3$ (20:1) as eluent to afford 33% of compound **51** as pale yellow foam.

^{1}H NMR (300 MHz, CDCl$_3$): δ = 8.26 (br s, 1H), 7.91 (d, J = 5.7 Hz, 1H), 7.64 (m, 4H), 7.14 (m, 5H), 6.17 (br s, 1H), 5.64 (m, 1H, NCHCO), 4.23 (br s, 1H, CH$_2$), 3.44 (m, 3H, CH$_2$), 3.27 (m, 2H, CH$_2$), 3.05 (s, 1H, CH$_2$), 1.04 (t, J = 7.1 Hz, 3H, CH$_3$).

^{13}C NMR (75 MHz, CDCl$_3$): δ = 167.8 (C=O), 166.7 (C=O), 147.8, 143.0, 136.8, 133.8, 131.3, 129.2, 128.4, 126.7, 123.1, 120.9, 104.7, 53.2, 47.3, 44.1, 40.1, 35.0, 10.9.

HRMS (ESI): m/z calcd for C$_{26}$H$_{25}$N$_4$O$_3$ [M+H]$^+$: 441.1927; found: 441.1917.

(S)-2-Amino-1-(1-ethyl-2,3-dihydropyrido[3,4-b]pyrazin-4(1H)-yl)-3-phenylpropan-1-one, 52

Hydrazine hydrate (4.0 mmol, 200 mg, 2 equiv.) was added to a solution of **51** (2.0 mmol, 880 mg, 1 equiv.) in MeOH (10 mL). After stitrring for 48 h at RT 5 ml of CH$_2$Cl$_2$ was added to the reaction mixture and the precipitate was filtered off. The solvent was removed under

Experimental part

reduced pressure and the residue was purified through column chromatography on SiO$_2$ with EtOAc/TEA/MeOH (10:1:1) as eluent to afford 83% of compound **52** as yellow oil.

$[\alpha]_D^{22}$ = +50.2° (c = 1, CHCl$_3$).

^{1}H NMR (300 MHz, CDCl$_3$): δ = 8.06 (br s, 1H), 7.96 (m, 1H), 7.03 (m, 4H), 6.84 (m, 1H), 6.36 (br s, 1H), 5.91 (s, 2H, NH$_2$), 4.38 (t, J = 7.4 Hz, 1H), 3.01 (m, 9H), 1.28 (m, 3H).

^{13}C NMR (75 MHz, CDCl$_3$): δ = 175.2, 167.8, 146.1, 146.1, 136.7, 129.0, 128.9, 128.7, 128.5, 128.3, 128.2, 126.5, 120.7, 105.0, 52.2, 47.2, 47.1, 44.4, 42.7, 34.2, 23.1, 22.2, 14.7, 10.9.

IR (neat): 2972 (w), 2932 (w), 2361 (w), 2342 (w), 1653 (m), 1597 (s), 1526 (s), 1432 (m), 1349(s), 1288 (m), 1236 (s), 1199 (s), 1170 (s), 1119 (m), 1075 (m), 1042 (m), 943 (w), 878 (w), 803 (m), 746 (s), 668 (m) cm^{-1}.

HRMS (ESI): m/z calcd for C$_{18}$H$_{23}$N$_4$O [M+H]$^+$: 311.1872; found: 311.1863.

(S)-1-(3,5-Bis(trifluoromethyl)phenyl)-3-(1-(1-ethyl-2,3-dihydropyrido[3,4-b]pyrazin-4(1H)-yl)-1-oxo-3-phenylpropan-2-yl)urea, 53

The compound **53** was prepared from 3,5-bis(trifluoromethyl)phenylisocyanate and amine **52** according to general procedure III. The crude product was purified by flash-chromatography on SiO$_2$ with EtOAc as eluent to afford 79% of compound **53** as yellow solid with m.p. 102-104°C.

$[\alpha]_D^{22}$ = +41.0° (c = 1, CHCl$_3$).

^{1}H NMR (400 MHz, C$_6$D$_6$, 60 °C): δ = 8.78 (br s, 1H, NH), 7.94 (m, 3H), 7.33 (s, 1H), 6.94 (m, 5H), 6.00 (m, 1H), 5.47 (br s, 1H, NH), 3.51 (m, 1H), 3.03 (m, 3H), 2.60 (m, 5H), 0.62 (m, 3H).

^{13}C NMR (101 MHz, C$_6$D$_6$, 60 °C): δ = 170.83, 154.64, 154.57, 154.47, 144.77, 144.72, 144.66, 144.64, 143.41, 143.35, 141.88, 131.89, 131.57, 124.95, 122.24, 117.76, 117.73, 114.45, 114.40, 105.36, 57.58, 51.47, 46.74, 46.29, 43.96, 43.71, 39.46, 39.39, 39.28, 39.26, 37.59, 37.56, 26.29, 10.35, 10.06.

IR (neat): 3338 (w), 2937 (w), 2362 (w), 1702 (w), 1655 (w), 1599 (m), 1528 (m), 1474 (m), 1386 (m), 1348 (w), 1275 (s), 1228 (m), 1170 (s), 1123 (s), 1043 (w), 940 (w), 878 (m), 804 (m), 731 (w), 700 (m) cm^{-1}.

HRMS (ESI): m/z calcd for C$_{27}$H$_{26}$F$_6$N$_5$O$_2$ [M+H]$^+$: 566.1991; found: 566.1979.

Experimental part

E4.4 Kinetic measurements

E4.4.1 General procedure for kinetic measurements.

Three stock solutions were prepared in dry calibrated 5 mL flasks. Stock solution A: 1.2 M in acetic anhydride (612.5 mg) and 0.6 M in 1,4-dioxane. Stock solution B: 1.8 M in NEt$_3$ (910.7 mg) and 0.6 M in ethynylcyclohexanol (372.5 mg). Stock solution C: catalyst (x mol% relative to alcohol, x = 2.5 –15). A nitrogen-flushed dry NMR tube was filled with 0.2 mL each of stock solution A, B and C. The NMR tube was fused with a gas burner and immediately injected into a Varian Mercury 200 MHz NMR. FIDs were collected in defined time intervals and the obtained multiple FID ^{1}H NMRs analyzed by integration of the product peaks with a self-written subroutine using the VNMR software package. The conversion was calculated directly from the intensities of dioxane (an internal standard) and product peaks:

$$\text{conv.} = \frac{(I_{Ester}/3)}{(I_{Dioxane}/8)} \cdot 100\% \tag{1}$$

Dependence of the conversion vs. time t was fitted by equation (2):

$$\text{conv.} = y_0\left(1 - \frac{1}{2e^{k(t-t_0)} - 1}\right) \tag{2}$$

$$k = k_2[ROH]_0 \tag{3}$$

E4.4.2 Initial rate measurements with catalyst cat11ur3.

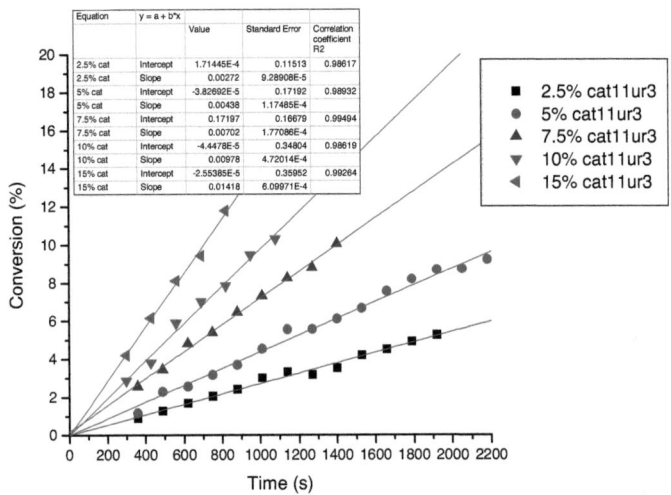

E4.4.3 Kinetic measurements with 10% catalyst loading.

Table E4.1. Catalytic activity of the 3,4-diamino and 4-aminopyridines.[a]

36a → (10 mol% catalyst, Ac_2O, NEt_3, $CDCl_3$, 23 °C) → 36b

Catalyst	run	k_2, l mol^{-1} s^{-1}	Δk_2	c_1	t_0, min	$t_{1/2}$, min	$\Delta t_{1/2}$, min
cat81ur1	1	8.56E-05	3.24E-07	1.009	-26	**395**	1
cat81ur1f	1					**1200**[b]	
cat81ur3	1	4.67E-05	4.69E-07	0.947	-37	**723**	7
cat81ur5	1	4.49E-05	3.92E-07	0.871	-44	**753**	7
cat81ur5	2	4.85E-05	3.59E-07	0.899	-49	697	5
	averaged					**725**	20
cat11ur1	1	7.74E-05	3.27E-07	0.886	-41	**436**	2
cat11ur1	2	8.72E-05	7.10E-07	0.995	-41	387	3
	averaged					**412**	17
cat11ur2	1	8.55E-05	6.42E-07	0.911	-13	**395**	3
cat11ur3	1	1.32E-04	1.96E-06	0.896	-68	**256**	4
cat11ur3	2	1.40E-04	2.11E-06	0.898	-46	241	4
	averaged					**249**	5
cat11ur4	1	9.95E-05	1.01E-06	0.982	-4	**339**	3
PheOMe	1	2.84E-05	4.20E-07	0.990	-90	**1192**	18
ValPh$_2$OH	1	4.35E-05	3.93E-07	0.956	-78	**776**	7
CyPhePh$_2$OH	1	7.40E-05	3.43E-07	0.973	-13	**457**	2
53	1	2.13E-04	7.28E-06	0.582	-71	**159**	5

[a] Conditions: 0.2 M alcohol, 2.0 equiv of Ac_2O, 3.0 equiv of NEt_3, 0.1 equiv catalyst, $CDCl_3$, 23.0±1.0 °C.
[b] Time of 50% conversion (the reaction did not reach 100% conversion).

Experimental part

Fitted conversion plots.

Fitting parameters:

$ka = k_2[ROH]_0$, where k_2 is the rate-constant of the second-order reaction;

t0 has a meaning of time axis offset (t_0 in equation 2);

y0 allows rescaling of the conversion axis (y_0 in equation 2).

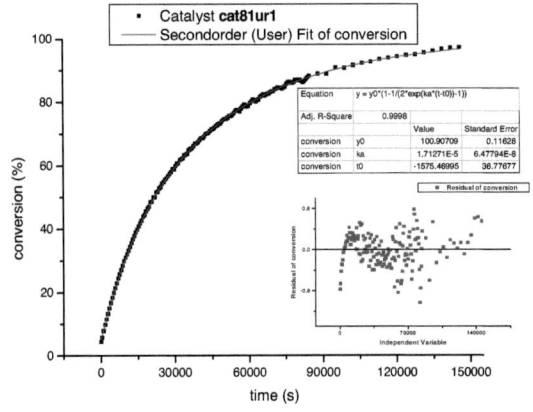

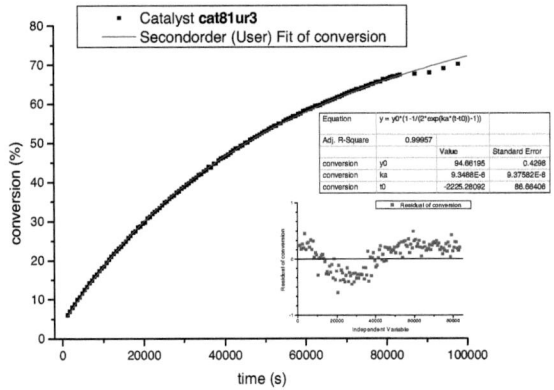

Experimental part

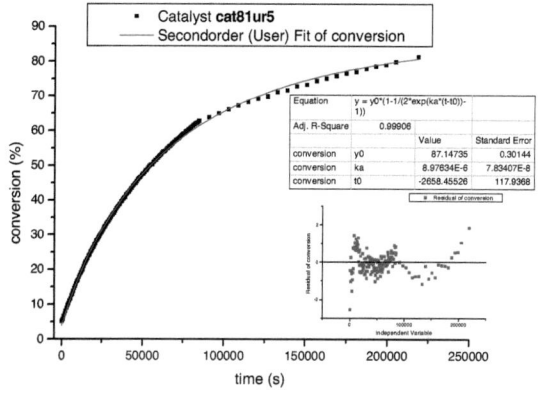

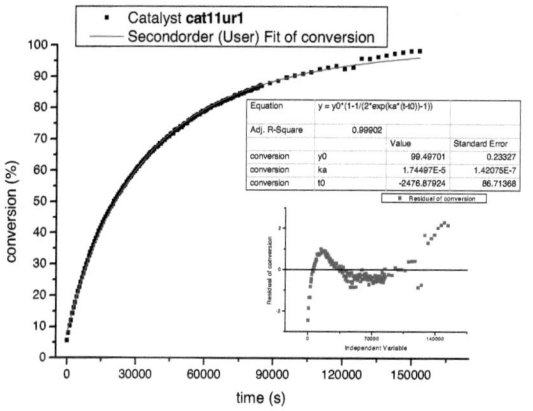

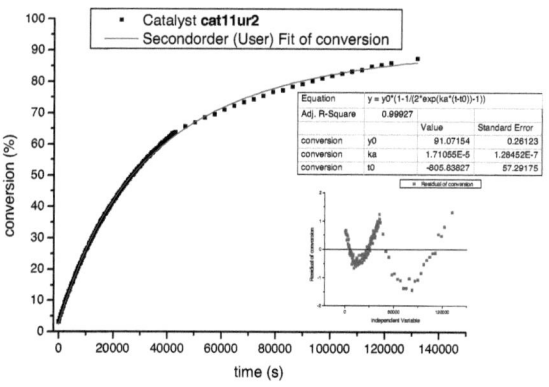

Experimental part

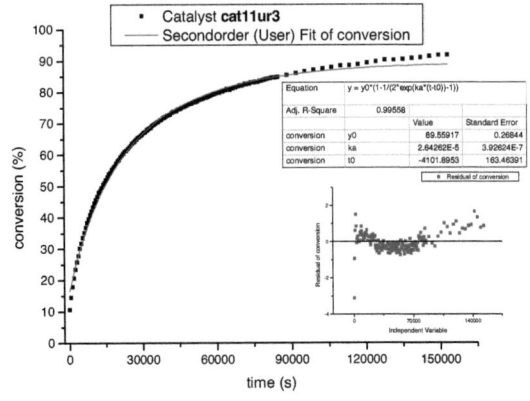

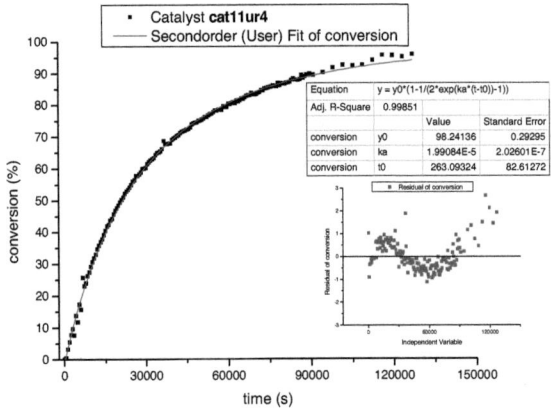

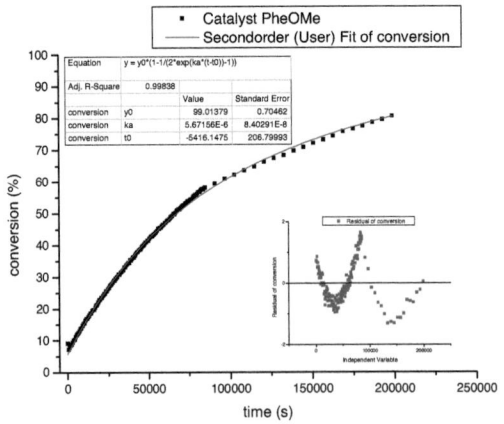

Experimental part

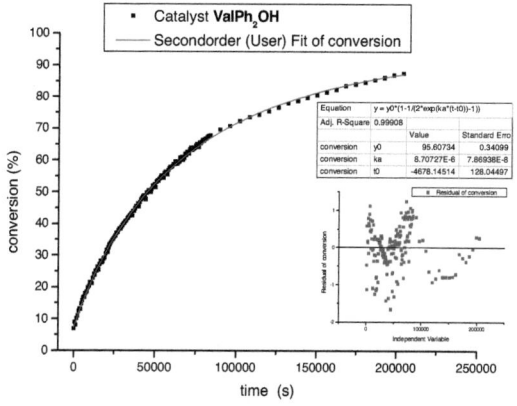

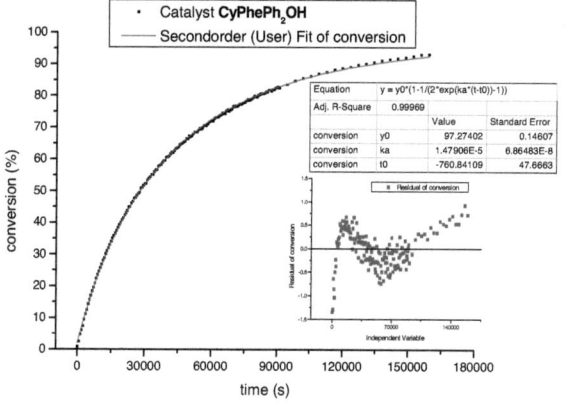

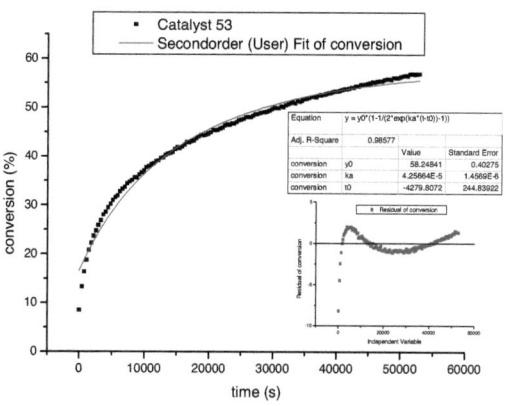

Experimental part

E4.5 Kinetic resolution experiments: general procedure.

A solution of alcohol (1 mmol), NEt$_3$ (0.75 mmol, 105 µL) and catalyst (0.01 mmol) in 3 mL of solvent (toluene or CH$_2$Cl$_2$) was cooled to –78 °C. Isobutyric anhydride (0.75 mmol, 124 µL) was added dropwise under vigorous stirring. After 9 h 100 µl of reaction mixture was taken off and quenched with 1 mL MeOH. The solvents were removed under reduced pressure and the crude mixture was analyzed by ^{1}H NMR and CSP HPLC.

HPLC retention times of alcohols and corresponding *iso*-butyrates.

1. (±)-1-(1-Naphthyl)ethanol (alc1).
Conditions for HPLC: Daicel Chiralpak OD-column analytical, hexane/*i*-propanol = 99/1, flow-rate 1.4 mL/min, detection at 254 nm, 25°C. Injection volumes: 5-10 µL. Retention times: 4.7 (*R*-ester), 6.0 (*S*-ester), 46.5 (*S*-alcohol), 78.8 (*R*-alcohol) min.

2. (*cis*)-4-Dimethylamino-benzoic acid 2-hydroxy-cyclohexyl ester (alc3).
Conditions for HPLC: Daicel Chiralpak OD-column analytical, hexane/*i*-propanol = 90/10, flow-rate 1.0 mL/min, detection at 254 nm, 25°C. Injection volume: 20 µL. Retention times: 7.2 (1*S*, 2*R*-ester), 9.1 (1*R*, 2*S* -ester), 16.5 (1*R*, 2*S*-alcohol), 29.6 (1*S*, 2*R*-alcohol) min.

3. *Trans*-2-Phenylcyclohexan-1-ol (alc2).
Conditions for HPLC: Daicel Chiralpak OD-column analytical, hexane/*i*-propanol = 99.5/0.5, flow-rate 1.0 mL/min, detection at 220 nm, 25°C. Injection volumes: 5-10 µL. Retention times: 48.8 (1*S*, 2*R* -alcohol), 54.3 (1*R*, 2*S* -alcohol) min.

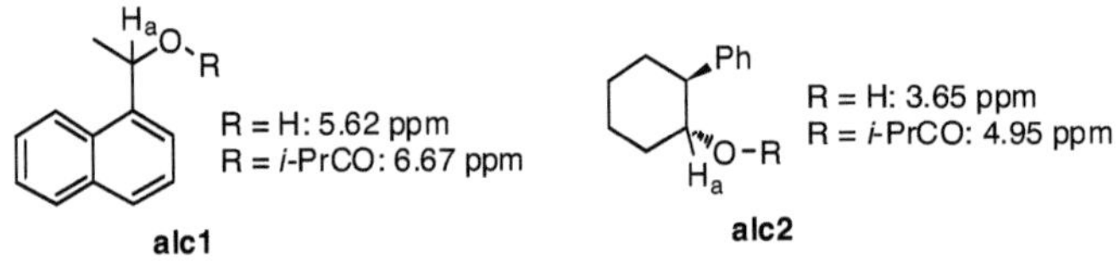

alc1: R = H: 5.62 ppm; R = *i*-PrCO: 6.67 ppm
alc2: R = H: 3.65 ppm; R = *i*-PrCO: 4.95 ppm

Scheme E4.1. Chemical shifts of hydrogens H$_a$ for alcohols and their *iso*-butyrates.

Conversion of alcohols **alc1** and **alc2** can be calculated from the integrals of α-hydrogen atoms in ^{1}H NMR spectra of reaction mixture:

$$y = \frac{I_{ester}}{I_{ester} + I_{ROH}} \cdot 100\% \qquad (1)$$

Experimental part

For alcohols **alc1** and **alc3** the conversion can be calculated from the enantiomeric excesses of alcohol and product:

$$y = \frac{ee_{ROH}}{ee_{ester} + ee_{ROH}} \cdot 100\% \qquad (2)^{[54]}$$

The selectivity s was then calculated by using equation (3) for **alc1** and **alc3** and equation (4) for alcohol **alc2**:

$$s = \frac{\ln[(1-ee_{ROH})/(1+ee_{ROH}/ee_{ester})]}{\ln[(1+ee_{ROH})/(1+ee_{ROH}/ee_{ester})]} \qquad (3)^{[54]}$$

$$s = \frac{\ln[(1-ee_{ROH})(1-y)]}{\ln[(1+ee_{ROH})(1-y)]} \qquad (4)^{[54]}$$

Chapter 5: Experimental details
E5.1 Determination of activation parameters.

General procedure for the isobutyrylation of 1-(1-naphthyl)ethanol.
A solution of 2 mmol (±)-1-(1-naphthyl)ethanol **60**, 6 mmol NEt$_3$, 0.33 mmol 1,3,5-trimethoxybenzene and 0.5 mol % PPY in 8 ml toluene was cooled to −78 °C. 4 mmol (i-PrCO)$_2$O was added dropwise with vigorous stirring. Every 10-30 minutes 50 μL of the reaction mixture was taken off and quenched with 1 mL MeOH. The solvents were distilled off under reduced pressure and ^{1}H NMR spectra were measured. Signals of ester (δ 6.64 ppm) and alcohol (δ 5.70 ppm) were integrated. Conversion y is given by equation 1:

$$y = \frac{I_{ester}}{I_{ester} + I_{ROH}} \cdot 100\% \qquad (1)$$

Dependence of the conversion y vs. time t was fitted by equation 2 for the second-order reaction kinetics:

$$y = y_0 \left(1 - \frac{1}{2e^{k(t-t_0)} - 1} \right) \qquad (2)$$

$$k = k_2 [ROH]_0 \qquad (3)$$

where k_2 is a rate-constant of the second-order reaction; t_0 has a meaning of time axis offset. With this parameter in the fitting process it's not necessary to measure the starting point of the reaction exactly. The variable y_0 allows for rescaling of the conversion axis.
A half-life can be calculated with equation (4):

$$\tau_{1/2} = \frac{\ln 1.5}{k_2 [ROH]_0} \qquad (4)$$

Table E5.1. Rate constants for **PPY**-catalyzed acylation, measured at different temperatures.

1/T, 1/K	0.005177	0.005047	0.00492247	0.00469	0.00448	0.004289	0.004113	0.00403
T, K	193.15	198.15	203.15	213.15	223.15	233.15	243.15	248.15
T, °C	−80	−75	−70	−60	−50	−40	−30	−25
k, min^{-1}	0.00489	0.00607	0.00778	0.01007	0.01533	0.02507	0.03164	0.03656
			0.00832	0.0115	0.01822	0.02245		
averaged k	0.00489	0.00607	0.00805	0.01079	0.01678	0.02376	0.03164	0.03656
ln(k)	−5.320563	−5.1044	−4.8220832	−4.5296	−4.08787	−3.73975	−3.45333	−3.3088
Absolute error k	1.70E-04	2.37E-04	2.80E-04	3.43E-04	1.02E-03	3.56E-04	4.53E-04	1.46E-03
Relative error k	0.035	0.039	0.035	0.032	0.061	0.015	0.014	0.040
k$_2$, M^{-1} s^{-1}	0.000408	0.000506	0.00067083	0.0009	0.0014	0.00198	0.002637	0.003047
ln(k$_2$/T)	−13.069	−12.878	−12.621	−12.377	−11.981	−11.676	−11.432	−11.308

Conversion *vs* time plots at different temperatures.

-80°C:

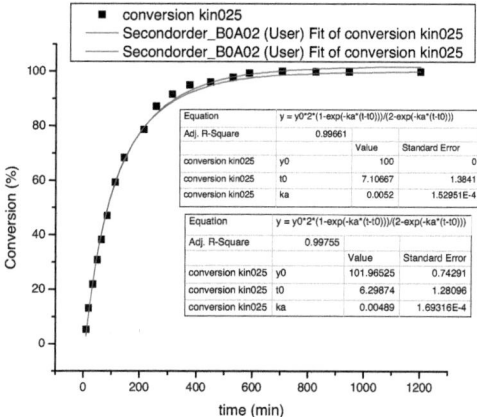

-75°C:

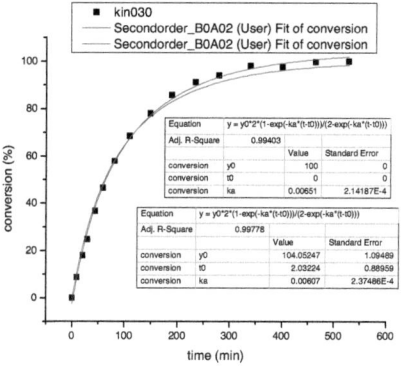

-70°C:

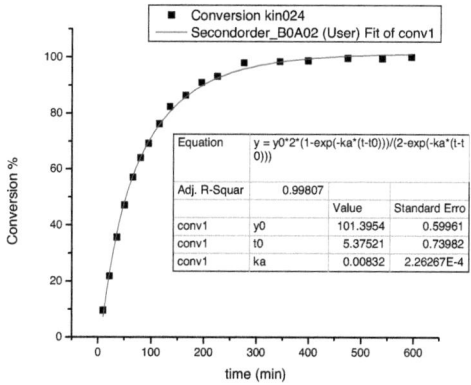

Experimental part

-60°C:

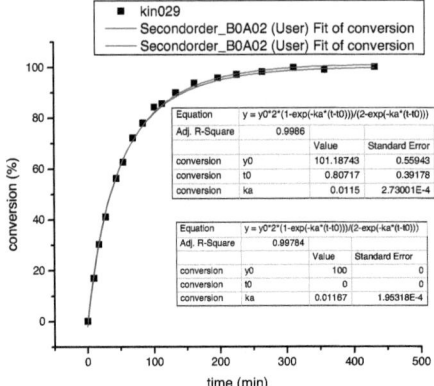

-50°C:

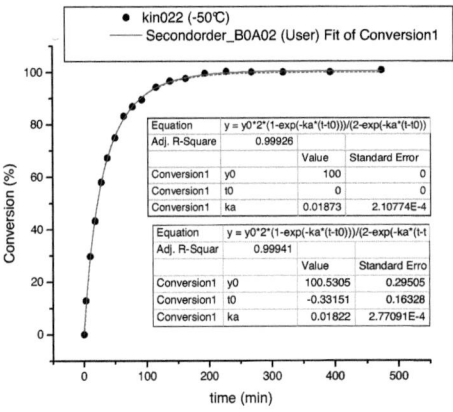

-40°C:

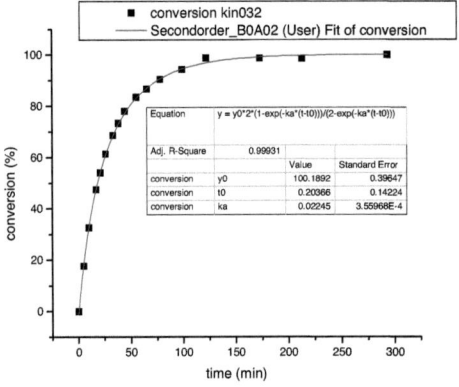

Experimental part

-30°C:

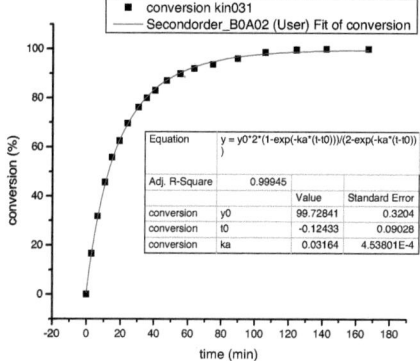

-25°C:

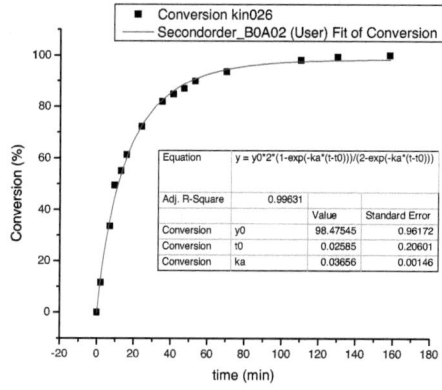

Arrhenius plot:

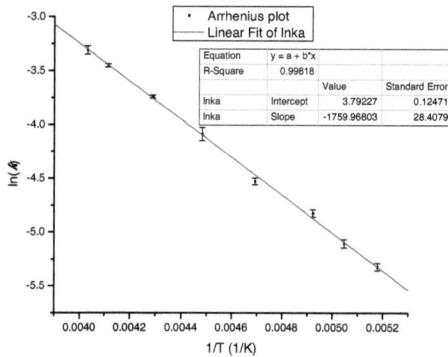

E_a = 14.6 kJ mol^{-1}

E5.2 Synthesis of the catalyst 5l-Me.

((5aS,6R,9S,9aR)-6,11,11-trimethyl-5a,6,7,8,9,9a-hexahydro-6,9-methanopyrido[3,4-b]quinoxaline-5,10-diyl)bis((3,5-dimethylphenyl)methanone), 72

The crude product was purified by column chromatography on SiO_2 with EtOAc/hexanes (1:3) as eluent to afford 44% of **72** as off-white oil.

^{1}H NMR (300 MHz, $CDCl_3$): δ = 7.78 (d, J = 5.5 Hz, 1H), 7.74 (d, J = 0.5 Hz, 1H), 7.10 (s, 3H), 7.06 (s, 3H), 6.34 (dd, J = 0.4 Hz, J = 5.5 Hz, 1H), 5.04 (d, J = 10.9 Hz, 1H, CH), 4.75 (d, J = 10.9, 1H, CH), 2.38 (m, 1H, CH), 2.31 (s, 6H, CH_3), 2.30 (s, 6H, CH_3), 1.95 (m, 1H, CH_2), 1.62 (m, 2H, CH_2), 1.47 (m, 1H, CH_2), 1.01 (s, 3H, CH_3), 0.85 (s, 3H, CH_3), 0.16 (s, 3H, CH_3).

^{13}C NMR (75 MHz, $CDCl_3$): δ = 170.9 (C=O), 170.7 (C=O), 147.8, 143.4, 138.8, 138.2, 138.2, 138.0, 135.2, 135.0, 134.9, 132.8, 132.7, 130.3, 127.7, 126.4, 126.3, 120.1, 67.6, 66.6, 52.5, 51.6, 44.9, 36.2, 27.6, 24.0, 21.3, 21.3, 21.1, 21.0, 11.3.

IR (neat): 2957 (w), 1662 (m), 1652 (m), 1576 (w), 1509 (w), 1446 (w), 1342 (m), 1318 (m), 1256 (w), 1134 (w), 836 (w), 779 (w), 695 (m) cm^{-1}.

HRMS (ESI): m/z calcd for $C_{33}H_{38}N_3O_2$ $[M+H]^+$: 508.2964; found: 508.2953.

(5aS,6R,9S,9aR)-5,10-bis(3,5-dimethylbenzyl)-6,11,11-trimethyl-5,5a,6,7,8,9,9a,10-octahydro-6,9-methanopyrido[3,4-b]quinoxaline, 5l-Me

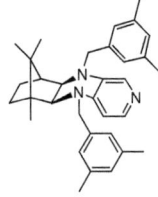

Reduction of the compound **72** (190 mg) was carried out according **general procedure II**. The crude product was purified by column chromatography on SiO_2 with EtOAc/TEA (40:1) as eluent to afford 40 % of **5l-Me** as off-white foam.

$[α]_D^{22}$ = +22.0° (c = 1, $CHCl_3$).

Experimental part

^{1}H NMR (300 MHz, CDCl$_3$) δ = 7.71 (s, 1H, CH$_{Py}$), 7.67 (d, J = 5.6 Hz, 1H, CH$_{Py}$), 6.95 (s, 2H, 2CH$_{Ar}$), 6.89 (s, 2H, 2CH$_{Ar}$), 6.80 (s, 2H, 2CH$_{Ar}$), 6.32 (d, J = 5.6 Hz, 1H, CH$_{Py}$), 4.70 (m, 2H, NCH$_2$), 4.53 (dd, J = 8.2 Hz, J = 17.3 Hz, 2H, NCH$_2$), 3.45 (d, J = 8.0 Hz, 1H, NCH), 3.40 (d, J = 8.0, 1H, NCH), 2.32 (s, 6H, ArCH$_3$), 2.30 (s, 6H, ArCH$_3$), 2.13 (d, J = 4.6 Hz, 1H, CH), 1.76 (m, 1H, CH$_2$), 1.56 (m, 1H, CH$_2$), 1.24 (m, 1H, CH$_2$), 1.13 (dd, J = 9.6 Hz, J = 13.5 Hz, 1H, CH$_2$), 1.06 (s, 6H, CH$_3$), 0.84 (s, 3H, CH$_3$).

^{13}C NMR (75 MHz, CDCl$_3$) δ = 139.1, 138.7, 138.2, 138.2, 138.1, 137.7, 131.5, 129.1, 128.6, 128.4, 124.3, 124.0, 106.9, 69.3, 64.2, 56.9, 54.6, 51.4, 48.6, 46.4, 36.0, 25.7, 22.2, 21.5, 21.4, 21.4, 20.5, 14.2.

IR (neat): 2951 (m), 2874 (w), 2360 (w), 1605 (m), 1579 (s), 1565 (m), 1516 (s), 1455 (m), 1390 (m), 1328 (s), 1274 (s), 1250 (m), 1204 (m), 1154 (m), 1118 (m), 1065 (m), 1036 (m), 922 (w), 864 (w), 830 (s), 802 (s), 688 (s) cm^{-1}.

HRMS (ESI): m/z calcd for C$_{33}$H$_{42}$N$_3$ [M+H]$^+$: 480.3379; found: 480.3370.

E5.3 Benchmark kinetics and KR with the catalyst 5l-Me.

Table E5.2. Catalytic activity of the catalyst **5l-Me**.[a]

catalyst	run	k_2, l mol^{-1} s^{-1}	Δk_2	c_1	t_0, min	$t_{1/2}$, min	Δ$t_{1/2}$, min
DMAP	1	2.18E-04	1.39E-06	1.010	-28	155.0	1.7
5b	1	1.90E-03	3.03E-05	1.088	-8	17.8	0.3
5l-Me	1	1.11E-4	2.75E-7	1.031	-9	61	1.0
5l-Me	2	1.01E-4	2.75E-7	1.011	-5	67	1.0
5l-Me	averaged					64	1.5

[a] Conditions: 0.2 M alcohol, 2.0 equiv of Ac$_2$O, 3.0 equiv of NEt$_3$, 0.1 equiv catalyst, CDCl$_3$, 23.0±1.0 °C.

Experimental part

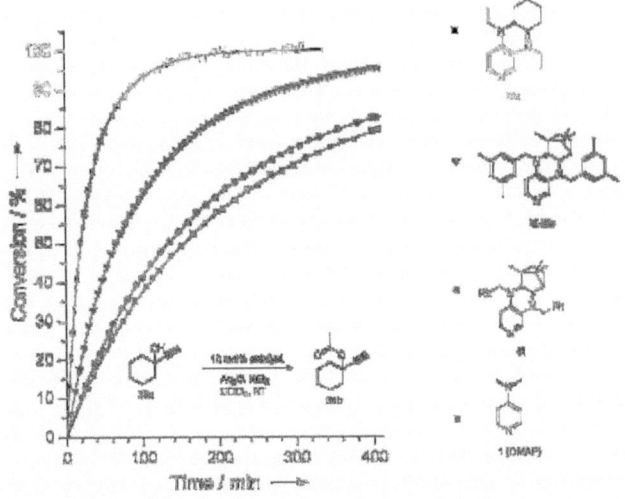

Figure E1.1. Conversion-time plots for the catalysts **5b**, **5l-Me**, **5l** and **DMAP**.

Table E5.3. KR of alcohols **alc1-3** using chiral 3,4-diaminopyridines.

Entry	Catalyst	mol% catalyst	Solvent	Time (h)	ee_A (%)[c]	ee_E (%)[d]	C (%)[a]	s [b]
1	**5b**	0.5	toluene	6.5	10.7	10.7	49	1.4
2	**5l**	0.5	toluene	3	5.3	8.7	38	1.3
3	**5l-Me**	1.0	toluene	9	32.4	13.6	70	1.7
4	**5l-Me**	0.5	toluene	7	7.0	15.4	31	1.5

[a] Conversion $C = 100*ee_A/(ee_A+ee_E)$.
[b] Selectivity factor s was calculated as described in ref. 54.
[c] ee of recovered alcohol, established by CSP-HPLC.
[d] ee of ester, established by CSP-HPLC.

Appendix

Chapter 2. Computational details.

Ground state model.

The relative acylation enthalpies were calculated at the optimized level of theory: MP2/6-31+G(2d,p)//B98/6-31G(d) with inclusion of solvent effects using PCM single point at RHF/6-31G(d) level with UAHF radii.

The conformational space of flexible 3,4-diaminopyridines has been searched using the MM3 force field or, for selected systems, also with the OPLS force field and the systematic search routine implemented in MACROMODEL 9.7. All stationary points located at force field level have then been reoptimized at B98/6-31G(d) level. Thermochemical corrections to 298.15 K have been calculated for all minima from unscaled vibrational frequencies obtained at the same level. The thermochemical corrections have been combined with single point energies calculated at the MP2(FC)/6-31+G(2d,p)//B98/6-31G(d) level to yield enthalpies H_{298} at 298.15 K:

$$H_{298}(\text{MP2(FC)}/6\text{-}31+\text{G(2d,p)}//\text{B98}/6-31\text{G(d)}) = \\ E_{tot}(\text{MP2(FC)}/6\text{-}31+\text{G(2d,p)}) + [H_{298}(\text{B98}/6-31\text{G(d)}) - E_{tot}(\text{B98}/6-31\text{G(d)})] \quad (1)$$

Inclusion of solvent effects effects using PCM single point at RHF/6-31G(d) level with UAHF radii gives the final enthalpies:

$$H_{298}(\text{MP2-5/solv}) = \\ H_{298}(\text{MP2(FC)}/6\text{-}31+\text{G(2d,p)}//\text{B98}/6-31\text{G(d)}) + \Delta G(\text{PCM/UAHF/RHF}/6-31\text{G(d)}) \quad (2)$$

In conformationally flexible systems enthalpies have been calculated as Boltzmann-averaged values over all available conformers:

$$\langle H_{298} \rangle = \sum_{i=1}^{n} w_i H_{298,i} \quad (3)$$

where Boltzmann-factors w_i were calculated with equation (4):

$$w_i = \frac{\exp(-H_{298,i}/RT)}{\sum_{i=1}^{n}\exp(-H_{298,i}/RT)} \tag{4}$$

Transition state model.

There are generally four possible orientations of the alcohol/anhydride part of the TS for each catalyst conformer: two orientations of the acetyl group and two variants of the alcohol attack on the reaction center (from the front face or the back face of the pyridine ring). For symmetrical aminopyridines such as DMAP (**3**) and PPY (**4**) this number reduces to one possible orientation. Up to four best conformations of the free catalyst were used in other cases to obtain the initial geometries of TSs, which were then optimized to energy minima with selected frozen bonds (marked bold in Scheme 2.6), followed by frequency analysis at the same level and single point calculations at MP2(FC)/6-31+G(2d,p) level.

Subsequent optimisation of the most stable conformations, obtained in the "frozen transition states" model, followed by frequency analysis at B98/6-31G(d) level and single point calculation at MP2 level as well as PCM solvation energy calculations, gave structures of the "optimized transition states".

The nature of all stationary points was verified through calculation of the vibrational frequency spectrum. All quantum mechanical calculations have been performed with Gaussian 03.

Appendix

A2.1 Relative acetylation enthalpies for 3,4-diamino and 4-aminopyridines.

Table A2.1. Calculated energies of conformers of 3,4-diaminopyridines and corresponding acetyl intermediates. Averaged enthalpies $<H_{298}>$ were calculated at B3LYP/6-311+G(d,p)// B3LYP/6-31G(d) level of theory in a gas phase.

Conformer	E_{tot} B3LYP/6-31G(d)	H_{298} B3LYP/6-31G(d)	E_{sp} B3LYP/6-311+G(d,p)	H_{298} B3LYP/6-311+G(d,p)// B3LYP/6-31G(d)	$<H_{298}>$ B3LYP/6-311+G(d,p)// B3LYP/6-31G(d)
Py					
py	-248.284973	-248.190708	-248.351162	-248.256898	-248.256898
py_ac	-401.299539	-401.151170	-401.401972	-401.253603	-401.253603
5c					-828.101443
4	-828.333870	-827.888867	-828.547489	-828.102486	
8	-828.333347	-827.888200	-828.546863	-828.101716	
19	-828.333103	-827.887884	-828.546600	-828.101381	
1	-828.332752	-827.887818	-828.546610	-828.101676	
2	-828.332627	-827.887663	-828.546281	-828.101317	
6	-828.332066	-827.887128	-828.545629	-828.100692	
3	-828.331729	-827.886850	-828.545306	-828.100427	
15	-828.331711	-827.886648	-828.545429	-828.100365	
14	-828.331518	-827.886378	-828.545085	-828.099945	
31	-828.331378	-827.886355	-828.545352	-828.100330	
35	-828.331252	-827.886285	-828.545189	-828.100222	
10	-828.331182	-827.886568	-828.544634	-828.100020	
9	-828.331067	-827.887065	-828.544818	-828.100816	
74	-828.330983	-827.885946	-828.544849	-828.099813	
32	-828.330710	-827.885310	-828.544241	-828.098840	
20	-828.330536	-827.885337	-828.544239	-828.099040	
28	-828.330456	-827.885512	-828.544136	-828.099193	
12	-828.330285	-827.885106	-828.543839	-828.098660	
5c_ac					-981.147416
4.ac2	-981.398200	-980.898023	-981.648702	-981.148525	
8.ac2	-981.397036	-980.896793	-981.647426	-981.147182	
4.ac1	-981.396749	-980.896636	-981.646942	-981.146829	
19.ac2	-981.396577	-980.896339	-981.646992	-981.146754	
15.ac2	-981.396071	-980.895960	-981.646652	-981.146541	
14.ac2	-981.395925	-980.895614	-981.646430	-981.146119	
8.ac1	-981.395559	-980.895355	-981.645624	-981.145421	
15.ac1	-981.394752	-980.894710	-981.645019	-981.144978	
19.ac1	-981.395212	-980.894840	-981.645316	-981.144944	
32.ac2	-981.395072	-980.894752	-981.645465	-981.145145	
1.ac2	-981.394685	-980.894733	-981.645496	-981.145544	
9.ac2	-981.394685	-980.894732	-981.645496	-981.145543	
25.ac2	-981.394557	-980.894348	-981.645067	-981.144858	
14.ac1	-981.394392	-980.894321	-981.644564	-981.144493	
18.ac2	-981.394319	-980.894027	-981.644779	-981.144487	
2.ac2	-981.394243	-980.894240	-981.644930	-981.144928	
6.ac2	-981.394241	-980.894205	-981.644740	-981.144704	
5e					-750.700337
018	-750.910344	-750.503666	-751.107744	-750.701066	
029	-750.909828	-750.503411	-751.107338	-750.700921	
019	-750.909733	-750.503224	-751.106973	-750.700463	
021	-750.909539	-750.503039	-751.106982	-750.700482	
012	-750.909236	-750.502918	-751.106739	-750.700421	

Appendix

Conformer	E_{tot} B3LYP/6-31G(d)	H_{298} B3LYP/6-31G(d)	E_{sp} B3LYP/6-311+G(d,p)	H_{298} B3LYP/6-311+G(d,p)// B3LYP/6-31G(d)	$<H_{298}>$ B3LYP/6-311+G(d,p)// B3LYP/6-31G(d)
011	-750.908823	-750.502469	-751.106108	-750.699754	
032	-750.908952	-750.502307	-751.106093	-750.699448	
010	-750.909034	-750.502423	-751.106604	-750.699993	
015	-750.908934	-750.502227	-751.106543	-750.699836	
036	-750.908337	-750.501876	-751.105704	-750.699243	
024	-750.908209	-750.501841	-751.105845	-750.699477	
027	-750.908197	-750.501772	-751.105660	-750.699235	
031	-750.908278	-750.501772	-751.105909	-750.699403	
023	-750.907536	-750.501152	-751.105033	-750.698649	
002	-750.907389	-750.500816	-751.104711	-750.698138	
5e_ac					-903.743174
04	-903.971708	-903.510209	-904.205721	-903.744221	
14	-903.971028	-903.509519	-904.205121	-903.743612	
12	-903.970979	-903.509510	-904.205066	-903.743597	
11	-903.970843	-903.509324	-904.204674	-903.743155	
13	-903.970853	-903.509363	-904.204986	-903.743496	
10	-903.970470	-903.508855	-904.204357	-903.742742	
25	-903.970119	-903.508596	-904.204032	-903.742509	
22	-903.970225	-903.508534	-903.970225	-903.508534	
21	-903.970141	-903.508575	-904.204018	-903.742453	
55	-903.969717	-903.508267	-904.203480	-903.742030	
53	-903.969544	-903.508094	-904.203350	-903.741899	
16	-903.969771	-903.508228	-904.203689	-903.742146	
29	-903.969683	-903.508132	-904.203643	-903.742092	
19	-903.969683	-903.508132	-904.203643	-903.742091	
52	-903.969520	-903.508040	-904.203009	-903.741529	
24	-903.969401	-903.508011	-904.203547	-903.742156	
28	-903.969401	-903.508011	-904.203547	-903.742156	
43	-903.969770	-903.508158	-904.204016	-903.742404	
20	-903.969309	-903.507635	-904.203473	-903.741799	
62	-903.968761	-903.507267	-904.202320	-903.740826	
02	-903.968732	-903.506944	-904.202568	-903.740780	
35	-903.968360	-903.506618	-904.202202	-903.740460	
03	-903.968360	-903.506619	-904.202201	-903.740460	
5f					-941.237860
13	-941.441168	-941.000048	-941.679325	-941.238205	
10	-941.440895	-940.999777	-941.679107	-941.237989	
4	-941.440911	-940.999909	-941.678960	-941.237959	
7	-941.441004	-941.000128	-941.678780	-941.237904	
6	-941.440604	-940.999799	-941.678645	-941.237840	
2	-941.440994	-941.000103	-941.678630	-941.237738	
1	-941.440452	-940.999271	-941.678237	-941.237056	
15	-941.439586	-940.998736	-941.677411	-941.236561	
5f_ac					-1094.282271
2.ac1	-1094.504581	-1094.008717	-1094.778484	-1094.282620	
1.ac2	-1094.504088	-1094.007995	-1094.778589	-1094.282496	
4.ac2	-1094.503737	-1094.007796	-1094.778423	-1094.282481	
4.ac1	-1094.504173	-1094.008173	-1094.778430	-1094.282430	
1.ac1	-1094.504379	-1094.008223	-1094.778515	-1094.282358	
6.ac2	-1094.502765	-1094.006845	-1094.777419	-1094.281499	

Appendix

Conformer	E_{tot} B3LYP/6-31G(d)	H_{298} B3LYP/6-31G(d)	E_{sp} B3LYP/6-311+G(d,p)	H_{298} B3LYP/6-311+G(d,p)// B3LYP/6-31G(d)	$<H_{298}>$ B3LYP/6-311+G(d,p)// B3LYP/6-31G(d)
6.ac1	-1094.503184	-1094.007274	-1094.777362	-1094.281451	
9.ac1	-1094.503161	-1094.007187	-1094.776996	-1094.281022	
9.ac2	-1094.502519	-1094.006632	-1094.776773	-1094.280886	
5g					-1211.548116
6_2	-1211.801513	-1211.244719	-1212.105737	-1211.548943	
4_1	-1211.801079	-1211.244044	-1212.105354	-1211.548320	
13_2	-1211.799741	-1211.242880	-1212.104198	-1211.547337	
10_2	-1211.799776	-1211.242678	-1212.104273	-1211.547175	
7_2	-1211.799188	-1211.242572	-1212.103190	-1211.546574	
14_1	-1211.798325	-1211.242090	-1212.102362	-1211.546127	
11_1	-1211.799388	-1211.242269	-1212.103106	-1211.545987	
7_1	-1211.798761	-1211.241876	-1212.102630	-1211.545745	
1_3	-1211.798604	-1211.241408	-1212.102824	-1211.545628	
14_2	-1211.797636	-1211.241075	-1212.101431	-1211.544870	
6_3	-1211.796911	-1211.240287	-1212.101305	-1211.544680	
2_1	-1211.796936	-1211.240044	-1212.101038	-1211.544146	
12_1	-1211.795729	-1211.238898	-1212.100541	-1211.543711	
2_2	-1211.796311	-1211.239462	-1212.100286	-1211.543437	
1_1	-1211.795842	-1211.239106	-1212.100162	-1211.543426	
11_2	-1211.795115	-1211.238410	-1212.099257	-1211.542552	
9_2	-1211.795435	-1211.238536	-1212.099379	-1211.542481	
6_1	-1211.795396	-1211.238462	-1212.099328	-1211.542393	
4_3	-1211.795500	-1211.238284	-1212.099324	-1211.542108	
5g_ac					-1364.592985
4_1_ac2	-1364.865089	-1364.253036	-1365.205655	-1364.593602	
10_2_ac2	-1364.864117	-1364.252150	-1365.205150	-1364.593183	
6_2_ac2	-1364.864254	-1364.252443	-1365.204982	-1364.593171	
4_1_ac1	-1364.862987	-1364.252093	-1365.203396	-1364.592502	
6_2_ac1	-1364.863746	-1364.251857	-1365.204279	-1364.592390	
10_2_ac1	-1364.863531	-1364.251562	-1365.204335	-1364.592367	
13_2_ac2	-1364.863200	-1364.251285	-1365.204167	-1364.592252	
13_2_ac1	-1364.862705	-1364.250687	-1365.203444	-1364.591426	
4_3_ac1	-1364.860981	-1364.248711	-1365.200843	-1364.588572	
4_3_ac2	-1364.859944	-1364.247747	-1365.200569	-1364.588372	
6_1_ac1	-1364.859136	-1364.246935	-1365.199714	-1364.587513	
5h					-786.364830
1	-786.527121	-786.160765	-786.732179	-786.365823	
16	-786.526243	-786.159865	-786.731243	-786.364865	
11	-786.526243	-786.159865	-786.731241	-786.364863	
2	-786.525182	-786.159317	-786.730119	-786.364253	
7	-786.525497	-786.159254	-786.730285	-786.364041	
13	-786.524594	-786.158573	-786.729975	-786.363954	
10	-786.524598	-786.158371	-786.729930	-786.363703	
14	-786.524937	-786.158863	-786.729705	-786.363631	
9	-786.524246	-786.158321	-786.729014	-786.363088	
12	-786.524020	-786.157456	-786.729574	-786.363010	
15	-786.523908	-786.157767	-786.729046	-786.362905	
4	-786.523666	-786.157448	-786.729003	-786.362785	
6	-786.523126	-786.157170	-786.728555	-786.362599	

Appendix

Conformer	E_{tot} B3LYP/6-31G(d)	H_{298} B3LYP/6-31G(d)	E_{sp} B3LYP/6-311+G(d,p)	H_{298} B3LYP/6-311+G(d,p)// B3LYP/6-31G(d)	$<H_{298}>$ B3LYP/6-311+G(d,p)// B3LYP/6-31G(d)
5h_ac					-939.408166
1_ac1	-939.588728	-939.167438	-939.830243	-939.408952	
1_ac2	-939.588379	-939.167144	-939.829601	-939.408366	
11_ac2	-939.587148	-939.165886	-939.828540	-939.407278	
11_ac1	-939.586928	-939.165589	-939.828036	-939.406697	
2_ac1	-939.586014	-939.164903	-939.827385	-939.406274	
2_ac2	-939.586001	-939.164875	-939.827064	-939.405938	
7_ac2	-939.585730	-939.164623	-939.827002	-939.405895	
7_ac1	-939.585677	-939.164462	-939.826689	-939.405474	
12_ac2	-939.585081	-939.163633	-939.826752	-939.405304	
14_ac2	-939.584606	-939.163617	-939.825755	-939.404766	
12_ac1	-939.584743	-939.163207	-939.826104	-939.404568	
10_ac1	-939.584105	-939.162900	-939.825762	-939.404557	
14_ac1	-939.584577	-939.163584	-939.825465	-939.404472	
13_ac1	-939.583198	-939.162175	-939.824815	-939.403792	
10_ac2	-939.582720	-939.161578	-939.824163	-939.403022	
15_ac1	-939.582831	-939.161680	-939.824164	-939.403013	
15_ac2	-939.582726	-939.161526	-939.823762	-939.402562	
13_ac2	-939.581881	-939.160863	-939.823294	-939.402276	
5i					-906.690099
7	-906.961129	-906.456105	-907.195945	-906.690921	
8	-906.960256	-906.455500	-907.195274	-906.690518	
1	-906.960018	-906.455260	-907.195139	-906.690381	
3	-906.959942	-906.455136	-907.194846	-906.690040	
23	-906.959926	-906.454745	-907.194748	-906.689567	
25	-906.959896	-906.454577	-907.194714	-906.689395	
28	-906.959743	-906.454389	-907.194458	-906.689104	
29	-906.959698	-906.454574	-907.194474	-906.689350	
5	-906.959160	-906.454106	-907.194062	-906.689008	
40	-906.959126	-906.453849	-907.193858	-906.688580	
2	-906.958975	-906.454191	-907.193780	-906.688996	
31	-906.958871	-906.454108	-907.193877	-906.689114	
15	-906.958843	-906.453673	-907.193623	-906.688453	
5i_ac					-1059.735847
7.ac1	-1060.025548	-1059.465249	-1060.297306	-1059.737007	
23.ac1	-1060.024192	-1059.464044	-1060.295957	-1059.735809	
29.ac1	-1060.024158	-1059.463980	-1060.295684	-1059.735506	
7.ac2	-1060.024153	-1059.463983	-1060.295582	-1059.735412	
28.ac1	-1060.024115	-1059.463905	-1060.295768	-1059.735558	
25.ac1	-1060.024055	-1059.464000	-1060.295810	-1059.735755	
8.ac1	-1060.023023	-1059.463171	-1060.294933	-1059.735081	
29.ac2	-1060.023003	-1059.462882	-1060.294234	-1059.734113	
23.ac2	-1060.022957	-1059.462824	-1060.294383	-1059.734251	
28.ac2	-1060.022769	-1059.462623	-1060.294107	-1059.733961	
46.ac1	-1060.022756	-1059.462555	-1060.293746	-1059.733545	
25.ac2	-1060.022738	-1059.462709	-1060.294175	-1059.734146	
56.ac2	-1060.022704	-1059.462579	-1060.294361	-1059.734235	
49.ac2	-1060.022667	-1059.462289	-1060.294220	-1059.733842	
40.ac1	-1060.022482	-1059.462277	-1060.294126	-1059.733921	
63.ac2	-1060.022255	-1059.462195	-1060.293867	-1059.733807	

Appendix

Conformer	E_{tot} B3LYP/6-31G(d)	H_{298} B3LYP/6-31G(d)	E_{sp} B3LYP/6-311+G(d,p)	H_{298} B3LYP/6-311+G(d,p)// B3LYP/6-31G(d)	$<H_{298}>$ B3LYP/6-311+G(d,p)// B3LYP/6-31G(d)
102.ac1	-1060.022121	-1059.461748	-1060.293989	-1059.733615	
5.ac1	-1060.021972	-1059.461879	-1060.293996	-1059.733903	
5j					-1221.581593
102	-1221.788419	-1221.273782	-1222.096927	-1221.582290	
82	-1221.788358	-1221.273724	-1222.096867	-1221.582233	
187	-1221.787692	-1221.272923	-1222.096127	-1221.581358	
158	-1221.787548	-1221.272832	-1222.095974	-1221.581258	
7	-1221.786251	-1221.271961	-1222.095034	-1221.580744	
195	-1221.785961	-1221.271262	-1222.094647	-1221.579947	
85	-1221.785999	-1221.271431	-1222.095038	-1221.580470	
71	-1221.785976	-1221.271220	-1222.094701	-1221.579945	
1	-1221.784814	-1221.270536	-1222.093249	-1221.578970	
214	-1221.785829	-1221.270879	-1222.094744	-1221.579794	
22	-1221.785954	-1221.271713	-1222.094824	-1221.580583	
135	-1221.785661	-1221.270823	-1222.094614	-1221.579775	
5j_ac					-1374.632494
102_ac1	-1374.857819	-1374.288066	-1375.202198	-1374.632444	
207_ac1	-1374.856493	-1374.287048	-1375.200809	-1374.631364	
106_ac1	-1374.856479	-1374.287018	-1375.200771	-1374.631310	
187_ac1	-1374.856372	-1374.286665	-1375.200594	-1374.630887	
22_ac2	-1374.855849	-1374.288993	-1375.200292	-1374.633436	
155_ac1	-1374.855432	-1374.286040	-1375.199728	-1374.630335	
33_ac1	-1374.854656	-1374.285227	-1375.198851	-1374.629422	
102_ac2	-1374.855228	-1374.285692	-1375.200446	-1374.630910	
46_ac1	-1374.854920	-1374.285445	-1375.199735	-1374.630260	
214_ac1	-1374.854447	-1374.284529	-1375.198907	-1374.628989	
82_ac2	-1374.854738	-1374.285149	-1375.199986	-1374.630397	
123_ac1	-1374.854302	-1374.284837	-1375.198763	-1374.629297	
187_ac2	-1374.853730	-1374.284096	-1375.198853	-1374.629219	
158_ac1	-1374.853121	-1374.283567	-1375.198325	-1374.628771	
22_ac1	-1374.852862	-1374.283411	-1375.197797	-1374.628346	
182_ac2	-1374.852862	-1374.283412	-1375.197796	-1374.628346	
5k					-1132.962481
13	-1133.176740	-1132.679811	-1133.459951	-1132.963021	
2	-1133.176526	-1132.679837	-1133.459580	-1132.962890	
15	-1133.176396	-1132.679542	-1133.459578	-1132.962724	
6	-1133.176317	-1132.679582	-1133.459404	-1132.962669	
10	-1133.175350	-1132.678143	-1133.459159	-1132.961952	
1	-1133.175398	-1132.678370	-1133.458791	-1132.961762	
4	-1133.175332	-1132.678206	-1133.458872	-1132.961746	
7	-1133.175241	-1132.678468	-1133.458500	-1132.961727	
12	-1133.174178	-1132.677206	-1133.457903	-1132.960932	
9	-1133.174447	-1132.677811	-1133.457277	-1132.960641	
16	-1133.174614	-1132.677437	-1133.457749	-1132.960572	
5k_ac					-1286.006517
1.ac2	-1286.239216	-1285.687367	-1286.559127	-1286.007278	
2.ac1	-1286.239661	-1285.687892	-1286.558775	-1286.007007	
2.ac2	-1286.239197	-1285.687440	-1286.558754	-1286.006997	
1.ac1	-1286.239374	-1285.687446	-1286.558718	-1286.006789	
4.ac2	-1286.238498	-1285.686583	-1286.558492	-1286.006577	

Conformer	E_{tot} B3LYP/6-31G(d)	H_{298} B3LYP/6-31G(d)	E_{sp} B3LYP/6-311+G(d,p)	H_{298} B3LYP/6-311+G(d,p)// B3LYP/6-31G(d)	$<H_{298}>$ B3LYP/6-311+G(d,p)// B3LYP/6-31G(d)
10.ac2	-1286.238548	-1285.686417	-1286.558543	-1286.006413	
4.ac1	-1286.238836	-1285.686857	-1286.558240	-1286.006261	
12.ac2	-1286.238058	-1285.686018	-1286.558010	-1286.005969	
6.ac2	-1286.238064	-1285.686287	-1286.557572	-1286.005794	
13.ac2	-1286.238234	-1285.686222	-1286.557785	-1286.005773	
12.ac1	-1286.238272	-1285.686313	-1286.557664	-1286.005705	
6.ac1	-1286.238498	-1285.686654	-1286.557497	-1286.005652	
15.ac2	-1286.238268	-1285.686371	-1286.557323	-1286.005426	
15.ac1	-1286.237592	-1285.685651	-1286.557114	-1286.005173	
16.ac2	-1286.237780	-1285.685944	-1286.556976	-1286.005141	
10.ac1	-1286.237429	-1285.685287	-1286.557101	-1286.004958	
13.ac1	-1286.237257	-1285.685252	-1286.556533	-1286.004528	
7.ac1	-1286.236155	-1285.684322	-1286.555457	-1286.003624	
7.ac2	-1286.235804	-1285.683846	-1286.555510	-1286.003553	
5l					-1288.919884
1	-1289.190133	-1288.598196	-1289.512328	-1288.920391	
2	-1289.189469	-1288.597319	-1289.511396	-1288.919246	
3	-1289.187642	-1288.595703	-1289.509658	-1288.917718	
4	-1289.188650	-1288.596397	-1289.510818	-1288.918566	
5l_ac					-1441.963936
3ac1	-1442.253201	-1441.606192	-1442.611644	-1441.964635	
1ac1	-1442.253098	-1441.606205	-1442.610722	-1441.963829	
1ac2	-1442.252637	-1441.605746	-1442.610610	-1441.963719	
2ac2	-1442.251977	-1441.604954	-1442.610530	-1441.963507	
4ac1	-1442.251930	-1441.604945	-1442.610037	-1441.963052	
2ac1	-1442.251938	-1441.604866	-1442.609947	-1441.962875	
4ac2	-1442.250842	-1441.603457	-1442.609277	-1441.961892	

Appendix

Table A2.2. Calculated energies of conformers of 3,4-diaminopyridines and corresponding acetyl intermediates. Averaged enthalpies <H$_{298}$> were calculated at MP2(FC)/6-31+G(2d,p)//B98/6-31G(d) level of theory with inclusion of solvent effects in chloroform at PCM/UAHF/RHF/6-31G(d) level.

Conformer	E$_{tot}$ B98/6-31G(d)	H$_{298}$ B98/6-31G(d)	E$_{tot}$ MP2(FC)/6-31+G(2d,p)	G$_{solv}$, kJ/mol	H$_{298}$ MP2-5 with solv	<H$_{298}$> MP2-5 with solv
Py						
Py	-248.181767	-248.087627	-247.589439	-9.00	-247.498727	
Py_ac	-401.140004	-400.991691	-400.215516	-142.55	-400.121498	
5a						-591.698885
1	-593.418065	-593.130880	-591.981245	-14.56	-591.699606	
2	-593.417088	-593.130024	-591.980224	-14.52	-591.698690	
3	-593.416648	-593.129499	-591.979898	-14.18	-591.698150	
4	-593.415838	-593.128722	-591.978867	-14.14	-591.697137	
5	-593.415488	-593.128357	-591.979020	-15.06	-591.697625	
6	-593.415277	-593.127870	-591.978978	-15.19	-591.697357	
7	-593.414082	-593.126927	-591.977706	-14.56	-591.696097	
5a_ac						-744.352678
1	-746.420416	-746.078238	-744.649502	-121.08	-744.353441	
2	-746.420285	-746.077937	-744.649302	-121.29	-744.353151	
3	-746.419224	-746.076995	-744.648332	-121.13	-744.352239	
4	-746.41917	-746.07687	-744.648337	-121.50	-744.352314	
5	-746.419086	-746.076798	-744.648046	-121.17	-744.351909	
6	-746.419086	-746.076799	-744.648047	-121.17	-744.351911	
7	-746.419131	-746.076892	-744.64761	-120.12	-744.351122	
8	-746.419131	-746.076893	-744.64761	-120.08	-744.351108	
9	-746.418954	-746.076873	-744.647375	-120.33	-744.351125	
10	-746.417975	-746.075703	-744.646556	-120.16	-744.350051	
11	-746.418056	-746.075775	-744.646790	-121.00	-744.350596	
5c						-825.516838
8	-828.001801	-827.557348	-825.963011	1.59	-825.517952	
4	-828.002033	-827.557927	-825.961857	-0.29	-825.517862	
2	-828.000917	-827.556782	-825.962331	1.59	-825.517591	
1	-828.000859	-827.556620	-825.961300	-0.50	-825.517252	
45	-827.998449	-827.554199	-825.961409	0.54	-825.517042	
7	-827.998623	-827.554291	-825.961033	-0.29	-825.516813	
28	-827.998803	-827.554666	-825.960348	-1.13	-825.516641	
31	-827.999580	-827.555168	-825.960747	-0.67	-825.516590	
11	-827.998462	-827.554037	-825.960445	-1.38	-825.516546	
14	-827.999940	-827.555453	-825.961145	0.38	-825.516515	
19	-828.001334	-827.556945	-825.961050	0.50	-825.516470	
35	-827.999401	-827.555214	-825.960485	-0.08	-825.516330	
24	-827.998368	-827.553719	-825.961522	1.51	-825.516300	
29	-827.997857	-827.553895	-825.961163	2.38	-825.516293	
74	-827.999336	-827.555047	-825.961453	2.59	-825.516176	
18	-827.998400	-827.554182	-825.960170	-0.50	-825.516143	
12	-827.998831	-827.554466	-825.960875	1.55	-825.515920	
10	-827.999587	-827.555798	-825.960577	2.72	-825.515752	
20	-827.998871	-827.554462	-825.959779	-0.63	-825.515609	
6	-828.000231	-827.556070	-825.959679	0.00	-825.515518	
5	-827.998289	-827.554336	-825.959602	0.63	-825.515410	
3	-827.999923	-827.555845	-825.959378	-0.13	-825.515349	
13	-827.997956	-827.553764	-825.959326	-0.50	-825.515326	
37	-827.997356	-827.553339	-825.960019	1.80	-825.515317	
22	-827.996082	-827.551755	-825.959506	0.25	-825.515083	
16	-827.998495	-827.554530	-825.959383	1.05	-825.515020	
15	-827.999816	-827.555598	-825.959146	-0.21	-825.515008	
32	-827.999030	-827.554465	-825.959228	-0.79	-825.514965	

Appendix

Conformer	E_{tot} B98/6-31G(d)	H_{298} B98/6-31G(d)	E_{tot} MP2(FC)/6-31+G(2d,p)	G_{solv}, kJ/mol	H_{298} MP2-5 with solv	$\langle H_{298}\rangle$ MP2-5 with solv
33	-827.997960	-827.553715	-825.958884	-0.84	-825.514958	
30	-827.996071	-827.551745	-825.958965	-0.46	-825.514814	
39	-827.996272	-827.552243	-825.958790	0.00	-825.514761	
54	-827.997251	-827.552678	-825.959050	-0.63	-825.514716	
21	-827.997575	-827.553365	-825.958567	-0.92	-825.514708	
40	-827.995914	-827.551958	-825.959180	1.92	-825.514491	
36	-827.996256	-827.551867	-825.958899	0.38	-825.514367	
23	-827.998327	-827.553927	-825.958346	-1.00	-825.514328	
27	-827.997507	-827.553932	-825.958726	2.51	-825.514196	
62	-827.996669	-827.552236	-825.958568	0.21	-825.514055	
25	-827.998959	-827.554428	-825.958275	-0.59	-825.513968	
44	-827.995953	-827.551966	-825.956942	-2.01	-825.513720	
9	-827.998539	-827.554271	-825.957891	0.00	-825.513622	
50	-827.995321	-827.551318	-825.958026	1.30	-825.513528	
61	-827.995667	-827.551628	-825.957431	-0.21	-825.513471	
56	-827.996281	-827.552182	-825.957329	-0.33	-825.513358	
84	-827.995191	-827.550835	-825.957461	-0.46	-825.513280	
71	-827.996169	-827.551664	-825.957109	-0.96	-825.512970	
43	-827.995723	-827.551388	-825.956643	-0.21	-825.512388	
70	-827.995463	-827.551442	-825.957181	2.30	-825.512284	
34	-827.996538	-827.552717	-825.956559	1.72	-825.512084	
52	-827.994934	-827.550548	-825.955925	-0.63	-825.511779	
51	-827.995061	-827.550749	-825.956102	0.92	-825.511439	
5c_ac						-978.172026
4_ac2	-981.009738	-980.510214	-978.635898	-97.11	-978.173362	
8_ac2	-981.008871	-980.509343	-978.636382	-95.14	-978.173092	
14_ac2	-981.007715	-980.508061	-978.635139	-96.73	-978.172330	
19_ac2	-981.008161	-980.508659	-978.634270	-98.37	-978.172234	
18_ac2	-981.006058	-980.506745	-978.633412	-97.57	-978.171261	
28_ac2	-981.005256	-980.505950	-978.632709	-99.29	-978.171219	
15_ac2	-981.007559	-980.508157	-978.633548	-97.07	-978.171118	
7_ac2	-981.005275	-980.505633	-978.633051	-98.62	-978.170970	
4_ac1	-981.008377	-980.508860	-978.633957	-94.98	-978.170614	
24_ac2	-981.005387	-980.505514	-978.634049	-95.56	-978.170573	
32_ac2	-981.006796	-980.506897	-978.633339	-97.36	-978.170523	
9_ac2	-981.006145	-980.506839	-978.632171	-98.74	-978.170474	
1_ac2	-981.006145	-980.506835	-978.632175	-98.74	-978.170474	
8_ac1	-981.007471	-980.507807	-978.634464	-93.22	-978.170306	
20_ac2	-981.005323	-980.506023	-978.632273	-97.99	-978.170294	
2_ac2	-981.005931	-980.506561	-978.632962	-96.36	-978.170292	
33_ac2	-981.003643	-980.504461	-978.631101	-100.42	-978.170165	
12_ac2	-981.005026	-980.505679	-978.633145	-95.40	-978.170133	
39_ac2	-981.001894	-980.502403	-978.631169	-100.83	-978.170084	
36_ac2	-981.002797	-980.503315	-978.631808	-99.12	-978.170078	
51_ac2	-981.005279	-980.505672	-978.631822	-99.33	-978.170047	
16_ac2	-981.004050	-980.504797	-978.631596	-98.95	-978.170033	
23_ac2	-981.005279	-980.505667	-978.631822	-99.29	-978.170027	
14_ac1	-981.006245	-980.506950	-978.633147	-94.93	-978.170012	
25_ac2	-981.006114	-980.506509	-978.632019	-98.58	-978.169959	
31_ac2	-981.005400	-980.505904	-978.631966	-97.70	-978.169681	
6_ac2	-981.005731	-980.506421	-978.631555	-98.11	-978.169615	
29_ac2	-981.002864	-980.503619	-978.631441	-98.24	-978.169614	
62_ac2	-981.003872	-980.504134	-978.632039	-97.65	-978.169496	
54_ac2	-981.004234	-980.504825	-978.631878	-97.15	-978.169472	
11_ac2	-981.003345	-980.503732	-978.630885	-100.29	-978.169471	
22_ac2	-981.002963	-980.503348	-978.631951	-97.45	-978.169451	

Appendix

Conformer	E_{tot} B98/6-31G(d)	H_{298} B98/6-31G(d)	E_{tot} MP2(FC)/6-31+G(2d,p)	G_{solv}, kJ/mol	H_{298} MP2-5 with solv	$<H_{298}>$ MP2-5 with solv
19_ac1	-981.006886	-980.507036	-978.632354	-96.61	-978.169300	
13_ac2	-981.004675	-980.505205	-978.631924	-96.15	-978.169075	
5_ac2	-981.003903	-980.504552	-978.631117	-97.82	-978.169024	
3_ac2	-981.005293	-980.506252	-978.630817	-97.36	-978.168859	
10_ac2	-981.003397	-980.504479	-978.630219	-98.58	-978.168846	
28_ac1	-981.004066	-980.504723	-978.631076	-97.36	-978.168816	
15_ac1	-981.006349	-980.507075	-978.631825	-95.19	-978.168806	
18_ac1	-981.004561	-980.505576	-978.631191	-96.02	-978.168779	
7_ac1	-981.003961	-980.504583	-978.631197	-96.86	-978.168711	
32_ac1	-981.005477	-980.505744	-978.631575	-96.15	-978.168463	
74_ac2	-981.003569	-980.504216	-978.631325	-95.35	-978.168290	
2_ac1	-981.004853	-980.505508	-978.631409	-95.10	-978.168286	
1_ac1	-981.005024	-980.505653	-978.630568	-96.99	-978.168137	
9_ac1	-981.005024	-980.505654	-978.630564	-96.99	-978.168134	
35_ac2	-981.004173	-980.504566	-978.630623	-97.32	-978.168083	
20_ac1	-981.004108	-980.504987	-978.630502	-96.15	-978.168002	
31_ac1	-981.004364	-980.505108	-978.630571	-96.32	-978.168000	
30_ac2	-981.001048	-980.501606	-978.629664	-98.99	-978.167926	
21_ac2	-981.002546	-980.503071	-978.629393	-99.79	-978.167925	
13_ac1	-981.003410	-980.504060	-978.630103	-97.24	-978.167788	
51_ac1	-981.003960	-980.504546	-978.630031	-97.53	-978.167765	
23_ac1	-981.003960	-980.504547	-978.630030	-97.53	-978.167765	
33_ac1	-981.002586	-980.503366	-978.629433	-98.32	-978.167662	
12_ac1	-981.003939	-980.504284	-978.631524	-93.93	-978.167645	
37_ac2	-981.000607	-980.501235	-978.628873	-100.08	-978.167620	
24_ac1	-981.003864	-980.503959	-978.631858	-93.55	-978.167586	
62_ac1	-981.002662	-980.503177	-978.630497	-96.02	-978.167585	
16_ac1	-981.002832	-980.503584	-978.629817	-97.15	-978.167572	
25_ac1	-981.004891	-980.505209	-978.630323	-96.78	-978.167501	
54_ac1	-981.003237	-980.503535	-978.630434	-95.65	-978.167162	
6_ac1	-981.004558	-980.505168	-978.629769	-96.40	-978.167096	
11_ac1	-981.002164	-980.502419	-978.629129	-98.53	-978.166914	
40_ac2	-981.000783	-980.501307	-978.628902	-98.03	-978.166764	
61_ac2	-981.001674	-980.502264	-978.628809	-97.99	-978.166721	
5_ac1	-981.002705	-980.503410	-978.629376	-96.15	-978.166702	
3_ac1	-981.004123	-980.504970	-978.629051	-96.57	-978.166678	
10_ac1	-981.002365	-980.503437	-978.628491	-95.52	-978.165945	
35_ac1	-981.002860	-980.503411	-978.628771	-95.44	-978.165672	
74_ac1	-981.002283	-980.502951	-978.629416	-93.30	-978.165621	
27_ac2	-981.001546	-980.502242	-978.627623	-97.61	-978.165498	
34_ac2	-981.001603	-980.502453	-978.627988	-95.40	-978.165172	
70_ac2	-981.000425	-980.501026	-978.627545	-96.65	-978.164959	
50_ac2	-980.998757	-980.499285	-978.626349	-99.29	-978.164693	
5e						-748.348429
029	-750.606553	-750.200824	-748.755443	1.00	-748.349332	
018	-750.607112	-750.201112	-748.755673	2.01	-748.348908	
021	-750.606375	-750.200488	-748.755172	1.13	-748.348854	
012	-750.605976	-750.200302	-748.754886	1.17	-748.348766	
010	-750.605854	-750.199821	-748.754930	0.46	-748.348722	
015	-750.605703	-750.199816	-748.754722	0.88	-748.348501	
019	-750.606566	-750.200615	-748.754755	1.72	-748.348151	
011	-750.605748	-750.200113	-748.754430	0.75	-748.348508	
002	-750.604419	-750.198622	-748.754286	0.54	-748.348282	
024	-750.604956	-750.199256	-748.754038	-0.29	-748.348450	
031	-750.605009	-750.199008	-748.754147	1.00	-748.347763	
032	-750.605803	-750.199873	-748.753254	-0.13	-748.347372	

197

Conformer	E_{tot} B98/6-31G(d)	H_{298} B98/6-31G(d)	E_{tot} MP2(FC)/6-31+G(2d,p)	G_{solv}, kJ/mol	H_{298} MP2-5 with solv	$\langle H_{298}\rangle$ MP2-5 with solv
027	-750.604951	-750.199015	-748.753218	0.63	-748.347044	
023	-750.604370	-750.198666	-748.752974	-0.25	-748.347366	
036	-750.605140	-750.199262	-748.753148	0.33	-748.347142	
014	-750.603217	-750.197343	-748.753121	-0.42	-748.347406	
003	-750.603738	-750.197727	-748.753217	0.71	-748.346934	
007	-750.602899	-750.197006	-748.752897	-0.25	-748.347100	
008	-750.602986	-750.197172	-748.752735	-0.29	-748.347033	
005	-750.602410	-750.196208	-748.752514	-0.79	-748.346615	
001	-750.601281	-750.195324	-748.752157	-0.84	-748.346519	
004	-750.602579	-750.196602	-748.752167	-0.50	-748.346381	
020	-750.601727	-750.195632	-748.751943	-0.33	-748.345975	
025	-750.602496	-750.196713	-748.751256	-0.29	-748.345584	
017	-750.602253	-750.195939	-748.751667	-0.67	-748.345608	
006	-750.600497	-750.194606	-748.750536	-0.54	-748.344852	
033	-750.601180	-750.195098	-748.750495	-0.63	-748.344652	
009	-750.600303	-750.194225	-748.750287	-0.84	-748.344528	
022	-750.601743	-750.195660	-748.750165	-0.46	-748.344258	
034	-750.601358	-750.195251	-748.749690	-0.75	-748.343870	
040	-750.601097	-750.194934	-748.749445	-0.17	-748.343346	
026	-750.599890	-750.193846	-748.748793	-1.00	-748.343132	
041	-750.599734	-750.193769	-748.748251	-1.13	-748.342716	
030	-750.599831	-750.193543	-748.748504	0.54	-748.342009	
5e_ac						-901.002684
04	-903.612034	-903.151115	-901.427163	-97.91	-901.003535	
13	-903.611153	-903.150164	-901.426547	-99.87	-901.003598	
12	-903.611208	-903.150313	-901.426323	-98.66	-901.003006	
14	-903.611285	-903.150338	-901.426274	-100.58	-901.003637	
10	-903.610912	-903.149873	-901.425934	-99.66	-901.002854	
11	-903.611202	-903.150244	-901.425667	-98.66	-901.002287	
22	-903.610538	-903.149472	-901.425300	-101.55	-901.002910	
16	-903.610131	-903.149100	-901.425047	-101.09	-901.002517	
02	-903.609317	-903.148009	-901.425265	-100.12	-901.002091	
53	-903.609918	-903.149110	-901.424762	-97.82	-901.001212	
19	-903.610056	-903.149049	-901.424942	-101.17	-901.002469	
29	-903.610056	-903.149052	-901.424937	-101.17	-901.002467	
21	-903.610418	-903.149444	-901.424886	-100.50	-901.002190	
20	-903.609568	-903.148619	-901.424844	-101.13	-901.002412	
25	-903.610415	-903.149507	-901.424794	-100.33	-901.002100	
43	-903.610048	-903.148798	-901.425134	-100.71	-901.002242	
24	-903.609600	-903.148883	-901.424547	-101.46	-901.002475	
28	-903.609600	-903.148880	-901.424546	-101.46	-901.002471	
05	-903.608313	-903.147341	-901.424639	-101.04	-901.002153	
55	-903.610063	-903.149151	-901.424559	-98.83	-901.001287	
35	-903.608866	-903.147812	-901.424466	-100.67	-901.001754	
03	-903.608866	-903.147810	-901.424466	-100.67	-901.001752	
07	-903.608521	-903.147446	-901.424428	-101.75	-901.002110	
52	-903.609957	-903.148966	-901.423780	-96.61	-900.999586	
15	-903.608067	-903.147197	-901.423405	-99.96	-901.000606	
33	-903.608109	-903.147038	-901.423473	-98.49	-900.999915	
08	-903.608109	-903.146985	-901.423304	-100.16	-901.000330	
47	-903.608109	-903.146984	-901.423305	-100.16	-901.000330	
45	-903.607707	-903.146490	-901.423247	-101.17	-901.000564	
62	-903.609133	-903.148272	-901.422877	-98.74	-900.999625	
09	-903.607344	-903.146331	-901.423012	-101.67	-901.000723	
46	-903.607344	-903.146327	-901.423013	-101.67	-901.000720	
64	-903.608309	-903.147318	-901.422950	-99.41	-900.999823	

Appendix

Conformer	E_{tot} B98/6-31G(d)	H_{298} B98/6-31G(d)	E_{tot} MP2(FC)/6-31+G(2d,p)	G_{solv}, kJ/mol	H_{298} MP2-5 with solv	$<H_{298}>$ MP2-5 with solv
44	-903.607030	-903.146207	-901.422725	-99.33	-900.999734	
58	-903.608369	-903.147403	-901.422741	-98.24	-900.999193	
49	-903.607280	-903.146400	-901.422600	-100.33	-900.999934	
40	-903.607274	-903.146155	-901.422809	-98.07	-900.999044	
59	-903.608786	-903.147659	-901.422802	-99.41	-900.999539	
17	-903.607456	-903.146299	-901.422574	-102.97	-901.000636	
01	-903.605978	-903.144756	-901.422600	-101.80	-901.000150	
23	-903.607291	-903.146298	-901.422016	-101.84	-900.999811	
18	-903.607291	-903.146289	-901.422017	-101.80	-900.999788	
27	-903.606929	-903.145717	-901.422139	-101.13	-900.999444	
26	-903.607397	-903.146025	-901.422258	-102.26	-900.999834	
56	-903.606854	-903.145870	-901.421662	-98.32	-900.998128	
50	-903.606398	-903.145217	-901.421140	-99.33	-900.997792	
51	-903.606943	-903.145885	-901.421008	-98.37	-900.997415	
34	-903.606565	-903.145688	-901.420714	-100.79	-900.998227	
63	-903.605809	-903.144761	-901.420641	-99.91	-900.997648	
37	-903.606520	-903.145356	-901.420540	-101.67	-900.998100	
42	-903.604988	-903.143725	-901.420616	-101.63	-900.998062	
06	-903.604988	-903.143722	-901.420616	-101.63	-900.998059	
32	-903.604763	-903.143401	-901.420700	-99.91	-900.997393	
57	-903.606160	-903.145004	-901.420214	-100.21	-900.997224	
61	-903.605095	-903.144112	-901.419945	-99.79	-900.996969	
60	-903.605814	-903.144336	-901.420185	-101.25	-900.997272	
5f						-938.344926
6	-941.062462	-940.622577	-938.783610	-5.06	-938.345653	
4	-941.062845	-940.622744	-938.783815	-4.48	-938.345420	
2	-941.062964	-940.622926	-938.784047	-2.80	-938.345077	
15	-941.061682	-940.621588	-938.783857	-2.59	-938.344751	
13	-941.063112	-940.622726	-938.784121	-1.38	-938.344262	
10	-941.062892	-940.622604	-938.784221	-0.75	-938.344219	
1	-941.062444	-940.622379	-938.783124	-2.38	-938.343968	
7	-941.063065	-940.623052	-938.783669	-0.29	-938.343768	
9	-941.062076	-940.622092	-938.782384	-2.97	-938.343532	
16	-941.061804	-940.621404	-938.782401	-3.47	-938.343323	
12	-941.060270	-940.619658	-938.782590	-2.64	-938.342982	
14	-941.062221	-940.622378	-938.782413	-0.67	-938.342826	
5f_ac						-1090.999723
2_ac1	-1094.070524	-1093.575470	-1091.459203	-95.48	-1091.000515	
2_ac2	-1094.069510	-1093.574222	-1091.458027	-98.74	-1091.000348	
4_ac2	-1094.069039	-1093.573899	-1091.456561	-99.91	-1090.999477	
9_ac1	-1094.069156	-1093.573911	-1091.457714	-97.11	-1090.999457	
9_ac2	-1094.068112	-1093.572822	-1091.456644	-100.00	-1090.999441	
6_ac2	-1094.068051	-1093.573005	-1091.455343	-101.88	-1090.999100	
4_ac1	-1094.069887	-1093.574522	-1091.457509	-96.52	-1090.998909	
6_ac1	-1094.068873	-1093.573644	-1091.456366	-98.99	-1090.998842	
1_ac2	-1094.069398	-1093.573913	-1091.456641	-97.40	-1090.998255	
1_ac1	-1094.070036	-1093.574785	-1091.457152	-94.14	-1090.997757	
15_ac2	-1094.062929	-1093.567514	-1091.453700	-103.60	-1090.997743	
16_ac2	-1094.067883	-1093.572376	-1091.455127	-99.41	-1090.997484	
16_ac1	-1094.068522	-1093.573165	-1091.455605	-96.40	-1090.996964	
7_ac1	-1094.067309	-1093.572188	-1091.454477	-98.28	-1090.996790	
10_ac1	-1094.068042	-1093.572954	-1091.454559	-95.65	-1090.995901	
13_ac1	-1094.067473	-1093.572262	-1091.453617	-97.82	-1090.995665	
12_ac1	-1094.066580	-1093.571072	-1091.454213	-94.68	-1090.994769	

Appendix

Conformer	E_{tot} B98/6-31G(d)	H_{298} B98/6-31G(d)	E_{tot} MP2(FC)/6-31+G(2d,p)	G_{solv}, kJ/mol	H_{298} MP2-5 with solv	$<H_{298}>$ MP2-5 with solv
5g						-1207.819503
4_1	-1211.312399	-1210.756414	-1208.381461	13.60	-1207.820297	
6_2	-1211.312697	-1210.756905	-1208.380677	12.47	-1207.820136	
10_2	-1211.311004	-1210.755073	-1208.380786	13.14	-1207.819851	
13_2	-1211.310918	-1210.754982	-1208.380161	12.43	-1207.819491	
7_2	-1211.310484	-1210.755045	-1208.378297	10.04	-1207.819033	
4_3	-1211.307244	-1210.751179	-1208.380018	13.93	-1207.818646	
6_1	-1211.306988	-1210.751164	-1208.379320	13.72	-1207.818269	
11_1	-1211.310772	-1210.754625	-1208.379032	12.97	-1207.817945	
14_1	-1211.309577	-1210.754309	-1208.376787	9.58	-1207.817869	
1_3	-1211.309836	-1210.753816	-1208.379027	13.72	-1207.817780	
12_1	-1211.306964	-1210.751178	-1208.379308	15.36	-1207.817673	
7_1	-1211.310182	-1210.754349	-1208.377495	10.50	-1207.817662	
2_1	-1211.308233	-1210.752278	-1208.377545	10.96	-1207.817414	
6_3	-1211.307907	-1210.752542	-1208.376699	10.50	-1207.817334	
2_2	-1211.307557	-1210.751815	-1208.376185	8.83	-1207.817080	
11_2	-1211.306443	-1210.750668	-1208.376622	11.17	-1207.816593	
1_1	-1211.307037	-1210.751253	-1208.377501	13.81	-1207.816458	
9_2	-1211.306770	-1210.750901	-1208.375445	8.45	-1207.816357	
14_2	-1211.309056	-1210.753429	-1208.376021	10.96	-1207.816218	
10_3	-1211.307435	-1210.751457	-1208.375572	10.46	-1207.815609	
2_3	-1211.301507	-1210.745834	-1208.374434	8.70	-1207.815446	
9_1	-1211.303419	-1210.747798	-1208.371177	9.00	-1207.812130	
5g_ac						-1360.472770
4_1_ac2	-1364.319820	-1363.708516	-1361.055082	-77.66	-1360.473355	
4_3_ac1	-1364.316651	-1363.705476	-1361.056803	-72.72	-1360.473325	
4_3_ac2	-1364.315092	-1363.703943	-1361.055270	-75.31	-1360.472806	
10_2_ac2	-1364.318656	-1363.707628	-1361.054256	-77.28	-1360.472662	
6_2_ac2	-1364.318791	-1363.707870	-1361.053435	-77.86	-1360.472172	
13_2_ac2	-1364.317774	-1363.706771	-1361.052658	-79.16	-1360.471806	
6_1_ac1	-1364.314227	-1363.703035	-1361.053552	-76.94	-1360.471666	
12_1_ac1	-1364.308866	-1363.697456	-1361.049916	-86.11	-1360.471303	
4_1_ac1	-1364.317938	-1363.706999	-1361.053512	-74.27	-1360.470859	
10_2_ac1	-1364.318127	-1363.707324	-1361.053333	-73.35	-1360.470466	
6_2_ac1	-1364.318393	-1363.707546	-1361.052601	-74.22	-1360.470025	
13_2_ac1	-1364.317343	-1363.706437	-1361.051833	-75.27	-1360.469596	
5h						-783.954589
1	-786.211003	-785.845253	-784.319500	-4.35	-783.955407	
11	-786.210144	-785.844351	-784.318658	-5.82	-783.955080	
7	-786.209518	-785.843845	-784.317937	-5.90	-783.954510	
14	-786.208937	-785.843458	-784.317282	-5.82	-783.954018	
12	-786.207914	-785.842026	-784.317745	-4.85	-783.953705	
2	-786.208914	-785.843775	-784.317084	-4.39	-783.953619	
10	-786.208438	-785.843093	-784.317385	-3.89	-783.953522	
13	-786.208216	-785.842967	-784.317316	-4.02	-783.953397	
6	-786.206906	-785.841416	-784.315807	-7.49	-783.953169	
15	-786.207729	-785.842314	-784.316707	-3.89	-783.952773	
4	-786.207467	-785.841833	-784.315847	-6.61	-783.952731	
9	-786.208078	-785.842878	-784.315466	-4.69	-783.952051	
5h_ac						-936.608509
1_ac1	-939.215919	-938.795126	-936.990756	-103.34	-936.609326	
11_ac2	-939.214397	-938.793616	-936.989281	-105.90	-936.608835	
1_ac2	-939.215697	-938.794921	-936.990057	-101.00	-936.607751	
7_ac2	-939.212999	-938.792484	-936.987507	-106.40	-936.607518	
11_ac1	-939.214299	-938.793614	-936.988645	-103.60	-936.607417	
2_ac1	-939.213085	-938.792539	-936.987561	-105.39	-936.607158	

Appendix

Conformer	E_{tot} B98/6-31G(d)	H_{298} B98/6-31G(d)	E_{tot} MP2(FC)/6-31+G(2d,p)	G_{solv}, kJ/mol	H_{298} MP2-5 with solv	$<H_{298}>$ MP2-5 with solv
14_ac2	-939.211901	-938.791668	-936.986127	-107.53	-936.606849	
12_ac2	-939.212274	-938.791416	-936.987575	-103.97	-936.606317	
7_ac1	-939.213104	-938.792579	-936.987082	-104.01	-936.606174	
2_ac2	-939.213226	-938.792736	-936.987278	-103.09	-936.606055	
14_ac1	-939.212078	-938.791542	-936.985822	-105.23	-936.605364	
12_ac1	-939.212080	-938.791160	-936.986777	-101.38	-936.604470	
5i						-903.840410
8	-906.595579	-906.091657	-904.345117	-1.42	-903.841737	
1	-906.595372	-906.091429	-904.344356	-3.43	-903.841719	
28	-906.595387	-906.090992	-904.345404	-0.29	-903.841120	
40	-906.594764	-906.090325	-904.344489	-1.97	-903.840800	
4	-906.594103	-906.090168	-904.343978	-1.80	-903.840729	
18	-906.594275	-906.090044	-904.343854	-2.22	-903.840467	
2	-906.594448	-906.090614	-904.343052	-2.68	-903.840238	
102	-906.594376	-906.089987	-904.343771	-2.09	-903.840179	
42	-906.593174	-906.088977	-904.345278	2.55	-903.840110	
5	-906.594709	-906.090505	-904.343251	-2.59	-903.840035	
9	-906.594123	-906.090280	-904.343095	-2.01	-903.840017	
29	-906.595359	-906.091171	-904.343975	-0.38	-903.839930	
23	-906.595495	-906.091246	-904.344081	0.08	-903.839800	
25	-906.595457	-906.091096	-904.344001	-0.25	-903.839736	
3	-906.595480	-906.091518	-904.343722	0.13	-903.839712	
59	-906.592955	-906.088865	-904.343993	0.67	-903.839648	
192	-906.592790	-906.088812	-904.342317	-3.18	-903.839550	
122	-906.592911	-906.088767	-904.342345	-3.14	-903.839396	
20	-906.594350	-906.090409	-904.343020	-0.79	-903.839382	
32	-906.594125	-906.090562	-904.343275	1.26	-903.839234	
13	-906.592865	-906.088903	-904.342899	-0.54	-903.839144	
7	-906.596591	-906.092456	-904.343974	1.84	-903.839137	
26	-906.593686	-906.089286	-904.345536	5.27	-903.839128	
19	-906.593695	-906.089603	-904.343544	0.92	-903.839101	
132	-906.593131	-906.088655	-904.343244	-0.59	-903.838991	
118	-906.593156	-906.088943	-904.342882	-0.79	-903.838971	
50	-906.593097	-906.088960	-904.343927	2.18	-903.838961	
131	-906.593090	-906.088647	-904.343140	-0.67	-903.838952	
31	-906.594329	-906.090437	-904.342542	-0.75	-903.838937	
6	-906.594190	-906.090147	-904.341988	-2.47	-903.838885	
63	-906.593588	-906.089351	-904.342678	-1.09	-903.838855	
56	-906.594069	-906.089777	-904.343037	-0.04	-903.838760	
95	-906.593168	-906.089336	-904.343008	1.26	-903.838698	
14	-906.592946	-906.088722	-904.342570	-0.92	-903.838697	
55	-906.593345	-906.089414	-904.343295	1.76	-903.838695	
30	-906.593579	-906.089335	-904.342547	-0.96	-903.838669	
62	-906.593096	-906.089263	-904.342464	0.00	-903.838631	
24	-906.592921	-906.088713	-904.342822	0.42	-903.838456	
49	-906.594088	-906.089945	-904.343268	1.80	-903.838440	
57	-906.593161	-906.089212	-904.343221	2.22	-903.838428	
65	-906.592825	-906.089232	-904.342512	1.30	-903.838425	
52	-906.593387	-906.089124	-904.343222	1.42	-903.838417	
11	-906.594155	-906.090204	-904.342898	1.55	-903.838358	
35	-906.594093	-906.090133	-904.342114	-0.29	-903.838265	
80	-906.592975	-906.089323	-904.343249	3.51	-903.838258	
82	-906.593151	-906.089370	-904.341125	-2.13	-903.838157	
54	-906.592780	-906.089255	-904.342615	2.51	-903.838134	
92	-906.593220	-906.088807	-904.342181	-0.88	-903.838103	
27	-906.594029	-906.089684	-904.341988	-1.13	-903.838074	

Appendix

Conformer	E_{tot} B98/6-31G(d)	H_{298} B98/6-31G(d)	E_{tot} MP2(FC)/6-31+G(2d,p)	G_{solv}, kJ/mol	H_{298} MP2-5 with solv	$<H_{298}>$ MP2-5 with solv
75	-906.593164	-906.088727	-904.342036	-1.09	-903.838013	
43	-906.593429	-906.089545	-904.341655	-0.33	-903.837898	
10	-906.593231	-906.089179	-904.341092	-1.92	-903.837773	
58	-906.592928	-906.088835	-904.342730	2.59	-903.837649	
12	-906.593168	-906.089258	-904.341086	-0.88	-903.837511	
46	-906.594240	-906.089811	-904.343016	2.85	-903.837504	
48	-906.593265	-906.088920	-904.341786	0.42	-903.837281	
74	-906.592991	-906.088684	-904.341182	-0.96	-903.837242	
72	-906.593022	-906.088749	-904.341417	0.54	-903.836937	
86	-906.592787	-906.088736	-904.340325	-0.79	-903.836577	
15	-906.594679	-906.090618	-904.341907	3.56	-903.836491	
5i_ac						-1056.495463
7_ac1	-1059.604369	-1059.044723	-1057.018841	-98.58	-1056.496740	
29_ac1	-1059.603285	-1059.044040	-1057.018766	-96.36	-1056.496222	
25_ac1	-1059.602986	-1059.043743	-1057.018176	-97.07	-1056.495905	
23_ac1	-1059.603144	-1059.043872	-1057.018425	-96.27	-1056.495821	
28_ac1	-1059.603134	-1059.043647	-1057.018231	-96.90	-1056.495652	
46_ac1	-1059.602284	-1059.042811	-1057.018814	-94.27	-1056.495245	
63_ac2	-1059.601250	-1059.042257	-1057.016767	-98.11	-1056.495144	
40_ac1	-1059.601422	-1059.042034	-1057.016358	-99.66	-1056.494931	
56_ac2	-1059.601853	-1059.042393	-1057.017464	-96.36	-1056.494705	
8_ac1	-1059.601652	-1059.042745	-1057.015841	-98.74	-1056.494542	
3_ac1	-1059.600451	-1059.041607	-1057.015860	-98.49	-1056.494530	
49_ac2	-1059.601931	-1059.042498	-1057.017937	-94.56	-1056.494519	
1_ac1	-1059.600597	-1059.041575	-1057.014617	-101.88	-1056.494399	
7_ac2	-1059.603076	-1059.043578	-1057.017045	-96.02	-1056.494121	
5_ac1	-1059.600773	-1059.041674	-1057.015189	-99.79	-1056.494098	
19_ac2	-1059.599848	-1059.040823	-1057.015965	-96.73	-1056.493784	
75_ac2	-1059.600556	-1059.040913	-1057.015954	-98.37	-1056.493777	
102_ac1	-1059.600964	-1059.041355	-1057.015841	-98.58	-1056.493777	
9_ac1	-1059.599530	-1059.040677	-1057.014189	-100.21	-1056.493503	
29_ac2	-1059.602431	-1059.043013	-1057.017507	-92.80	-1056.493435	
15_ac1	-1059.600505	-1059.041353	-1057.017247	-92.55	-1056.493346	
35_ac1	-1059.600644	-1059.041497	-1057.015707	-96.48	-1056.493308	
25_ac2	-1059.601773	-1059.042580	-1057.016478	-94.52	-1056.493285	
72_ac2	-1059.600694	-1059.041330	-1057.015790	-96.69	-1056.493253	
6_ac1	-1059.599409	-1059.040415	-1057.013836	-100.79	-1056.493232	
28_ac2	-1059.601883	-1059.042443	-1057.016524	-94.68	-1056.493147	
55_ac2	-1059.600230	-1059.040706	-1057.016342	-95.35	-1056.493137	
31_ac1	-1059.600491	-1059.041326	-1057.015108	-97.40	-1056.493042	
23_ac2	-1059.602089	-1059.042582	-1057.016951	-93.26	-1056.492965	
27_ac2	-1059.600384	-1059.040997	-1057.015152	-97.65	-1056.492960	
20_ac1	-1059.600315	-1059.041010	-1057.014946	-97.91	-1056.492931	
2_ac1	-1059.599960	-1059.040904	-1057.013451	-101.00	-1056.492865	
18_ac1	-1059.600217	-1059.040844	-1057.014491	-98.99	-1056.492823	
63_ac1	-1059.600060	-1059.041120	-1057.015123	-95.98	-1056.492740	
8_ac2	-1059.600814	-1059.041766	-1057.014840	-95.86	-1056.492301	
40_ac2	-1059.600221	-1059.040939	-1057.014609	-97.07	-1056.492298	
49_ac1	-1059.600777	-1059.041418	-1057.016378	-92.55	-1056.492270	
56_ac1	-1059.600666	-1059.041359	-1057.015789	-93.85	-1056.492226	
3_ac2	-1059.599401	-1059.040303	-1057.014508	-96.65	-1056.492222	
1_ac2	-1059.599513	-1059.040503	-1057.013142	-99.50	-1056.492028	
46_ac2	-1059.600781	-1059.041347	-1057.016457	-90.88	-1056.491636	
5_ac2	-1059.599582	-1059.040515	-1057.013521	-97.53	-1056.491600	
102_ac2	-1059.599492	-1059.040205	-1057.014292	-96.02	-1056.491579	
75_ac1	-1059.599607	-1059.039997	-1057.014592	-94.64	-1056.491029	

Appendix

Conformer	E_{tot} B98/6-31G(d)	H_{298} B98/6-31G(d)	E_{tot} MP2(FC)/6-31+G(2d,p)	G_{solv}, kJ/mol	H_{298} MP2-5 with solv	$<H_{298}>$ MP2-5 with solv
35_ac1	-1059.599912	-1059.040826	-1057.014568	-93.05	-1056.490924	
15_ac2	-1059.599459	-1059.040105	-1057.015652	-90.67	-1056.490832	
72_ac1	-1059.599700	-1059.040262	-1057.014359	-94.18	-1056.490793	
27_ac1	-1059.599613	-1059.040411	-1057.014072	-94.27	-1056.490774	
2_ac2	-1059.598807	-1059.039910	-1057.011813	-98.53	-1056.490446	
5j						-1217.910363
1	-1221.300472	-1220.787113	-1218.426019	2.43	-1217.911735	
102	-1221.302906	-1220.789129	-1218.420923	-7.53	-1217.910014	
187	-1221.302230	-1220.788302	-1218.420066	-8.79	-1217.909485	
46	-1221.299409	-1220.786171	-1218.417271	-14.02	-1217.909371	
7	-1221.300922	-1220.787456	-1218.420254	-5.61	-1217.908924	
85	-1221.300735	-1220.786939	-1218.421235	-3.60	-1217.908809	
28	-1221.299636	-1220.786175	-1218.420075	-5.44	-1217.908686	
82	-1221.302848	-1220.789187	-1218.421419	-2.43	-1217.908683	
214	-1221.300369	-1220.786137	-1218.419442	-8.74	-1217.908541	
61	-1221.299675	-1220.786121	-1218.418254	-9.67	-1217.908382	
158	-1221.302118	-1220.788136	-1218.420520	-4.56	-1217.908274	
195	-1221.300761	-1220.786865	-1218.420009	-5.65	-1217.908264	
71	-1221.300703	-1220.786764	-1218.419907	-6.02	-1217.908264	
207	-1221.299925	-1220.786709	-1218.417510	-10.38	-1217.908246	
106	-1221.299854	-1220.786331	-1218.418068	-9.67	-1217.908226	
17	-1221.299839	-1220.786345	-1218.419359	-5.86	-1217.908096	
31	-1221.299523	-1220.786194	-1218.418592	-7.41	-1217.908083	
22	-1221.300344	-1220.786825	-1218.418231	-8.37	-1217.907900	
112	-1221.299490	-1220.785759	-1218.419279	-5.94	-1217.907810	
217	-1221.299857	-1220.786633	-1218.418334	-6.69	-1217.907660	
126	-1221.299902	-1220.786623	-1218.418410	-6.49	-1217.907601	
137	-1221.300116	-1220.786807	-1218.418535	-6.23	-1217.907600	
135	-1221.300217	-1220.786237	-1218.420028	-3.22	-1217.907275	
160	-1221.299959	-1220.786488	-1218.418634	-5.52	-1217.907267	
155	-1221.299090	-1220.786023	-1218.416219	-10.67	-1217.907216	
227	-1221.300138	-1220.786156	-1218.419562	-4.23	-1217.907189	
277	-1221.299584	-1220.786379	-1218.417691	-6.61	-1217.907004	
172	-1221.299151	-1220.786123	-1218.417032	-7.03	-1217.906681	
123	-1221.299443	-1220.785987	-1218.416700	-8.70	-1217.906558	
182	-1221.299370	-1220.786085	-1218.415962	-10.08	-1217.906518	
32	-1221.295325	-1220.782134	-1218.417955	-3.60	-1217.906135	
255	-1221.299702	-1220.786054	-1218.418791	-1.80	-1217.905828	
84	-1221.298567	-1220.785130	-1218.416755	-6.44	-1217.905772	
152	-1221.297812	-1220.784179	-1218.415933	-8.70	-1217.905614	
5j_ac						-1370.564416
33_ac1	-1374.313136	-1373.744397	-1371.099821	-90.63	-1370.565599	
102_ac2	-1374.312839	-1373.744081	-1371.097101	-96.73	-1370.565187	
82_ac2	-1374.312327	-1373.743559	-1371.097518	-93.18	-1370.564240	
46_ac1	-1374.312730	-1373.743987	-1371.095838	-97.40	-1370.564194	
106_ac1	-1374.314494	-1373.745936	-1371.097840	-90.96	-1370.563927	
187_ac2	-1374.311409	-1373.742739	-1371.095803	-96.40	-1370.563850	
158_ac2	-1374.310765	-1373.741953	-1371.095977	-95.81	-1370.563659	
182_ac2	-1374.310528	-1373.741791	-1371.095227	-96.90	-1370.563397	
22_ac1	-1374.310528	-1373.741791	-1371.095227	-96.90	-1370.563397	
102_ac1	-1374.315742	-1373.746520	-1371.099042	-88.07	-1370.563366	
187_ac1	-1374.314317	-1373.745353	-1371.097370	-90.75	-1370.562971	
7_ac2	-1374.309841	-1373.741193	-1371.095886	-93.55	-1370.562871	
207_ac1	-1374.314521	-1373.745965	-1371.096363	-91.38	-1370.562612	
155_ac1	-1374.313468	-1373.744718	-1371.096410	-91.76	-1370.562608	
135_ac1	-1374.309488	-1373.740839	-1371.096361	-91.17	-1370.562437	

Appendix

Conformer	E_{tot} B98/6-31G(d)	H_{298} B98/6-31G(d)	E_{tot} MP2(FC)/6-31+G(2d,p)	G_{solv}, kJ/mol	H_{298} MP2-5 with solv	$<H_{298}>$ MP2-5 with solv
227_ac2	-1374.309488	-1373.740848	-1371.096370	-91.09	-1370.562423	
123_ac2	-1374.309098	-1373.740367	-1371.093731	-98.11	-1370.562370	
22_ac2	-1374.313638	-1373.744899	-1371.097535	-87.45	-1370.562102	
137_ac2	-1374.308448	-1373.739844	-1371.094872	-93.93	-1370.562044	
71_ac1	-1374.309452	-1373.741294	-1371.094126	-94.56	-1370.561983	
214_ac2	-1374.309310	-1373.740279	-1371.093798	-97.45	-1370.561882	
160_ac2	-1374.308411	-1373.739702	-1371.094651	-94.35	-1370.561878	
123_ac1	-1374.312210	-1373.743568	-1371.095873	-89.83	-1370.561445	
85_ac2	-1374.309378	-1373.740538	-1371.093119	-97.19	-1370.561298	
106_ac2	-1374.308588	-1373.740205	-1371.090645	-101.67	-1370.560987	
135_ac2	-1374.308657	-1373.739794	-1371.093967	-94.10	-1370.560945	
227_ac1	-1374.308716	-1373.739743	-1371.094001	-94.27	-1370.560932	
255_ac1	-1374.308955	-1373.740132	-1371.094958	-91.17	-1370.560860	
17_ac1	-1374.307978	-1373.739909	-1371.092086	-95.69	-1370.560462	
214_ac1	-1374.312368	-1373.743183	-1371.095843	-88.58	-1370.560395	
195_ac2	-1374.308094	-1373.739561	-1371.091520	-96.78	-1370.559847	
255_ac2	-1374.308871	-1373.740232	-1371.093167	-91.88	-1370.559523	
217_ac2	-1374.308068	-1373.739442	-1371.091340	-96.06	-1370.559303	
85_ac1	-1374.307492	-1373.738716	-1371.090503	-95.60	-1370.558141	
1_ac1	-1374.306489	-1373.737992	-1371.089571	-93.89	-1370.556834	
5k						-1129.494665
5k_6	-1132.719571	-1132.223683	-1129.991588	0.59	-1129.495476	
5k_2	-1132.719743	-1132.224095	-1129.990748	1.21	-1129.494637	
5k_16	-1132.717991	-1132.222095	-1129.989202	1.38	-1129.492781	
5k_9	-1132.717803	-1132.222100	-1129.988772	2.05	-1129.492289	
5k_4	-1132.718542	-1132.222495	-1129.990984	2.51	-1129.493981	
5k_14	-1132.717516	-1132.222752	-1129.987748	3.39	-1129.491693	
5k_13	-1132.720046	-1132.223977	-1129.992325	4.06	-1129.494711	
5k_15	-1132.719620	-1132.223722	-1129.991243	4.18	-1129.493751	
5k_1	-1132.718643	-1132.222572	-1129.990748	5.31	-1129.492652	
5k_10	-1132.718590	-1132.222209	-1129.991153	5.98	-1129.492492	
5k_7	-1132.718577	-1132.222727	-1129.990500	6.86	-1129.492037	
5k_12	-1132.717365	-1132.221306	-1129.990284	9.00	-1129.490799	
5k_ac						-1282.147863
5k_2_ac1	-1285.726811	-1285.175964	-1282.666285	-86.73	-1282.148474	
5k_6_ac2	-1285.724706	-1285.173928	-1282.663656	-93.43	-1282.148463	
5k_2_ac2	-1285.725949	-1285.175058	-1282.665334	-89.04	-1282.148355	
5k_9_ac1	-1285.725437	-1285.174418	-1282.665419	-88.32	-1282.148040	
5k_6_ac1	-1285.725512	-1285.174914	-1282.664508	-89.04	-1282.147821	
5k_9_ac2	-1285.722401	-1285.173515	-1282.661184	-92.68	-1282.147596	
5k_16_ac1	-1285.724411	-1285.173429	-1282.663035	-90.71	-1282.146602	
5k_4_ac2	-1285.724972	-1285.173806	-1282.663722	-88.99	-1282.146452	
5k_1_ac2	-1285.725819	-1285.174714	-1282.664002	-87.86	-1282.146363	
5k_13_ac2	-1285.724837	-1285.173744	-1282.662904	-90.63	-1282.146328	
5k_15_ac2	-1285.725375	-1285.174421	-1282.665132	-83.89	-1282.146130	
5k_4_ac1	-1285.725693	-1285.174551	-1282.664360	-86.23	-1282.146063	
5k_16_ac2	-1285.724828	-1285.173855	-1282.663331	-88.41	-1282.146031	
5k_15_ac1	-1285.723602	-1285.173469	-1282.662132	-89.04	-1282.145910	
5k_1_ac1	-1285.726252	-1285.175297	-1282.664111	-84.18	-1282.145219	
5k_10_ac2	-1285.724986	-1285.173959	-1282.662922	-86.86	-1282.144978	
5k_12_ac2	-1285.724473	-1285.173331	-1282.662928	-86.02	-1282.144551	
5k_13_ac1	-1285.723877	-1285.172770	-1282.661463	-89.62	-1282.144491	
5k_7_ac2	-1285.722445	-1285.171115	-1282.660277	-92.80	-1282.144294	
5k_14_ac1	-1285.721651	-1285.171621	-1282.659651	-90.29	-1282.144011	
5k_7_ac1	-1285.723188	-1285.171928	-1282.661620	-88.12	-1282.143921	
5k_12_ac1	-1285.725071	-1285.174042	-1282.663278	-82.42	-1282.143642	

Appendix

Conformer	E_{tot} B98/6-31G(d)	H_{298} B98/6-31G(d)	E_{tot} MP2(FC)/6-31+G(2d,p)	G_{solv}, kJ/mol	H_{298} MP2-5 with solv	$<H_{298}>$ MP2-5 with solv
5k_14_ac2	-1285.720973	-1285.170353	-1282.658527	-91.71	-1282.142839	
5k_10_ac1	-1285.723876	-1285.172589	-1282.661382	-85.19	-1282.142540	
5l						-1284.973428
1	-1288.674630	-1288.083688	-1285.568297	9.71	-1284.973658	
2	-1288.673958	-1288.082890	-1285.567733	13.05	-1284.971692	
3	-1288.672207	-1288.081221	-1285.565655	20.92	-1284.966701	
4	-1288.672877	-1288.081718	-1285.564962	16.40	-1284.967556	
5l_ac						-1437.624915
1ac1	-1441.681515	-1441.035695	-1438.241929	-75.90	-1437.625017	
1ac2	-1441.680630	-1441.034903	-1438.240962	-78.78	-1437.625243	
2ac1	-1441.680095	-1441.033915	-1438.240092	-75.14	-1437.622534	
2ac2	-1441.679775	-1441.033834	-1438.240106	-77.19	-1437.623567	
3ac1	-1441.680924	-1441.034925	-1438.240386	-67.24	-1437.619996	
3ac2	-1441.677032	-1441.032019	-1438.235746	-69.58	-1437.617235	
4ac1	-1441.679966	-1441.033950	-1438.238863	-70.33	-1437.619635	
4ac2	-1441.678282	-1441.031988	-1438.236760	-75.65	-1437.619278	

Table A2.3. Calculated energies of conformers of 4-amino- and 4-guanidinylpyridines and corresponding acetyl intermediates. Averaged enthalpies <H$_{298}$> were calculated at MP2(FC)/6-31+G(2d,p)//B98/6-31G(d) level of theory with inclusion of solvent effects in chloroform at PCM/UAHF/RHF/6-31G(d) level.

Conformer	E$_{tot}$ B98/6-31G(d)	H$_{298}$ B98/6-31G(d)	E$_{tot}$ MP2(FC)/6-31+G(2d,p)	G$_{solv}$, kJ/mol	H$_{298}$ MP2-5 with solv	<H$_{298}$> MP2-5 with solv
Py						
Py	-248.181767	-248.087627	-247.589439	-9.00	-247.498727	-247.498727
Py_ac	-401.140004	-400.991691	-400.215516	-142.55	-400.121498	-400.121498
DMAP						
DMAP	-382.100962	-381.928959	-381.179977	-13.68	-381.013184	-381.013184
DMAP_ac	-535.091159	-534.864305	-533.836116	-131.34	-533.659287	-533.659287
PPY						
PPY	-459.499042	-459.289458	-458.383907	-16.90	-458.1807599	-458.1807599
PPY_ac	-612.493384	-612.228939	-611.043991	-130.58	-610.8292813	-610.8292813
6a						-535.362849
TCAP_b	-536.905604	-536.658613	-535.602351	-20.33	-535.3631033	
TCAP_a	-536.904889	-536.657992	-535.601147	-20.75	-535.3621533	
6a_ac						-688.016947
TCAP_ac2	-689.905522	-689.603408	-688.268359	-133.80	-688.0172068	
TCAP_ac1	-689.904728	-689.602521	-688.267441	-133.72	-688.0161653	
6b						-458.1883116
catMe	-459.503374	-459.293804	-458.39138	-17.07	-458.1883116	
6b_ac						-610.838384
ac1	-612.49845	-612.234055	-611.052154	-132.51	-610.8382294	
ac2	-612.498722	-612.234281	-611.052535	-132.34	-610.8384997	
7a						-607.721061
3a	-609.437278	-609.163890	-607.990831	-9.50	-607.721061	
7a_ac						-760.369530
ac2	-762.440075	-762.111592	-760.655492	-112.05	-760.369687	
ac1	-762.439141	-762.110580	-760.654585	-113.60	-760.369292	
7a_Nac[a]						-760.372579
Nac2	-762.419100	-762.091098	-760.650873	-130.54	-760.3725911	
Nac1	-762.417554	-762.090243	-760.647493	-122.26	-760.3667484	
7b						-646.889338
1	-648.739813	-648.437021	-647.190473	-5.02	-646.889593	
2	-648.737139	-648.434075	-647.189498	-5.77	-646.888632	
7b_ac						-799.539929
Ac2	-801.745225	-801.387184	-799.857844	-106.15	-799.5402334	
Ac1	-801.744402	-801.386219	-799.857176	-106.15	-799.5394234	
Ac3	-801.740146	-801.382030	-799.854065	-106.02	-799.5363299	
Ac4	-801.739262	-801.381071	-799.853080	-106.94	-799.5356203	
7b_Nac[a]						-799.542958
Nac2	-801.726548	-801.368811	-799.856508	-116.78	-799.5432502	
Nac3	-801.726255	-801.368531	-799.856334	-116.44	-799.5429597	
Nac4	-801.721772	-801.364235	-799.852345	-123.89	-799.5419952	
Nac1	-801.718232	-801.360445	-799.848735	-125.69	-799.5388208	
7c						-842.696676
2	-845.225804	-844.773120	-843.151656	4.35	-842.6973152	
1	-845.226266	-844.773664	-843.151150	3.89	-842.6970664	
8	-845.224920	-844.772280	-843.151380	5.52	-842.6966375	
3	-845.225141	-844.772591	-843.150273	3.43	-842.6964166	
7	-845.224944	-844.772307	-843.150164	3.51	-842.6961901	
9	-845.224639	-844.771780	-843.151589	6.82	-842.6961324	
6	-845.225043	-844.772409	-843.150398	4.77	-842.6959472	
4	-845.225399	-844.772609	-843.150866	5.69	-842.6959088	
5	-845.225251	-844.772595	-843.150973	6.57	-842.6958146	

Appendix

Conformer	E_{tot} B98/6-31G(d)	H_{298} B98/6-31G(d)	E_{tot} MP2(FC)/6-31+G(2d,p)	G_{solv}, kJ/mol	H_{298} MP2-5 with solv	$<H_{298}>$ MP2-5 with solv
7c_ac						-995.347569
2	-998.231706	-997.724119	-995.818900	-97.11	-995.3483003	
1	-998.232565	-997.724897	-995.819661	-94.89	-995.3481347	
9	-998.230948	-997.723126	-995.819330	-95.73	-995.3479697	
3	-998.231953	-997.724053	-995.820285	-92.38	-995.3475707	
10	-998.230527	-997.722839	-995.818092	-97.03	-995.3473608	
11	-998.230519	-997.722723	-995.818036	-96.27	-995.3469073	
12	-998.230476	-997.722694	-995.818050	-96.02	-995.3468401	
4	-998.231443	-997.723537	-995.818836	-94.14	-995.3467861	
14	-998.230740	-997.722574	-995.818674	-95.19	-995.346764	
8	-998.231579	-997.723492	-995.819356	-92.97	-995.3466794	
6	-998.231366	-997.723577	-995.818758	-93.64	-995.3466346	
13	-998.230521	-997.722593	-995.818395	-94.85	-995.3465935	
5	-998.231417	-997.723514	-995.818758	-93.55	-995.3464863	
7	-998.231393	-997.723380	-995.819127	-92.55	-995.3463645	
7d						-606.5703889
3d	-608.248922	-607.997703	-606.815152	-16.95	-606.5703889	
7d_ac						-759.219846
ac1	-761.253972	-760.947690	-759.480966	-118.83	-759.2199440	
ac2	-761.253112	-760.946828	-759.480190	-120.29	-759.2197221	
7e						-645.739143
3	-647.550815	-647.269999	-646.015591	-12.09	-645.7393798	
2	-647.551603	-647.270936	-646.014767	-12.80	-645.7389753	
1	-647.549715	-647.268968	-646.012418	-12.76	-645.7365310	
7e_ac						-798.390625
1	-800.558357	-800.222515	-798.682759	-115.02	-798.3907258	
2	-800.559119	-800.223296	-798.683274	-113.39	-798.390639	
4	-800.554459	-800.218868	-798.679798	-112.34	-798.3869951	
3	-800.553670	-800.217939	-798.679003	-113.22	-798.3863953	
7e_Nac[a]						-798.387359
Nac1	-800.534459	-800.198985	-798.675115	-126.19	-798.3877043	
Nac4	-800.534032	-800.198552	-798.674792	-125.56	-798.3871353	
Nac3	-800.528229	-800.193138	-798.669482	-134.56	-798.3856422	
Nac2	-800.525202	-800.190078	-798.666451	-135.69	-798.3830086	
7f						-841.545967
5	-844.036400	-843.605886	-841.976620	-1.92	-841.546837	
4	-844.036216	-843.605834	-841.977251	2.01	-841.546103	
7	-844.035767	-843.605382	-841.975710	-1.92	-841.546056	
8	-844.036125	-843.605687	-841.976356	-0.08	-841.545948	
6	-844.036255	-843.605695	-841.975293	-2.89	-841.545934	
2	-844.037044	-843.606530	-841.975303	-1.97	-841.545539	
3	-844.037044	-843.606530	-841.975303	-1.97	-841.545539	
1	-844.037144	-843.606699	-841.975565	-0.17	-841.545185	
9	-844.035455	-843.605001	-841.974594	-2.59	-841.545126	
10	-844.035456	-843.604966	-841.974249	-2.80	-841.544825	
11	-844.035576	-843.604960	-841.973940	-2.47	-841.544265	
12	-844.035410	-843.604857	-841.973606	-2.47	-841.543994	
13	-844.034772	-843.604406	-841.973025	-2.47	-841.543600	
14	-844.033715	-843.603332	-841.972631	-1.09	-841.542663	
7f_ac						-994.197488
1	-997.046466	-996.560962	-994.645085	-102.26	-994.198530	
6	-997.044593	-996.559081	-994.643719	-103.76	-994.197727	
3	-997.044744	-996.559383	-994.642986	-105.02	-994.197625	
9	-997.044342	-996.558805	-994.643316	-104.01	-994.197394	
7	-997.044580	-996.559039	-994.643666	-103.01	-994.197359	
2	-997.045386	-996.559941	-994.644275	-101.09	-994.197333	

Appendix

Conformer	E_{tot} B98/6-31G(d)	H_{298} B98/6-31G(d)	E_{tot} MP2(FC)/6-31+G(2d,p)	G_{solv} kJ/mol	H_{298} MP2-5 with solv	$\langle H_{298}\rangle$ MP2-5 with solv
4	-997.043785	-996.558338	-994.642910	-103.64	-994.196937	
13	-997.044277	-996.559610	-994.642851	-101.17	-994.196718	
5	-997.044780	-996.559270	-994.643901	-100.16	-994.196540	
15	-997.043032	-996.557678	-994.642475	-102.47	-994.196150	
8	-997.044453	-996.558957	-994.642756	-102.01	-994.196114	
14	-997.043336	-996.557947	-994.642060	-103.09	-994.195936	
10	-997.043432	-996.557812	-994.642227	-103.22	-994.195921	
12	-997.044284	-996.558670	-994.642550	-102.13	-994.195835	
11	-997.044204	-996.558647	-994.642787	-100.96	-994.195684	
16	-997.041856	-996.556543	-994.641937	-101.34	-994.195222	
17	-997.041856	-996.556543	-994.641936	-101.34	-994.195221	
19	-997.040761	-996.555274	-994.642548	-94.47	-994.193043	
18	-997.040348	-996.554837	-994.641995	-95.69	-994.192930	
7g						-953.218523
py3g_z2	-955.903158	-955.499018	-953.621331	-5.06	-953.219118	
py3g_z3	-955.901028	-955.496897	-953.621346	-3.97	-953.218727	
h3g_1	-955.903126	-955.499183	-953.619806	-5.86	-953.218095	
h3g_3	-955.900586	-955.497080	-953.617966	-7.36	-953.217264	
h3g_4	-955.901229	-955.497343	-953.619450	-3.97	-953.217076	
h3g_2	-955.902205	-955.498497	-953.617240	-7.74	-953.216481	
h3g_6	-955.897733	-955.494139	-953.616828	-4.27	-953.214860	
h3g_7	-955.897976	-955.494064	-953.617214	-4.06	-953.214849	
py3g_z4	-955.897697	-955.494087	-953.616685	-4.35	-953.214732	
7g_ac						-1105.870495
3	-1108.912657	-1108.454069	-1106.292039	-98.11	-1105.870819	
7	-1108.911987	-1108.453133	-1106.291541	-99.62	-1105.870630	
2	-1108.913135	-1108.454221	-1106.291777	-97.40	-1105.869961	
5	-1108.912567	-1108.453657	-1106.291288	-98.62	-1105.869940	

[a] '7*_Nac' indicates acetylation at the nitrogen atom in 4-position.

Table A2.4. Acetylation enthalpies and structure parameters of 4-amino- and 4-guanidinylpyridines at the MP2/6-31+G(2d,p)//B98/6-31G(d) level of theory. Values in brackets indicate acetylation at the nitrogen atom in 4-position.

Catalyst	ΔH_{ac} B3LYP/6-311+G(d,p)//B3LYP/6-31G(d) [kJ/mol]	On the basis of the energetically lowest conformer					Averaged
		q_{NPA} (Ac)[a,b]	r(C-N)[pm][a]	ΔH_{ac} (MP2-5) [kJ/mol]	$\Delta\Delta G_{solv}$ [kJ/mol]	ΔH_{ac} (MP2-5/solv) [kJ/mol]	ΔH_{ac} (MP2-5/solv) [kJ/mol]
py	0.0	0.368	153.3	0.0	0.00	0.0	0.0
DMAP	-82.1	0.302	148.3	-77.2	15.86	-61.3	-61.3
7a	-113.1	0.273	146.8	-98.9 (-88.0)	31.00 (12.51)	-67.9 (-75.5)	-67.5 (-75.5)
PPY	-93.1	0.295	147.9	-87.5	19.87	-67.6	-67.6
7d	-118.9	0.270	146.7	-102.0	31.67	-70.3	-70.1
7b	-120.5	0.271	146.8	-105.6 (-102.9)	32.43 (21.80)	-73.2 (-81.1)	-73.0 (-81.0)
7c	-123.1	0.268	146.7	-109.0	34.77	-74.1	-73.8
7e	-126.7	0.268	146.6	-107.0 (-86.5)	32.26 (19.46)	-75.0 (-67.1)	-75.4 (-66.8)
7f	-130.1	0.266	146.6	-109.1	33.22	-75.9	-75.5
7g	-133.1	0.266	146.6	-116.4	40.5	-76.0	-76.7
6a	-108.9	0.283	147.3	-102.3	20.08	-82.3	-82.3
6b	-96.0	0.292	147.8	-90.3	18.28	-72.0	-71.7

[a] Charge and distance parameters of the most favorable conformer
[b] In units of elemental charge e

Appendix

Table A2.5. Acetylation enthalpies and structure parameters for 3,4-diaminopyridines, as calculated at MP2/6-31+G(2d,p)//B98/6-31G(d) level with inclusion of solvent effects at PCM/UAHF/RHF/6-31G(d) level. Data for 4-aminopyridines **DMAP**, **PPY** and **6a** are also shown.

Catalyst	On the basis of the energetically lowest conformer					Averaged
	q_{NPA} (Ac)[a,b]	r(C-N)[pm][a]	ΔH_{ac} (MP2-5) [kJ/mol]	$\Delta\Delta G_{solv}$ [kJ/mol]	ΔH_{ac} (MP2-5/solv) [kJ/mol]	ΔH_{ac} (MP2-5/solv) [kJ/mol]
py	0.368	153.3	0.0	0.0	0.0	0.0
DMAP	0.302	148.3	-77.2	15.9	-61.3	-61.3
PPY	0.295	147.9	-87.5	19.9	-67.6	-67.6
6a	0.283	147.3	-102.3	20.1	-82.3	-82.3
5a	0.282	147.4	-108.6	27.0	-81.6	-81.4
5b	0.276	147.1	-119.6	33.3	-86.3	-85.2
5c	0.274	147.0	-121.8	36.1	-85.7	-85.1
5e	0.280	147.3	-117.2	34.4	-82.8	-82.7
5f	0.279	147.4	-127.4	43.1	-84.3	-84.1
5g	0.273	147.0	-121.8	42.3	-79.5	-80.1
5h	0.280	147.2	-116.3	34.5	-81.8	-81.8
5i	0.275	147.0	-121.9	37.3	-84.6	-84.8
5j	0.268	146.9	-122.1	40.5	-81.6	-82.1
5k	0.276	147.4	-124.1	44.7	-79.4	-79.9
5l	0.282	147.4	-123.0	47.3	-75.7	-75.4
5a	0.282	147.4	-108.6	27.0	-81.6	-81.4

[a] Charge and distance parameters of the most favorable conformer
[b] In units of elemental charge e
[c] $\Delta\Delta G_{solv} = \Delta H_{ac}$ (MP2-5/solv) $- \Delta H_{ac}$ (MP2-5)

Table A2.6. Comparison of different basis sets for the MP2 single points

Conformer	E_{tot} B98/6-31G(d)	H_{298} B98/6-31G(d)	E_{tot} MP2/6-311+G(2d,p)	<H_{298}> "MP2-6/solv"[a]	E_{tot} MP2/6-311+G(3df,2p)	<H_{298}> "MP2-7/solv"[b]
Py						
Py	-248.181767	-248.087627	-247.656777	-247.562637	-247.754374	-247.660234
Py_ac	-401.140004	-400.991691	-400.330778	-400.182464	-400.484202	-400.335889
DMAP						
DMAP	-382.100962	-381.928959	-381.289295	-381.117291	-381.442355	-381.270352
DMAP_ac	-535.091159	-534.864305	-533.993468	-533.766615	-534.202318	-533.975465
PPY						
PPY	-459.499042	-459.289458	-458.513262	-458.303677	-458.697778	-458.488194
PPY_ac	-612.493384	-612.228939	-611.221396	-610.956952	-611.461694	-611.197249
6a						
TCAP_b	-536.905604	-536.658613	-535.752726	-535.505735	-535.966387	-535.719396
TCAP_ac2	-689.905522	-689.603408	-688.466596	-688.164481	-688.736063	-688.433948
7a						
7a	-609.437278	-609.163890	-608.168122	-607.894735	-608.412135	-608.138748
7a_ac2	-762.440075	-762.111592	-760.881481	-760.552998	-761.181388	-760.852905
7b						
7b_1	-648.739813	-648.437021	-647.379176	-647.076384	-647.638806	-647.336014
7b_ac2	-801.745225	-801.387183	-800.095240	-799.737198	-800.410741	-800.052699
5a						
5a_1	-593.418065	-593.130880	-592.151774	-591.864589	-592.391322	-592.104137
5a_ac1	-746.420416	-746.078238	-744.868480	-744.526302	-745.163729	-744.821551
5b						
5b_6	-749.405682	-749.021776	-747.789867	-747.405960	-748.092581	-747.708675
5b_ac2	-902.413587	-901.973819	-900.511669	-900.071901	-900.870467	-900.430699
5k						
5k_13	-1132.720046	-1132.223977	-1130.301147	-1129.805079	-1130.751595	-1130.255527
5k_2_ac1	-1285.726811	-1285.175964	-1283.023612	-1282.472765	-1283.529626	-1282.978779

[a] "MP2-6/solv" = MP2/6-311+G(2d,p)//B98/6-31G(d) with solvation energies calculated at PCM/UAHF/RHF/6-31G(d) level
[b] "MP2-7/solv" = MP2/6-311+G(3df,2p)//B98/6-31G(d) with solvation energies calculated at PCM/UAHF/RHF/6-31G(d) level

A2.2 Relative activation enthalpies for 3,4-diamino and 4-aminopyridines.

Table A2.7. Calculated energies of conformers of 3,4-diaminopyridines and corresponding transition states. Final enthalpies "H₂₉₈ MP2-5 with solv" were calculated at MP2(FC)/6-31+G(2d,p)//B98/6-31G(d) level of theory with inclusion of solvent effects in chloroform at chosen level. Methods for calculation PCM single point energies in chloroform: "solv1": RHF/6-31G(d) with UAHF radii; "solv2": RHF/6-31G(d) with Pauling radii; "solv3": B98/6-31G(d) with Pauling radii.

Conformer	E_{tot} B98/6-31G(d)	H_{298} B98/6-31G(d)	E_{tot} MP2(FC)/6-31+G(2d,p)	solv1 G_{solv} kJ/mol	solv2 G_{solv} kJ/mol	solv3 G_{solv} kJ/mol	solv1 H_{298} MP2-5 with solv	solv2 H_{298} MP2-5 with solv	solv3 H_{298} MP2-5 with solv
alc									
alc	-233.576452	-233.432655	-233.020561	-7.78	-0.63	2.30	-232.879728	-232.877003	-232.875888
anh									
anh	-381.583992	-381.475889	-380.793597	-6.65	-34.73	-20.92	-380.688028	-380.698721	-380.693462
anh2	-381.583860	-381.475922	-380.794157	-8.33	-36.61	-22.89	-380.689391	-380.700163	-380.694936
py									
py	-248.181767	-248.087627	-247.589439	-9.00	-5.27	-1.92	-247.498726	-247.497308	-247.496033
frozen_ts									
py_ts optimized ts	-863.330376	-862.984096	-861.407258	1.67	-7.41	5.44	-861.060340	-861.063799	-861.058906
tsopt	-863.330413	-862.983065	-861.407201	2.64	-8.95	3.72	-861.058849	-861.063263	-861.058435
3 (DMAP)									
DMAP	-382.100962	-381.928959	-381.179977	-13.68	-11.88	-8.49	-381.013184	-381.012499	-381.011208
frozen_ts									
DMAP_ts optimized ts	-997.259012	-996.834800	-995.006218	-2.93	3.14	13.72	-994.583122	-994.580811	-994.576779
tsopt	-997.259033	-996.834636	-995.006254	-3.26	0.17	10.84	-994.583100	-994.581793	-994.577730
4 (PPY)									
PPY	-459.499042	-459.289458	-458.383907	-16.90	-5.27	-1.92	-458.180761	-458.176331	-458.175056
frozen_ts									
PPY_ts optimized ts	-1074.658021	-1074.196281	-1072.211182	-6.11	*	*	-1071.751769	*	*
tsopt	-1074.658051	-1074.196091	-1072.211254	-6.40	1.67	12.55	-1071.751732	-1071.748656	-1071.744513
7a									
7a	-609.437278	-609.163890	-607.990831	-9.50	-2.93	8.49	-607.721060	-607.718559	-607.714208
frozen_ts									
ts1	-1224.596611	-1224.070835	-1221.817596	3.60	20.38	32.59	-1221.290450	-1221.284060	-1221.279406

Appendix

Conformer	E_{tot} B98/6-31G(d)	H_{298} B98/6-31G(d)	E_{tot} MP2(FC)/6-31+G(2d,p)	solv1 G_{solv} kJ/mol	solv2 G_{solv} kJ/mol	solv3 G_{solv} kJ/mol	solv1 H_{298} MP2-5 with solv	solv2 H_{298} MP2-5 with solv	solv3 H_{298} MP2-5 with solv
ts2	-1224.595822	-1224.070121	-1221.816900	1.88	18.58	31.46	-1221.290482	-1221.284124	-1221.279215
ts3	-1224.595121	-1224.069546	-1221.815904	-0.54	15.48	28.33	-1221.290536	-1221.284433	-1221.279540
ts4	-1224.597176	-1224.071446	-1221.818658	6.49	21.13	33.93	-1221.290457	-1221.284880	-1221.280003
optimized ts									
ts2opt	-1224.595818	-1224.069937	-1221.816846	1.51	16.86	29.96	-1221.290392	-1221.284543	-1221.279555
ts3opt	-1224.595134	-1224.069252	-1221.815909	-1.09	15.31	28.12	-1221.290442	-1221.284195	-1221.279318
ts4opt	-1224.597204	-1224.071399	-1221.818683	8.33	21.59	34.35	-1221.289707	-1221.284655	-1221.279795
7b									
1	-648.739813	-648.437021	-647.190473	-5.02	15.90	19.79	-646.889594	-646.881626	-646.880144
2	-648.737139	-648.434075	-647.189498	-5.77	11.92	16.74	-646.888633	-646.881893	-646.880060
frozen_ts									
7b_1_ts2	-1263.900062	-1263.344925	-1261.019353	10.71	37.15	48.58	-1260.460137	-1260.450065	-1260.445715
7b_1_ts1	-1263.899612	-1263.344451	-1261.018317	8.33	34.18	44.56	-1260.459985	-1260.450136	-1260.446184
7b_1_ts4	-1263.899366	-1263.343952	-1261.018287	7.82	33.43	44.64	-1260.459894	-1260.450141	-1260.445870
7b_1_ts3	-1263.900231	-1263.344812	-1261.019441	11.25	37.15	48.03	-1260.459735	-1260.449871	-1260.445727
7b_2_ts2	-1263.897147	-1263.342127	-1261.017918	10.71	34.27	45.81	-1260.458818	-1260.449846	-1260.445448
7b_2_ts4	-1263.897899	-1263.342544	-1261.018974	13.72	35.82	47.91	-1260.458392	-1260.449978	-1260.445372
7b_2_ts1	-1263.895771	-1263.340508	-1261.016559	8.12	29.75	41.97	-1260.458204	-1260.449965	-1260.445312
7b_2_ts3	-1263.894986	-1263.339595	-1261.015462	5.44	27.20	39.33	-1260.457999	-1260.449712	-1260.445091
optimized ts									
1_ts2opt	-1263.900085	-1263.344874	-1261.019372	10.42	37.07	48.37	-1260.460193	-1260.450041	-1260.445739
1_ts1opt	-1263.899655	-1263.344289	-1261.018453	8.16	35.56	45.98	-1260.459980	-1260.449542	-1260.445574
1_ts4opt	-1263.899412	-1263.343862	-1261.018402	7.41	33.05	*	-1260.460031	-1260.450263	*
6a									
6a_a	-536.904889	-536.657992	-535.601147	-20.75	2.59	4.60	-535.362154	-535.353262	-535.352497
6a_b	-536.905604	-536.658613	-535.602351	-20.33	2.43	4.69	-535.363105	-535.354436	-535.353575
frozen_ts									
a_ts2	-1152.066869	-1151.567550	-1149.433194	-5.65	22.93	31.97	-1148.936027	-1148.925142	-1148.921700
a_ts1	-1152.066916	-1151.567448	-1149.433173	-5.23	23.81	32.80	-1148.935697	-1148.924638	-1148.921211
b_ts3	-1152.066653	-1151.567301	-1149.432888	-5.52	23.43	32.64	-1148.935640	-1148.924612	-1148.921106
b_ts2	-1152.066666	-1151.567293	-1149.433011	-4.85	22.80	32.01	-1148.935487	-1148.924953	-1148.921447
optimized									

Appendix

Conformer	E_{tot} B98/6-31G(d)	H_{298} B98/6-31G(d)	E_{tot} MP2(FC)/6-31+G(2d,p)	solv1 G_{solv} kJ/mol	solv2 G_{solv} kJ/mol	solv3 G_{solv} kJ/mol	solv1 H_{298} MP2-5 with solv	solv2 H_{298} MP2-5 with solv	solv3 H_{298} MP2-5 with solv
ts									
a ts2opt	-1152.066909	-1151.567461	-1149.433307	-5.98	22.34	31.46	-1148.936138	-1148.925350	-1148.921876
5a									
1	-593.418065	-593.130880	-591.981245	-14.56	5.69	8.08	-591.699606	-591.691893	-591.690984
2	-593.417088	-593.130024	-591.980224	-14.52	6.15	8.33	-591.698690	-591.690817	-591.689989
3	-593.416648	-593.129499	-591.979898	-14.18	6.69	8.45	-591.698151	-591.690199	-591.689529
4	-593.415838	-593.128722	-591.978867	-14.14	6.69	8.45	-591.697137	-591.689201	-591.688532
5	-593.415488	-593.128357	-591.979020	-15.06	4.23	6.65	-591.697626	-591.690279	-591.689355
6	-593.415277	-593.127870	-591.978978	-15.19	4.39	6.61	-591.697356	-591.689898	-591.689053
7	-593.414082	-593.126927	-591.977706	-14.56	6.95	8.41	-591.696097	-591.687905	-591.687348
frozen ts									
1 ts4	-1208.582930	-1208.043150	-1205.818760	6.82	38.03	47.57	-1205.276383	-1205.264494	-1205.260861
2 ts2	-1208.582625	-1208.042984	-1205.818421	5.06	38.20	47.70	-1205.276852	-1205.264230	-1205.260613
1 ts1	-1208.582055	-1208.042450	-1205.817825	8.79	40.46	49.33	-1205.274873	-1205.262809	-1205.259431
2 ts4	-1208.581606	-1208.041903	-1205.817318	4.77	39.08	47.86	-1205.275798	-1205.262730	-1205.259384
1 ts3	-1208.578252	-1208.038635	-1205.811364	0.59	26.07	35.06	-1205.271524	-1205.261819	-1205.258393
3 ts4	-1208.580190	-1208.040780	-1205.815293	5.61	40.12	48.37	-1205.273748	-1205.260601	-1205.257461
3 ts2	-1208.580852	-1208.041126	-1205.816281	5.48	*	*	-1205.274467	*	*
1 ts2	-1208.578588	-1208.038935	-1205.811847	0.08	*	*	-1205.272163	*	*
optimized ts									
1 ts4opt	-1208.582983	-1208.043047	-1205.818948	4.64	37.95	47.45	-1205.277243	-1205.264558	-1205.260940
2 ts4opt	-1208.581766	-1208.041791	-1205.817607	4.14	39.20	47.95	-1205.276054	-1205.262700	-1205.259369
2 ts3opt	-1208.578330	-1208.038788	-1205.811568	0.04	24.89	34.56	-1205.272010	-1205.262544	-1205.258863
2 ts2opt	-1208.582671	-1208.042830	-1205.818582	4.52	*	*	-1205.277019	*	*
5b									
1	-749.407023	-749.022449	-747.576974	-8.03	27.11	28.58	-747.195459	-747.182073	-747.181515
2	-749.405930	-749.021754	-747.576291	-7.87	30.75	32.13	-747.195111	-747.180402	-747.179876
4	-749.406234	-749.021772	-747.576561	-7.99	28.03	29.08	-747.195142	-747.181421	-747.181023
6	-749.405682	-749.021776	-747.576377	-8.24	28.79	30.25	-747.195610	-747.181507	-747.180949
7	-749.405742	-749.021319	-747.575522	-8.37	26.94	28.07	-747.194286	-747.180836	-747.180405
9	-749.405012	-749.020692	-747.574665	-8.20	28.49	29.96	-747.193468	-747.179492	-747.178935
10	-749.404995	-749.020441	-747.576374	-8.16	25.82	27.70	-747.194928	-747.181988	-747.181270
11	-749.406385	-749.021639	-747.576067	-8.74	25.82	27.41	-747.194652	-747.181488	-747.180883

Appendix

Conformer	E_{tot} B98/6-31G(d)	H_{298} B98/6-31G(d)	E_{tot} MP2(FC)/6-31+G(2dp)	solv1 G_{solv}, kJ/mol	solv2 G_{solv}, kJ/mol	solv3 G_{solv}, kJ/mol	solv1 H_{298} MP2-5 with solv	solv2 H_{298} MP2-5 with solv	solv3 H_{298} MP2-5 with solv
12	-749.404828	-749.020084	-747.576198	-8.20	25.98	27.57	-747.194577	-747.181557	-747.180952
13	-749.404877	-749.020263	-747.576944	-8.33	26.82	28.91	-747.195501	-747.182114	-747.181318
14	-749.404725	-749.020437	-747.574364	-8.28	28.62	30.04	-747.193232	-747.179176	-747.178634
15	-749.404598	-749.020095	-747.576157	-7.66	27.78	29.46	-747.194570	-747.181072	-747.180435
16	-749.406393	-749.021713	-747.576140	-8.79	25.73	27.36	-747.194807	-747.181660	-747.181038
frozen_ts									
1_ts2	-1364.573141	-1363.935995	-1361.416481	13.14	61.09	69.71	-1360.774331	-1360.756069	-1360.752786
11_ts2	-1364.572026	-1363.935017	-1361.415342	12.22	58.37	67.07	-1360.773680	-1360.756103	-1360.752788
1_ts4	-1364.572125	-1363.935075	-1361.415374	13.97	62.84	70.84	-1360.773001	-1360.754388	-1360.751344
11_ts4	-1364.571028	-1363.934298	-1361.414326	13.14	63.43	71.59	-1360.772592	-1360.753437	-1360.750330
6_ts2	-1364.569908	-1363.933251	-1361.413216	12.47	60.92	69.25	-1360.771811	-1360.753357	-1360.750185
4_ts2	-1364.570886	-1363.934056	-1361.414105	14.48	61.00	69.33	-1360.771761	-1360.754040	-1360.750869
6_ts4	-1364.569156	-1363.932413	-1361.412601	13.64	63.35	71.71	-1360.770663	-1360.751731	-1360.748544
2_ts2	-1364.568399	-1363.931902	-1361.411564	12.22	60.96	69.50	-1360.770414	-1360.751849	-1360.748598
4_ts4	-1364.570163	-1363.933406	-1361.413240	16.36	63.26	71.21	-1360.770252	-1360.752388	-1360.749360
13_ts2	-1364.568126	-1363.931144	-1361.411620	12.68	60.67	69.25	-1360.769810	-1360.751531	-1360.748264
1_ts3	-1364.568905	-1363.932425	-1361.409456	8.62	49.33	57.91	-1360.769693	-1360.754187	-1360.750920
13_ts4	-1364.568427	-1363.931502	-1361.412079	14.90	59.04	68.24	-1360.769480	-1360.752668	-1360.749162
2_ts4	-1364.567886	-1363.930832	-1361.410835	11.80	64.14	71.80	-1360.769287	-1360.749351	-1360.746435
1_ts1	-1364.568512	-1363.931490	-1361.408878	8.28	49.25	57.45	-1360.768701	-1360.753100	-1360.749976
11_ts3	-1364.567997	-1363.930928	-1361.408409	7.03	*	*	-1360.768663	*	*
11_ts1	-1364.567541	-1363.930297	-1361.407917	7.99	*	*	-1360.767629	*	*
optimized ts									
1_ts2opt	-1364.573153	-1363.935988	-1361.416613	12.64	59.96	68.45	-1360.774636	-1360.756612	-1360.753377
11_ts2opt	-1364.572050	-1363.934964	-1361.415460	12.22	60.46	68.99	-1360.773720	-1360.755346	-1360.752095
1_ts4opt	-1364.572257	-1363.935082	-1361.415669	13.39	62.68	70.54	-1360.773395	-1360.754622	-1360.751626
1_ts3opt	-1364.568935	-1363.932278	-1361.409530	8.16	50.58	59.16	-1360.769765	-1360.753606	-1360.750339
5j									
1	-1221.300472	-1220.787113	-1218.426019	2.43	35.15	41.13	-1217.911735	-1217.899273	-1217.896994
102	-1221.302906	-1220.789129	-1218.420923	-7.53	20.63	25.94	-1217.910014	-1217.899289	-1217.897265
187	-1221.302230	-1220.788302	-1218.420066	-8.79	19.87	25.31	-1217.908811	-1217.900493	-1217.898421
46	-1221.299409	-1220.786171	-1218.417271	-14.02	18.74	*	-1217.909371	-1217.896893	*
7	-1221.300922	-1220.787456	-1218.420254	-5.61	26.36	31.30	-1217.908924	-1217.896748	-1217.894868

Appendix

Conformer	E_{tot} B98/6-31G(d)	H_{298} B98/6-31G(d)	E_{tot} MP2(FC)/6-31+G(2d,p)	solv1 G_{solv} kJ/mol	solv2 G_{solv} kJ/mol	solv3 G_{solv} kJ/mol	solv1 H_{298} MP2-5 with solv	solv2 H_{298} MP2-5 with solv	solv3 H_{298} MP2-5 with solv
85	-1221.300735	-1220.786939	-1218.421235	-3.60	27.61	33.47	-1217.908809	-1217.896921	-1217.894690
frozen ts									
33_ts1	-1836.473923	-1835.707423	-1832.269217	29.50	79.87	*	-1831.491483	-1831.472296	*
158_ts1	-1836.472722								
102_ts1	-1836.472973								
102_ts2	-1836.473093								
33_ts2	-1836.472785								
85_ts2	-1836.471401								
158_ts2	-1836.471722								
85_ts1	-1836.471491								
1_ts1	-1836.469635								
106_ts2	-1836.470443								
207_ts1	-1836.469182								
155_ts2	-1836.469421								
optimized ts									
33_ts1	-1836.473927	-1835.707440	-1832.269541	29.25	79.58	89.08	-1831.491914	-1831.472743	-1831.469126
158_ts1	-1836.472718	-1835.706546	-1832.268247	27.45	77.45	87.45	-1831.491621	-1831.472578	-1831.468769
102_ts2	-1836.472742	-1835.706082	-1832.268925	29.12	75.73	86.19	-1831.491174	-1831.473421	-1831.469437
33_ts2	-1836.472582	-1835.706038	-1832.268708	34.06	78.87	89.29	-1831.489192	-1831.472125	-1831.468156
85_ts2	-1836.471338	-1835.705063	-1832.266938	31.42	74.77	85.69	-1831.488695	-1831.472186	-1831.468026
158_ts2	-1836.471600	-1835.704589	-1832.267210	31.46	75.56	86.06	-1831.488214	-1831.471418	-1831.467418
102_ts1	-1836.473163	-1835.706720	-1832.26545	24.31	74.48	*	-1831.489744	-1831.470636	*
85_ts1	-1836.471625	-1835.704943	-1832.262886	25.40	*	*	-1831.486531	*	*
1_ts1	-1836.469945	-1835.703777	-1832.263680	29.33	*	*	-1831.486340	*	*
106_ts2	-1836.470615	-1835.704426	-1832.260658	23.60	*	*	-1831.485481	*	*
207_ts1	-1836.469471	-1835.703334	-1832.259291	23.01	*	*	-1831.484390	*	*
155_ts2	-1836.469614	-1835.702959	-1832.259531	24.56	*	*	-1831.483522	*	*
5k									
5k_6	-1132.71957	-1132.22368	-1129.99159	0.59	37.82	41.51	-1129.495476	-1129.481293	-1129.479891
5k_13	-1132.72005	-1132.22398	-1129.99233	4.06	40.88	44.98	-1129.494711	-1129.480687	-1129.479125
5k_2	-1132.71974	-1132.22410	-1129.99075	1.21	38.24	41.88	-1129.494637	-1129.480534	-1129.479147
5k_4	-1132.71854	-1132.22250	-1129.99098	2.51	35.82	39.33	-1129.493981	-1129.481296	-1129.479957
5k_15	-1132.71962	-1132.22372	-1129.99124	4.18	43.47	47.24	-1129.493751	-1129.478787	-1129.477353

214

Appendix

Conformer	E_{tot} B98/6-31G(d)	H_{298} B98/6-31G(d)	E_{tot} MP2(FC)/6-31+G(2d,p)	solv1 G_{solv} kJ/mol	solv2 G_{solv} kJ/mol	solv3 G_{solv} kJ/mol	solv1 H_{298} MP2-5 with solv	solv2 H_{298} MP2-5 with solv	solv3 H_{298} MP2-5 with solv
5k_16	-1132.71799	-1132.22210	-1129.98920	1.38	36.74	40.46	-1129.492781	-1129.479315	-1129.477897
5k_1	-1132.71864	-1132.22257	-1129.99075	5.31	38.16	41.84	-1129.492652	-1129.480143	-1129.478740
5k_10	-1132.71859	-1132.22221	-1129.99115	5.98	39.16	42.97	-1129.492492	-1129.479855	-1129.478405
5k_9	-1132.71780	-1132.22210	-1129.98877	2.05	39.25	43.05	-1129.492289	-1129.478122	-1129.476671
5k_7	-1132.71858	-1132.22273	-1129.99050	6.86	44.52	49.08	-1129.492037	-1129.477694	-1129.475957
5k_14	-1132.71752	-1132.22275	-1129.98775	3.39	39.62	43.39	-1129.491693	-1129.477892	-1129.476458
5k_12	-1132.71736	-1132.22131	-1129.99028	9.00	41.59	45.61	-1129.490799	-1129.478385	-1129.476855
frozen_ts									
2_ts4	-1747.880519	-1747.132325	-1743.831159	29.20	77.74	88.62	-1743.071841	-1743.053355	-1743.049212
15_ts4	-1747.879917	-1747.131619	-1743.830837	32.09	79.71	90.96	-1743.070316	-1743.052181	-1743.047894
4_ts4	-1747.881367	-1747.133047	-1743.829639	34.14	78.20	88.87	-1743.068314	-1743.051534	-1743.047470
6_ts4	-1747.881486	-1747.133197	-1743.829398	29.16	79.08	89.79	-1743.070001	-1743.050989	-1743.046910
13_ts4	-1747.882464	-1747.134185	-1743.828136	30.63	77.70	88.20	-1743.068192	-1743.050264	-1743.046264
6_ts2	-1747.878594	-1747.130399	-1743.824327	21.76	67.95	78.53	-1743.067845	-1743.050252	-1743.046220
4_ts2	-1747.882272	-1747.133570	-1743.828156	34.73	80.54	90.04	-1743.066227	-1743.048777	-1743.045160
6_ts1	-1747.882385	-1747.134297	-1743.828111	32.09	*	*	-1743.067800	*	*
2_ts2	-1747.878208	-1747.130341	-1743.826701	29.79	*	*	-1743.067487	*	*
13_ts2	-1747.882612	-1747.134304	-1743.828501	33.93	*	*	-1743.067268	*	*
15_ts2	-1747.877628	-1747.129289	-1743.826485	31.46	*	*	-1743.066162	*	*
6_ts3	-1747.878813	-1747.130595	-1743.823137	23.05	*	*	-1743.066138	*	*
13_ts1	-1747.879267	-1747.130979	-1743.822186	21.80	*	*	-1743.065595	*	*
13_ts3	-1747.879313	-1747.130911	-1743.822685	27.74	*	*	-1743.063717	*	*
optimized ts									
2_ts4opt	-1747.880519	-1747.132325	-1743.831159	29.20	77.74	88.62	-1743.071841	-1743.053355	-1743.049212
6_ts4opt	-1747.881953	-1747.132894	-1743.829966	28.58	79.12	89.70	-1743.070023	-1743.050772	-1743.046740
15_ts4opt	-1747.879940	-1747.131692	-1743.830756	32.68	79.71	90.88	-1743.070062	-1743.052150	-1743.047895
13_ts3opt	-1747.879331	-1747.130841	-1743.822725	27.36	62.89	74.35	-1743.063813	-1743.050283	-1743.045917
13_ts2opt	-1747.882726	-1747.134099	-1743.828704	34.69	77.28	88.16	-1743.066867	-1743.050644	-1743.046500
6_ts1opt	-1747.882437	-1747.134116	-1743.828315	30.88	83.18	92.72	-1743.068233	-1743.048313	-1743.044680
4_ts2opt	-1747.882338	-1747.133553	-1743.828482	34.94	80.50	89.91	-1743.066391	-1743.049037	-1743.045451
5l									
1	-1288.674630	-1288.083688	-1285.568297	9.71	56.36	59.96	-1284.973658	-1284.955890	-1284.954519
2	-1288.673958	-1288.082890	-1285.567733	13.05	57.24	61.04	-1284.971692	-1284.954864	-1284.953414

Appendix

Conformer	E_{tot} B98/6-31G(d)	H_{298} B98/6-31G(d)	E_{tot} MP2(FC)/6-31+G(2d,p)	solv1 G_{solv} kJ/mol	solv2 G_{solv} kJ/mol	solv3 G_{solv} kJ/mol	solv1 H_{298} MP2-5 with solv	solv2 H_{298} MP2-5 with solv	solv3 H_{298} MP2-5 with solv
3	-1288.672207	-1288.081221	-1285.565655	20.92	64.14	67.99	-1284.966701	-1284.950239	-1284.948773
4	-1288.672877	-1288.081718	-1285.564962	16.40	60.04	63.81	-1284.967556	-1284.950935	-1284.949501
frozen ts									
2_ts2	-1903.835205	-1902.993930	-1899.407900	41.42	*	*	-1898.550848	*	*
1_ts2	-1903.835868	-1902.992951	-1899.408837	42.38	*	*	-1898.549777	*	*
1_ts4	-1903.834904	-1902.991323	-1899.404425	37.66	*	*	-1898.546501	*	*
2_ts4	-1903.835205	-1902.991744	-1899.404675	42.26	*	*	-1898.545119	*	*
1_ts3	-1903.832939	-1902.989490	-1899.400697	34.48	*	*	-1898.544116	*	*
1_ts1	-1903.833103	-1902.989984	-1899.399492	33.22	*	*	-1898.543719	*	*
2_ts3	-1903.832420	-1902.989028	-1899.398396	34.23	*	*	-1898.541968	*	*
2_ts1	-1903.831909	-1902.988305	-1899.397270	35.56	*	*	-1898.540120	*	*
optimized ts									
1_ts2opt	-1903.835843	-1902.992792	-1899.408762	42.38	103.64	114.31	-1898.549569	-1898.526238	-1898.522175
2_ts2opt	-1903.837941	-1902.993738	-1899.407946	41.51	*	*	-1898.547935	*	*

* PCM SCRF can not be converged with the chosen parameters (i.e. Pauling radii).
** For catalysts **5a**, **5b**, **5k** and **5l** conformers ts2 and ts4 contain the acetate-*ortho*-hydrogen contact on the side of 3N-substituent.

Appendix

Table A2.8. Calculated free energies of conformers of 3,4-diaminopyridines and corresponding transition states. Final free energies "G_{298} MP2-5 with solv" were calculated at MP2(FC)/6-31+G(2d,p)//B98/6-31G(d) level of theory with inclusion of solvent effects in chloroform at chosen level. Methods for calculation PCM single point energies in chloroform: "solv1": RHF/6-31G(d) with UAHF radii; "solv2": RHF/6-31G(d) with Pauling radii; "solv3": B98/6-31G(d) with Pauling radii.

Conformer	E_{tot} B98/6-31G(d)	G_{298} B98/6-31G(d)	E_{tot} MP2(FC)/6-31+G(2d,p)	solv1 G_{solv} kJ/mol	solv2 G_{solv} kJ/mol	solv3 G_{solv} kJ/mol	solv1 G_{298} MP2-5 with solv	solv2 G_{298} MP2-5 with solv	solv3 G_{298} MP2-5 with solv
alc									
alc	-233.576452	-233.469331	-233.020561	-7.78	-0.63	2.30	-232.916404	-232.913679	-232.912564
anh									
anh	-381.583992	-381.518074	-380.793597	-6.65	-34.73	-20.92	-380.730213	-380.740906	-380.735647
anh2	-381.583860	-381.518359	-380.794157	-8.33	-36.61	-22.89	-380.731828	-380.742600	-380.737373
py									
py optimized ts	-248.181767	-248.120261	-247.589439	-9.00	-5.27	-1.92	-247.531360	-247.529942	-247.528667
tsopt									
3 (DMAP)									
DMAP	-863.330413	-863.052429	-861.407201	2.64	-8.95	3.72	-861.128213	-861.132627	-861.127799
optimized ts	-382.100962	-381.972152	-381.179977	-13.68	-11.88	-8.49	-381.056377	-381.055692	-381.054401
tsopt	-997.259033	-996.914872	-995.006254	-3.26	0.17	10.84	-994.663336	-994.662029	-994.657966
4 (PPY)									
PPY	-459.499042	-459.334344	-458.383907	-16.90	-5.27	-1.92	-458.225647	-458.221217	-458.219942
optimized ts									
tsopt	-1074.658051	-1074.277708	-1072.211254	-6.40	1.67	12.55	-1071.833349	-1071.830273	-1071.826130
6a									
6a_a	-536.904889	-536.704231	-535.601147	-20.75	2.59	4.60	-535.408393	-535.399501	-535.398736
6a_b	-536.905604	-536.705023	-535.602351	-20.33	2.43	4.69	-535.409515	-535.400846	-535.399985
optimized ts									
a_ts2opt	-1152.066909	-1151.650866	-1149.433307	-5.98	22.34	31.46	-1149.019543	-1149.008755	-1149.005281
7a									
7a	-609.437278	-609.221210	-607.990831	-9.50	-2.93	8.49	-607.778380	-607.775879	-607.771528
optimized ts									
ts2opt	-1224.595818	-1224.163862	-1221.816846	1.51	16.86	29.96	-1221.384317	-1221.378468	-1221.373480

217

Appendix

Conformer	E_{tot} B98/6-31G(d)	G_{298} B98/6-31G(d)	E_{tot} MP2(FC)/6-31+G(2d,p)	solv1 G_{solv} kJ/mol	solv2 G_{solv} kJ/mol	solv3 G_{solv} kJ/mol	solv1 G_{298} MP2-5 with solv	solv2 G_{298} MP2-5 with solv	solv3 G_{298} MP2-5 with solv
ts3opt	-1224.595134	-1224.163996	-1221.815909	-1.09	15.31	28.12	-1221.385186	-1221.378939	-1221.374062
ts4opt	-1224.597204	-1224.165708	-1221.818683	8.33	21.59	34.35	-1221.384016	-1221.378964	-1221.374104
7b									
1	-648.739813	-648.497957	-647.190473	-5.02	15.90	19.79	-646.9505296	-646.9425616	-646.9410795
2	-648.737139	-648.494272	-647.189498	-5.77	11.92	16.74	-646.9488305	-646.9420895	-646.9402569
optimized ts									
1_ts2opt	-1263.900085	-1263.441597	-1261.019372	10.42	37.07	48.37	-1260.556916	-1260.546764	-1260.542462
1_ts1opt	-1263.899655	-1263.441251	-1261.018453	8.16	35.56	45.98	-1260.556942	-1260.546504	-1260.542536
1_ts4opt	-1263.899412	-1263.441526	-1261.018402	7.41	33.05	*	-1260.557695	-1260.547927	
5a									
1	-593.418065	-593.184844	-591.981245	-14.56	5.69	8.08	-591.753570	-591.745857	-591.744948
2	-593.417088	-593.184505	-591.980224	-14.52	6.15	8.33	-591.753171	-591.745298	-591.744470
3	-593.416648	-593.184065	-591.979898	-14.18	6.69	8.45	-591.752717	-591.744765	-591.744095
4	-593.415838	-593.183378	-591.978867	-14.14	6.69	8.45	-591.751793	-591.743857	-591.743188
5	-593.415488	-593.182805	-591.979020	-15.06	4.23	6.65	-591.752074	-591.744727	-591.743803
6	-593.415277	-593.181700	-591.978978	-15.19	4.39	6.61	-591.751186	-591.743728	-591.742883
7	-593.414082	-593.181914	-591.977706	-14.56	6.95	8.41	-591.751084	-591.742892	-591.742335
optimized ts									
1_ts4opt	-1208.582983	-1208.132786	-1205.818948	4.64	37.95	47.45	-1205.366982	-1205.354297	-1205.350679
2_ts2opt	-1208.582671	-1208.132235	-1205.818582	4.52	*	*	-1205.366424		
2_ts3opt	-1208.578330	-1208.129381	-1205.811568	0.04	24.89	34.56	-1205.362603	-1205.353137	-1205.349456
2_ts4opt	-1208.581766	-1208.131805	-1205.817607	4.14	39.20	47.95	-1205.366068	-1205.352714	-1205.349383
5b									
1	-749.407023	-749.083230	-747.576974	-8.03	27.11	28.58	-747.256240	-747.242854	-747.242296
2	-749.405930	-749.082046	-747.576291	-7.87	30.75	32.13	-747.255403	-747.240694	-747.240168
4	-749.406234	-749.082709	-747.576561	-7.99	28.03	29.08	-747.256079	-747.242358	-747.241960
6	-749.405682	-749.082275	-747.576377	-8.24	28.79	30.25	-747.256109	-747.242006	-747.241448
7	-749.405742	-749.081909	-747.575522	-8.37	26.94	28.07	-747.254876	-747.241426	-747.240995
9	-749.405012	-749.080802	-747.574665	-8.20	28.49	29.96	-747.253578	-747.239602	-747.239045
10	-749.404995	-749.080845	-747.576374	-8.16	25.82	27.70	-747.255332	-747.242392	-747.241674
11	-749.406385	-749.080802	-747.576067	-8.74	25.82	27.41	-747.253815	-747.240651	-747.240046
12	-749.404828	-749.080142	-747.576198	-8.20	25.98	27.57	-747.254635	-747.241615	-747.241010

Appendix

Conformer	E_{tot} B98/6-31G(d)	G_{298} B98/6-31G(d)	E_{tot} MP2(FC)/6-31+G(2d,p)	solv1 G_{solv} kJ/mol	solv2 G_{solv} kJ/mol	solv3 G_{solv} kJ/mol	solv1 G_{298} MP2-5 with solv	solv2 G_{298} MP2-5 with solv	solv3 G_{298} MP2-5 with solv
14	-749.404725	-749.080568	-747.574364	-8.28	28.62	30.04	-747.253363	-747.239307	-747.238765
15	-749.404598	-749.080109	-747.576157	-7.66	27.78	29.46	-747.254584	-747.241086	-747.240449
16	-749.406393	-749.081996	-747.576140	-8.79	25.73	27.36	-747.255090	-747.241943	-747.241321
optimized ts									
1 ts2opt	-1364.573153	-1364.031977	-1361.416613	12.64	59.96	68.45	-1360.870625	-1360.852601	-1360.849366
11 ts2opt	-1364.572050	-1364.031426	-1361.415460	12.22	60.46	68.99	-1360.870182	-1360.851808	-1360.848557
1 ts4opt	-1364.572257	-1364.031677	-1361.415669	13.39	62.68	70.54	-1360.869990	-1360.851217	-1360.848221
1 ts3opt	-1364.568935	-1364.029792	-1361.409530	8.16	50.58	59.16	-1360.867279	-1360.851120	-1360.847853
5j									
1	-1221.300472	-1220.869017	-1218.426019	2.43	35.15	41.13	-1217.993638	-1217.981176	-1217.978898
102	-1221.302906	-1220.872804	-1218.420923	-7.53	20.63	25.94	-1217.993689	-1217.982963	-1217.980941
187	-1221.302230	-1220.871238	-1218.420066	-8.79	19.87	25.31	-1217.992422	-1217.981506	-1217.979434
46	-1221.299409	-1220.868699	-1218.417271	-14.02	18.74	*	-1217.991901	-1217.979423	
7	-1221.300922	-1220.869851	-1218.420254	-5.61	26.36	31.30	-1217.99192	-1217.979143	-1217.977261
85	-1221.300735	-1220.870339	-1218.421235	-3.60	27.61	33.47	-1217.99221	-1217.980323	-1217.978091
optimized ts									
33 ts1	-1836.473927	-1835.824366	-1832.269541	29.25	79.58	89.08	-1831.608839	-1831.58967	-1831.586051
158 ts1	-1836.472718	-1835.824618	-1832.268247	27.45	77.45	87.45	-1831.609692	-1831.590648	-1831.586839
102 ts2	-1836.472742	-1835.822502	-1832.268925	29.12	75.73	86.19	-1831.607594	-1831.589841	-1831.585857
33 ts2	-1836.472582	-1835.823229	-1832.268708	34.06	78.87	89.29	-1831.606382	-1831.589315	-1831.585346
85 ts2	-1836.471338	-1835.823331	-1832.266938	31.42	74.77	85.69	-1831.606964	-1831.590453	-1831.586293
158 ts2	-1836.471600	-1835.820827	-1832.267210	31.46	75.56	86.06	-1831.604455	-1831.587658	-1831.583658
5k									
5k_6	-1132.71957	-1132.30075	-1129.99159	0.59	37.82	41.51	-1129.572547	-1129.558364	-1129.556962
5k_13	-1132.72005	-1132.30096	-1129.99233	4.06	40.88	44.98	-1129.571690	-1129.557666	-1129.556104
5k_2	-1132.71974	-1132.30147	-1129.99075	1.21	38.24	41.88	-1129.572011	-1129.557908	-1129.556521
5k_4	-1132.71854	-1132.30024	-1129.99098	2.51	35.82	39.33	-1129.571729	-1129.559044	-1129.557705
5k_15	-1132.71962	-1132.30102	-1129.99124	4.18	43.47	47.24	-1129.571044	-1129.556080	-1129.554646
5k_16	-1132.71799	-1132.29999	-1129.98920	1.38	36.74	40.46	-1129.570671	-1129.557205	-1129.555787
5k_1	-1132.71864	-1132.30050	-1129.99075	5.31	38.16	41.84	-1129.570575	-1129.558066	-1129.556663
5k_10	-1132.71859	-1132.30019	-1129.99115	5.98	39.16	42.97	-1129.570472	-1129.557835	-1129.556385
5k_9	-1132.71780	-1132.30003	-1129.98877	2.05	39.25	43.05	-1129.570219	-1129.556052	-1129.554601

219

Appendix

Conformer	E_{tot} B98/6-31G(d)	G_{298} B98/6-31G(d)	E_{tot} MP2(FC)/6-31+G(2d,p)	solv1 G_{solv} kJ/mol	solv2 G_{solv} kJ/mol	solv3 G_{solv} kJ/mol	solv1 G_{298} MP2-5 with solv	solv2 G_{298} MP2-5 with solv	solv3 G_{298} MP2-5 with solv
5k_7	-1132.71858	-1132.30111	-1129.99050	6.86	44.52	49.08	-1129.570416	-1129.556073	-1129.554336
optimized ts									
2_ts4opt	-1747.880519	-1747.245129	-1743.831159	29.20	77.74	88.62	-1743.184645	-1743.166159	-1743.162016
6_ts4opt	-1747.881953	-1747.245528	-1743.829966	28.58	79.12	89.70	-1743.182657	-1743.163406	-1743.159374
15_ts4opt	-1747.879940	-1747.245916	-1743.830756	32.68	79.71	90.88	-1743.184286	-1743.166374	-1743.162119
5l									
1	-1288.674630	-1288.168666	-1285.568297	9.71	56.36	59.96	-1285.058636	-1285.040868	-1285.039497
2	-1288.673958	-1288.168161	-1285.567733	13.05	57.24	61.04	-1285.056963	-1285.040135	-1285.038685
3	-1288.672207	-1288.168075	-1285.565655	20.92	64.14	67.99	-1285.053555	-1285.037093	-1285.035627
4	-1288.672877	-1288.167743	-1285.564962	16.40	60.04	63.81	-1285.053581	-1285.036960	-1285.035526
optimized ts									
1_ts2opt	-1903.835843	-1903.113899	-1899.408762	42.38	103.64	114.31	-1898.670676	-1898.647345	-1898.643282
2_ts2opt	-1903.837941	-1903.113558	-1899.407946	41.51	*	*	-1898.667755		

* PCM SCRF can not be converged with the chosen parameters (i.e. Pauling radii).

Appendix

Table A2.9. Relative activation enthalpies ΔH$_{act}$ as calculated at MP2(FC)/6-31+G(2d,p)//B98/6-31G(d) level of theory in gas phase ("MP2-5") or with inclusion of solvent effects in chloroform at chosen level. Methods for calculation PCM single point energies in chloroform: "solv1": RHF/6-31G(d) with UAHF radii; "solv2": RHF/6-31G(d) with Pauling radii; "solv3": B98/6-31G(d) with Pauling radii. Correlations of the activation enthalpies with relative reaction rates ln(1/t$_{1/2}$) give correlation coefficients R^2, which are shown in the last three rows.

Catalyst	ln(1/t$_{1/2}$)	Boltzmann averaged enthalpies ΔH$_{act}$ [kJ/mol]				Boltzmann averaged enthalpies ΔH$_{act}$ [kJ/mol]				Activation enthalpies ΔH$_{act}$, based on the best conformers [kJ/mol]			
		frozen transition states				optimized transition states				optimized transition states			
		MP2-5	solv1	solv2	solv3	MP2-5	solv1	solv2	solv3	MP2-5	solv1	solv2	solv3
py	-5.6870	0.00	0.00	0.00	0.00	0.00	0.00	0.00	0.00	0.00	0.00	0.00	0.00
7a	-5.0173	-24.40	-20.50	1.46	-6.71	-27.70	-23.89	0.03	-7.84	-27.35	-24.18	-0.08	-7.73
3 (DMAP)	-4.6444	-21.94	-21.86	-4.78	-7.08	-24.50	-25.71	-8.77	-10.82	-24.50	-25.71	-8.77	-10.82
7b	-4.2047	-27.84	-23.01	-4.50	-7.17	-30.88	-27.87	-6.05	-8.29	-33.05	-28.68	-5.90	-8.78
4 (PPY)	-2.7081	-24.79	-24.67	*	*	-27.35	-28.49	-16.72	-18.52	-36.62	-28.48	-16.72	-18.52
6a	-4.9273	-33.94	-29.65	-11.09	-13.75	-37.30	-34.57	-13.71	-16.17	-23.55	-33.90	-13.02	-15.49
5l	-4.7536	-62.25	-40.69	*	*	-62.88	-41.41	-12.29	-14.55	-58.88	-41.45	-10.13	-13.79
5k	-3.9318	-55.75	-39.09	-14.36	-16.52	-59.16	-43.47	-15.99	-18.00	-59.63	-42.64	-14.62	-17.99
5a	-2.8904	-51.23	-41.46	-16.79	-19.09	-54.71	-47.20	-18.51	-20.72	-53.56	-45.98	-17.62	-19.83
5b	-2.7726	-54.86	-44.40	-19.37	-21.82	-59.17	-49.99	-22.36	-24.84	-62.49	-49.63	-21.02	-24.84
5j		-64.23	-49.28	-15.41	*	-66.44	-53.07	-18.03	-22.64	-67.85	-52.66	-18.31	-22.62
Correlation coefficients R^2													
all catalysts		0.1653	0.3361	0.5682	0.4874	0.1828	0.4024	0.5434	0.5791	0.1166	0.3974	0.5848	0.5788
4-aminopyridines		0.7176	0.9911	0.9089	0.9201	0.6691	0.9895	0.5628	0.5186	0.0231	0.9652	0.5193	0.4838
3,4-diaminopyridines		0.0061	0.7760	0.3333	0.9973	0.0568	0.9510	0.6657	0.8709	0.3753	0.9636	0.7765	0.8637

* PCM SCRF cannot be converged with the chosen parameters (i.e. Pauling radii).

Appendix

Table A2.10. Relative activation free energies ΔG_{act} as calculated at MP2(FC)/6-31+G(2d,p)//B98/6-31G(d) level of theory in gas phase ("MP2-5*") or with inclusion of solvent effects in chloroform at chosen level. Methods for calculation PCM single point energies in chloroform: "solv1": RHF/6-31G(d) with UAHF radii; "solv2": RHF/6-31G(d) with Pauling radii; "solv3": B98/6-31G(d) with Pauling radii. Correlations of the activation free energies with relative reaction rates $\ln(1/t_{1/2})$ give correlation coefficients R^2, which are shown in the last three rows.

Catalyst	$\ln(1/t_{1/2})$	Boltzmann averaged ΔG_{act} [kJ/mol]			ΔG_{act}, based on the best conformers [kJ/mol]
		optimized transition states			optimized transition states
		solv1	solv2	solv3	solv1
py		0.00	0.00	0.00	0
3 (DMAP)	-5.0173	-26.53	-9.59	-11.64	-26.53
4 (PPY)	-4.2047	-28.48	-16.73	-18.53	-28.48
6a	-2.7081	-35.28	-14.40	-16.87	-34.59
7a	-5.6870	-25.07	-0.73	-8.66	-23.85
7b	-4.6444	-26.76	-6.37	-6.65	-28.71
5a	-3.9318	-44.12	-15.69	-17.82	-43.48
5b	-2.8904	-46.71	-19.08	-21.38	-46.37
5j	-2.7726	-49.76	-13.38	-17.76	-50.28
5k	-4.7536	-41.36	-13.85	-15.67	-40.03
5l	-4.9273	-40.28	-10.78	-13.06	-39.87
Correlation coefficients R^2					
all catalysts		0.4208	0.5191	0.5500	0.4498
4-aminopyridines		0.9516	0.5807	0.4567	0.9737
3,4-diaminopyridines		0.9419	0.3830	0.7073	0.9226

Appendix

Table A2.11. Numbers of transition states conformers used for Boltzmann-averaging to obtain relative activation enthalpies ΔH_{act} as calculated at MP2(FC)/6-31+G(2d,p)/B98/6-31G(d) level of theory in gas phase ("MP2-5") or with inclusion of solvent effects in chloroform at chosen level. Methods for calculation PCM single point energies in chloroform: "solv1": RHF/6-31G(d) with UAHF radii; "solv2": RHF/6-31G(d) with Pauling radii; "solv3": B98/6-31G(d) with Pauling radii.

Catalyst	$\ln(1/t_{1/2})$	frozen transition states				optimized transition states				free catalyst
		MP2-5	solv1	solv2	solv3	MP2-5	solv1	solv2	solv3	
py										
3 (DMAP)	-5.0173	1	1	1	1	1	1	1	1	1
4 (PPY)	-4.2047	1	1	*	1	1	1	1	1	1
6a	-2.7081	4	4	4	4	1	1	1	1	2
7a	-5.6870	4	4	4	4	3	3	3	3	1
7b	-4.6444	8	8	8	8	3	3	3	2	2
5a	-3.9318	8	8	6	6	4	4	3	3	7
5b	-2.8904	16	16	14	14	4	4	4	4	13
5k	-4.7536	14	14	7	7	7	7	7	7	12
5l	-4.9273	8	8	*	*	2	2	1	1	4
5j	-2.7726	1	1	1	*	12	12	7	6	6

* PCM SCRF cannot be converged with the chosen parameters (i.e. Pauling radii).

A2.4. Influence of the explicit solvation.

Table A2.12. Calculated energies of pyridines and corresponding transition states, solvated by chloroform. Methods for calculation PCM single point energies in chloroform: "solv1": RHF/6-31G(d) with UAHF radii; "solv2": RHF/6-31G(d) with Pauling radii; "solv3": B98/6-31G(d) with Pauling radii.

Conformer	E_{tot} B98/6-31G(d)	H_{298} B98/6-31G(d)	E_{tot} MP2(FC)/6-31+G(2d,p)	$<H_{298}>$ MP2(FC)/6-31+G(2d,p)//B98/6-31G(d)	solv1 G_{solv}, kJ/mol	solv2 G_{solv}, kJ/mol	solv3 G_{solv}, kJ/mol
chloroform	-1419.094429	-1419.068994	-1417.478989	-1417.453554	0.54	-11.25	-9.04
py							
py_alc	-481.770379	-481.530830	-480.623500	-480.383951	-2.26	4.52	
cat_CHCl₃	-1667.286776	-1667.164971	-1665.079878	-1664.958073	8.45	0.46	2.85
ts_CHCl₃	-2282.440733	-2282.066001	-2278.905033	-2278.530301	30.29	29.25	38.58
3 (DMAP)							
DMAP.alc	-615.691516	-615.374106	-614.215625	-613.898214	-5.98	9.58	
DMAP.alc.2	-615.691872	-615.373393	-614.215168	-613.896689	-4.81	14.43	
cat_CHCl₃.1	-1801.207922	-1801.008289	-1798.671846	-1798.472213	5.06	6.07	7.74
cat_CHCl₃.2	-1801.200451	-1801.000781					
ts_CHCl₃							
ts_CHCl₃_3	-2416.370091	-2415.918107	-2412.505641	-2412.053656	30.08	38.62	46.82
4 (PPY)							
PPY.alc	-693.089778	-692.734773	-691.419919	-691.064914	-9.00	16.32	
PPY.alc.2	-693.089199	-692.734290	-691.418990	-691.064080	-8.70	16.02	
cat_CHCl₃	-1878.606247	-1878.368964	-1875.876311	-1875.639028	2.09	12.51	14.35
ts_CHCl₃							
ts_CHCl₃.1	-2493.769008	-2493.279459	-2489.710663	-2489.221115	27.70	46.69	55.15
6a							
6a_b.alc.1	-770.496548	-770.104082	-768.638892	-768.246426	-11.46	25.36	
6a_b.alc.2	-770.496593	-770.104194	-768.638950	-768.246551	-11.42	25.40	
cat_CHCl₃_1	-1956.012971	-1955.738290	-1953.095233	-1952.820552	-0.38	21.67	22.47
cat_CHCl₃_2	-1956.005629						
ts_CHCl₃	-1152.066909	-1151.567461	-1149.433307	-1148.933860	-5.98	0.00	31.46
ts_CHCl₃_1	-2571.176550	-2570.649703	-2566.931649	-2566.404802	24.81		
7a	-609.437278	-609.163890	-607.990831	-607.717443	-9.50	-2.93	8.49
7a.alc.1	-843.028110	-842.608168	-841.026825	-840.606883	-1.26	25.86	-1.26
7a.alc.2	-843.027933	-842.608060	-841.026721	-840.606848	-0.75	26.65	-0.75
7a.Nalc.1	-843.028012	-842.608076	-841.029063	-840.609127	7.53	35.52	7.53
7a.Nalc.2	-843.026872	-842.607024	-841.028316	-840.608468	7.99	39.29	7.99
cat_CHCl₃_1	-2028.544445	-2028.243409	-2025.483018	-2025.181982	9.96	21.46	25.65
cat_CHCl₃_2	-2028.542546	-2028.241421	-2025.487454	-2025.186329	15.94	25.23	30.00
cat_CHCl₃_3	-2028.535796	-2028.234814					
ts_CHCl₃	-1224.595818	-1224.069937	-1221.816846	-1221.290965	1.51		
ts_CHCl₃_1	-2643.705649	-2643.152255	-2639.314376	-2638.760983	33.14		
5a							
5a_1.alc.1	-827.010237	-826.576631	-825.018453	-824.584847	-1.13	35.56	
5a_1.alc.2	-827.010136	-826.576078	-825.018519	-824.584460	-0.75	36.32	
cat_CHCl₃.1	-2012.525707	-2012.210613	-2009.474695	-2009.159601	7.49	27.57	28.87
cat_CHCl₃.3	-2012.518774	-2012.203954					
ts_CHCl₃							
ts_CHCl₃_1	-2627.690987	-2627.123439	-2623.315872	-2622.748324	34.39	68.45	75.98
5b							
5b_6.alc.1	-982.997653	-982.466821	-980.613639	-980.082807	6.49	61.04	

Appendix

Conformer	E_{tot} B98/6-31G(d)	H_{298} B98/6-31G(d)	E_{tot} MP2(FC)/6-31+G(2d,p)	$<H_{298}>$ MP2(FC)/6-31+G(2d,p)//B98/6-31G(d)	solv1 G_{solv}, kJ/mol	solv2 G_{solv}, kJ/mol	solv3 G_{solv}, kJ/mol
5b_6.alc.2	-982.997908	-982.467058	-980.613887	-980.083037	5.44	58.74	
cat_CHCl₃.1	-2168.513355	-2168.101290	-2165.069737	-2164.657671	14.02	50.00	50.46
cat_CHCl₃.2	-2168.506508	-2168.094645					
cat_CHCl₃.3	-2168.513418	-2168.101291	-2165.070070	-2164.657943	14.06	50.71	51.21
ts_CHCl₃							
ts_CHCl₃_3	-2783.681530	-2783.017748	-2778.915576	-2778.251795	46.94		
5k							
5k_6.alc.1	-1366.310732	-1365.668329	-1363.029180	-1362.386777	19.33	74.73	
5k_6.alc.2	-1366.311755	-1365.669281	-1363.031429	-1362.388955	19.33	77.53	
5k_6.alc.3	-1366.310720	-1365.668340	-1363.029300	-1362.386920	18.91	73.72	
5k_6.alc.4	-1366.310704	-1365.668452	-1363.028649	-1362.386396	14.85	67.66	
5k_6_CHCl₃	-2551.826952	-2551.303491	-2547.486284	-2546.962823	26.99	64.18	66.86
ts_CHCl₃							
2_ts4opt_CHCl₃	-3166.989287	-3166.213770	-3161.331034	-3160.555517	64.22	114.43	123.01
5l							
alc.1	-1522.266836	-1521.529166	-1518.608130	-1517.870460	28.74	91.76	
alc.2	-1522.265687	-1521.528069	-1518.606133	-1517.868516	28.87	93.55	
alc.3	-1522.265579	-1521.528003	-1518.606197	-1517.868620	26.40	87.82	
alc.4	-1522.266686	-1521.529949	-1518.608225	-1517.871488	28.12	92.22	
cat_CHCl₃_1	-2707.782096	-2707.163411	-2703.063430	-2702.444745	35.56	82.38	84.98
cat_CHCl₃_2	-2707.782096	-2707.163426	-2703.063434	-2702.444764	35.56	82.38	84.98
ts_CHCl₃							
1_ts2_CHCl₃	-3322.944773	-3322.073934	-3316.908431	-3316.037592	76.36		

A2.5. Single point calculations at B3LYP-D level.

Table A2.13. Calculated energies of conformers of **6a** and **5b** and corresponding transition states. Final enthalpies "H_{298} B3LYP-D with solv" were calculated at B3LYP-D/6-311+G(d,p)//B98/6-31G(d) level of theory with inclusion of solvent effects in chloroform at PCM/UAHF/RHF/6-31G(d) level. $<H_{298}>$ are Boltzmann-averaged enthalpies.

Conformer	E_{tot} B98/6-31G(d)	H_{298} B98/6-31G(d)	E_{tot} B3LYP-D/6-311+G(d,p)	G_{solv}, kJ/mol	"H_{298} B3LYP-D with solv"	$<H_{298}>$ B3LYP-D with solv
py						
py	-248.181767	-248.087627	-248.357210	-9.00	-248.266497	-248.266497
tsopt	-863.330413	-862.983065	-863.986592	2.64	-863.638241	-863.638241
6a						-537.0505796
6a_a	-536.904889	-536.657992	-537.289168	-20.75	-537.050175	
6a_b	-536.905604	-536.658613	-537.290037	-20.33	-537.050790	
6a_ts						-1152.435922
a_ts2opt	-1152.066909	-1151.567461	-1152.933091	-5.98	-1152.435922	
5b						-749.587190
1	-749.407023	-749.022449	-749.969255	-8.03	-749.587741	
2	-749.405930	-749.021754	-749.967880	-7.87	-749.586699	
4	-749.406234	-749.021772	-749.968358	-7.99	-749.586940	
6	-749.405682	-749.021776	-749.968007	-8.24	-749.587240	
7	-749.405742	-749.021319	-749.968241	-8.37	-749.587005	
9	-749.405012	-749.020692	-749.967260	-8.20	-749.586063	
10	-749.404995	-749.020441	-749.967658	-8.16	-749.586211	
11	-749.406385	-749.021639	-749.969006	-8.74	-749.587591	
12	-749.404828	-749.020084	-749.967695	-8.20	-749.586074	
13	-749.404877	-749.020263	-749.967834	-8.33	-749.586391	
14	-749.404725	-749.020437	-749.967087	-8.28	-749.585954	
15	-749.404598	-749.020095	-749.967367	-7.66	-749.585781	
16	-749.406393	-749.021713	-749.969079	-8.79	-749.587746	
5b_ts						-1364.976473
1_ts2opt	-1364.573153	-1363.935988	-1365.618759	12.64	-1364.976782	
1_ts4opt	-1364.572257	-1363.935082	-1365.617776	13.39	-1364.975502	
11_ts2opt	-1364.572050	-1363.934964	-1365.618247	12.22	-1364.976507	
1_ts3opt	-1364.568935	-1363.932278	-1365.613155	8.16	-1364.973390	

Appendix

Chapter 3. Computational details.
A3.1 Relative acetylation enthalpies for potential photoswitchable 3,4-diaminopyridines.

Table A3.1. Calculated energies of conformers for potential photoswitchable 3,4-diaminopyridines, as calculated at MP2/6-31+G(2d,p)//B98/6-31G(d) level with inclusion of solvent effects at PCM/UAHF/RHF/6-31G(d) level.

Confor mer	E_{tot} B98/6-31G(d)	H_{298} B98/6-31G(d)	E_{tot} MP2(FC)/6-31+G(2d,p)	G_{solv}, kJ/mol	H_{298} MP2-5 with solv	$<H_{298}>$ MP2-5 with solv
			Py			
Py	-248.181767	-248.087627	-247.589439	-9.00	-247.498727	-247.498727
Py_ac	-401.140004	-400.991691	-400.215516	-142.55	-400.121498	-400.121498
			diaza1			
diaza1tr						-1044.022505
3	-1046.857690	-1046.478012	-1044.402223	-11.63	-1044.026975	
1	-1046.857580	-1046.477989	-1044.402053	-11.67	-1044.026908	
diaza1tr_ac						-1196.675477
3_ac1	-1199.858325	-1199.424128	-1197.068686	-108.28	-1196.675732	
1_ac2	-1199.858318	-1199.424100	-1197.068697	-108.24	-1196.675706	
3_ac2	-1199.856957	-1199.422878	-1197.066994	-108.20	-1196.674126	
1_ac1	-1199.856955	-1199.422802	-1197.066869	-108.28	-1196.673958	
diaza1ci						-1044.008449
1	-1046.833133	-1046.454151	-1044.383069	-12.18	-1044.008724	
4	-1046.833002	-1046.453791	-1044.382675	-13.26	-1044.008515	
7	-1046.832807	-1046.453629	-1044.382354	-13.85	-1044.008451	
5	-1046.832796	-1046.453467	-1044.382303	-13.56	-1044.008137	
4new	-1046.831620	-1046.45269	-1044.382978	-11.25	-1044.008335	
2	-1046.831344	-1046.452302	-1044.382393	-13.01	-1044.008307	
3	-1046.831558	-1046.452591	-1044.382727	-12.26	-1044.008430	
6	-1046.831275	-1046.452370	-1044.382436	-12.97	-1044.008471	
diaza1ci_ac						-1196.656734
7_ac1	-1199.832729	-1199.399018	-1197.047782	-113.30	-1196.657226	
5_ac2	-1199.832665	-1199.398953	-1197.047786	-113.22	-1196.657197	
4_ac2	-1199.831533	-1199.397899	-1197.046766	-113.97	-1196.656542	
1_ac2	-1199.831564	-1199.397960	-1197.047009	-112.80	-1196.656368	
5_ac1	-1199.831165	-1199.397592	-1197.045821	-113.51	-1196.655482	
7_ac2	-1199.831221	-1199.397592	-1197.045806	-113.47	-1196.655396	
1_ac1	-1199.830601	-1199.397022	-1197.045616	-112.68	-1196.654953	
4_ac1	-1199.830668	-1199.396959	-1197.045708	-112.51	-1196.654851	
			diaza1cn			
diaza1tr						-1136.056814
3	-1139.064106	-1138.684114	-1136.431421	-14.43	-1136.056927	
1	-1139.063984	-1138.683898	-1136.431237	-14.48	-1136.056666	
diaza1tr_ac						-1288.703967
1_ac2	-1292.058452	-1291.623979	-1289.092961	-120.04	-1288.704208	
3_ac1	-1292.058464	-1291.624024	-1289.092943	-120.00	-1288.704207	
1_ac1	-1292.057152	-1291.622859	-1289.091178	-119.96	-1288.702574	
3_ac2	-1292.057158	-1291.622754	-1289.091143	-119.96	-1288.702428	
diaza1ci						-1136.037495
7	-1139.039005	-1138.659639	-1136.410931	-16.15	-1136.037716	
4	-1139.039185	-1138.659680	-1136.411005	-16.15	-1136.037651	
5	-1139.039209	-1138.659623	-1136.411170	-15.73	-1136.037576	
1	-1139.039202	-1138.659538	-1136.411299	-14.77	-1136.037261	
4new	-1139.037014	-1138.657520	-1136.410914	-14.39	-1136.036901	
2	-1139.036955	-1138.657669	-1136.410733	-15.82	-1136.037471	
3	-1139.037093	-1138.657654	-1136.410975	-15.40	-1136.037400	
6	-1139.037144	-1138.657915	-1136.410933	-15.36	-1136.037552	

Appendix

Confor mer	E_{tot} B98/6-31G(d)	H_{298} B98/6-31G(d)	E_{tot} MP2(FC)/6-31+G(2d,p)	G_{solv}, kJ/mol	H_{298} MP2-5 with solv	$<H_{298}>$ MP2-5 with solv
diaza1ci_ac						-1288.686840
4ac1new**	-1292.036108	-1291.602523	-1289.082969	-100.08	-1288.687503	
7_ac1	-1292.033216	-1291.599308	-1289.072491	-123.22	-1288.685515	
5_ac2	-1292.033155	-1291.599259	-1289.072377	-123.34	-1288.685460	
4_ac2	-1292.033005	-1291.599221	-1289.073934	-116.19	-1288.684404	
1_ac2	-1292.032999	-1291.599160	-1289.073762	-116.57	-1288.684320	
5_ac1	-1292.031797	-1291.597913	-1289.070576	-123.22	-1288.683623	
7_ac2	-1292.031694	-1291.597709	-1289.070512	-123.39	-1288.683522	
1_ac1	-1292.031456	-1291.597746	-1289.070768	-120.62	-1288.683002	
4_ac1	-1292.031474	-1291.597804	-1289.070752	-120.50	-1288.682978	
diaza1ome						
diaza1tr						-1158.237020
3_2	-1161.339892	-1160.924702	-1158.647079	-14.06	-1158.237244	
1_1	-1161.339746	-1160.924617	-1158.646918	-14.18	-1158.237192	
3_1	-1161.339527	-1160.924391	-1158.646518	-14.06	-1158.236736	
1_2	-1161.339460	-1160.924225	-1158.646432	-14.10	-1158.236568	
diaza1tr_ac						-1310.886271
3_2_ac2	-1314.343692	-1313.873929	-1311.315562	-107.40	-1310.886707	
1_1_ac2	-1314.343654	-1313.873816	-1311.315529	-107.36	-1310.886583	
1_2_ac2	-1314.343207	-1313.873467	-1311.315060	-107.74	-1310.886356	
3_1_ac2	-1314.343189	-1313.873439	-1311.315014	-107.61	-1310.886252	
1_1_ac1	-1314.342298	-1313.872577	-1311.313747	-107.49	-1310.884965	
3_2_ac1	-1314.342319	-1313.872598	-1311.313702	-107.40	-1310.884888	
1_2_ac1	-1314.341844	-1313.872277	-1311.313201	-107.74	-1310.884669	
3_1_ac1	-1314.341816	-1313.872144	-1311.313141	-107.57	-1310.884440	
diaza1ci						-1158.217408
4_2	-1161.313891	-1160.899264	-1158.626205	-15.69	-1158.217554	
7_1	-1161.313736	-1160.899143	-1158.626025	-16.07	-1158.217551	
5_1	-1161.313722	-1160.899125	-1158.626041	-15.90	-1158.217500	
1_2	-1161.314173	-1160.899492	-1158.626777	-14.18	-1158.217498	
4_1	-1161.313883	-1160.899105	-1158.626425	-14.98	-1158.217352	
5_2	-1161.313392	-1160.898853	-1158.625741	-15.56	-1158.217131	
1_1	-1161.314072	-1160.899382	-1158.626714	-12.93	-1158.216948	
7_2	-1161.313416	-1160.898867	-1158.625743	*		
4new_1	-1161.312917	-1160.898455	-1158.626881	-11.84	-1158.216929	
4new_2	-1161.312893	-1160.898548	-1158.626689	-13.35	-1158.217427	
2_1	-1161.312411	-1160.897865	-1158.626105	-14.98	-1158.217264	
3_1	-1161.312806	-1160.898579	-1158.626613	-13.31	-1158.217454	
6_1	-1161.312577	-1160.898111	-1158.626108	-15.19	-1158.217427	
2_2	-1161.312626	-1160.898111	-1158.626250	-15.27	-1158.217551	
3_2	-1161.312756	-1160.898420	-1158.626372	-14.60	-1158.217598	
6_2	-1161.312213	-1160.897954	-1158.625897	-14.94	-1158.217327	
diaza1ci_ac						-1310.865945
7_1_ac2	-1314.316788	-1313.847750	-1311.292828	-112.59	-1310.866675	
5_1_ac2	-1314.316776	-1313.847540	-1311.292817	-112.68	-1310.866497	
7_2_ac2	-1314.315546	-1313.846437	-1311.291902	-114.06	-1310.866235	
5_2_ac2	-1314.315537	-1313.846448	-1311.291749	-114.18	-1310.866150	
1_2_ac2	-1314.315644	-1313.846431	-1311.292130	-112.34	-1310.865705	
4_2_ac2	-1314.315615	-1313.846476	-1311.291860	-112.21	-1310.865462	
1_1_ac2	-1314.313707	-1313.844687	-1311.290228	-115.10	-1310.865048	
4_1_ac2	-1314.313672	-1313.844656	-1311.290061	-115.27	-1310.864948	
5_1_ac1	-1314.315226	-1313.846079	-1311.290829	-112.84	-1310.864662	
7_1_ac1	-1314.315255	-1313.846004	-1311.290868	-112.84	-1310.864597	
5_2_ac1	-1314.313951	-1313.844872	-1311.289728	-114.47	-1310.864250	
4_2_ac1	-1314.314620	-1313.845519	-1311.290791	-111.71	-1310.864240	
7_2_ac1	-1314.313968	-1313.844718	-1311.289891	-114.35	-1310.864195	

Appendix

Conformer	E_{tot} B98/6-31G(d)	H_{298} B98/6-31G(d)	E_{tot} MP2(FC)/6-31+G(2d,p)	G_{solv}, kJ/mol	H_{298} MP2-5 with solv	$<H_{298}>$ MP2-5 with solv
1_2_ac1	-1314.314591	-1313.845488	-1311.290483	-112.30	-1310.864153	
1_1_ac1	-1314.313019	-1313.844021	-1311.288948	-113.64	-1310.863232	
4_1_ac1	-1314.312990	-1313.843908	-1311.288886	-113.72	-1310.863118	
diaza3						
diaza3tr						-1044.026911
3	-1046.858831	-1046.479035	-1044.401895	-12.72	-1044.026943	
1	-1046.858788	-1046.479001	-1044.401914	-12.47	-1044.026876	
diaza3tr_ac						-1196.675234
3_ac2	-1199.856353	-1199.421957	-1197.066593	-113.93	-1196.675591	
3_ac1	-1199.856291	-1199.421917	-1197.066551	-113.43	-1196.675379	
1_ac2	-1199.856964	-1199.422639	-1197.067069	-109.20	-1196.674337	
1_ac1	-1199.856840	-1199.422570	-1197.066913	-108.95	-1196.674140	
diaza3ci						-1044.008394
1	-1046.834230	-1046.455465	-1044.382535	-13.05	-1044.008742	
4	-1046.834105	-1046.454748	-1044.382233	-14.06	-1044.008230	
7	-1046.833798	-1046.454738	-1044.381630	-14.73	-1044.008179	
5	-1046.833785	-1046.454669	-1044.381789	-14.35	-1044.008139	
diaza3ci_ac						-1196.657594
1_ac1	-1199.830973	-1199.397156	-1197.047564	-116.69	-1196.658192	
4_ac1	-1199.830933	-1199.397049	-1197.047590	-116.27	-1196.657992	
7_ac1	-1199.830633	-1199.396905	-1197.046074	-117.28	-1196.657015	
1_ac2	-1199.830632	-1199.396894	-1197.047204	-113.55	-1196.656717	
5_ac2	-1199.830683	-1199.396877	-1197.045795	-117.40	-1196.656706	
4_ac2	-1199.830308	-1199.396600	-1197.046945	-111.55	-1196.655723	
7_ac2	-1199.831315	-1199.397614	-1197.046529	-112.42	-1196.655649	
5_ac1	-1199.831405	-1199.397645	-1197.046266	-112.80	-1196.655470	
diaza2						
diaza2ci						-1504.646119
7	-1508.737017	-1508.187430	-1505.202219	15.90	-1504.646576	
6	-1508.737030	-1508.187354	-1505.201915	15.86	-1504.646200	
4	-1508.736456	-1508.186795	-1505.199878	13.97	-1504.644895	
12	-1508.736311	-1508.186624	-1505.199552	13.60	-1504.644686	
8	-1508.739247	-1508.189284	-1505.197318	13.64	-1504.642160	
11	-1508.738758	-1508.188864	-1505.196532	12.55	-1504.641857	
10	-1508.739087	-1508.189058	-1505.197057	13.60	-1504.641849	
9	-1508.738807	-1508.188770	-1505.196897	14.77	-1504.641235	
diaza2ci_ac						-1657.292099
7_ac2	-1661.739770	-1661.135585	-1657.868256	-75.77	-1657.292931	
6_ac2	-1661.739713	-1661.135305	-1657.868442	-73.72	-1657.292114	
12_ac2	-1661.738235	-1661.134169	-1657.865534	-77.28	-1657.290901	
7_ac1	-1661.738459	-1661.134140	-1657.866495	-74.27	-1657.290463	
6_ac1	-1661.738415	-1661.134093	-1657.866680	-73.76	-1657.290453	
4_ac2	-1661.738125	-1661.134033	-1657.865666	-75.69	-1657.290402	
9_ac2	-1661.741785	-1661.137188	-1657.865667	-74.10	-1657.289293	
11_ac2	-1661.741977	-1661.137259	-1657.865439	-74.43	-1657.289071	
10_ac2	-1661.741049	-1661.136758	-1657.864756	-75.10	-1657.289069	
8_ac2	-1661.740984	-1661.136532	-1657.864698	-75.14	-1657.288867	
4_ac1	-1661.737403	-1661.133245	-1657.864332	-75.31	-1657.288859	
12_ac1	-1661.737424	-1661.133267	-1657.863985	-76.19	-1657.288847	
9_ac1	-1661.740458	-1661.136108	-1657.863928	-73.64	-1657.287626	
8_ac1	-1661.740221	-1661.135821	-1657.863520	-74.48	-1657.287486	
10_ac1	-1661.740290	-1661.135686	-1657.863471	-74.39	-1657.287201	
11_ac1	-1661.740648	-1661.136093	-1657.863685	-73.64	-1657.287178	
diaza2tr						-1504.662688
1	-1508.760339	-1508.210318	-1505.218216	13.64	-1504.663000	
2	-1508.760247	-1508.210055	-1505.218110	13.68	-1504.662707	

Appendix

Confor mer	E_{tot} B98/6-31G(d)	H_{298} B98/6-31G(d)	E_{tot} MP2(FC)/6-31+G(2d,p)	G_{solv}, kJ/mol	H_{298} MP2-5 with solv	$<H_{298}>$ MP2-5 with solv
8	-1508.763667	-1508.213344	-1505.216447	13.72	-1504.660897	
4	-1508.763555	-1508.212975	-1505.216075	13.72	-1504.660268	
diaza2tr_ac						-1657.309450
2_ac2	-1661.764638	-1661.159654	-1657.886974	-73.35	-1657.309925	
1_ac2	-1661.764678	-1661.159820	-1657.886926	-73.14	-1657.309924	
8_ac2	-1661.767432	-1661.162421	-1657.886049	-72.13	-1657.308512	
4_ac2	-1661.767357	-1661.162365	-1657.886070	-71.76	-1657.308409	
2_ac1	-1661.763456	-1661.158710	-1657.885301	-72.51	-1657.308172	
1_ac1	-1661.763511	-1661.158765	-1657.885234	-72.38	-1657.308057	
8_ac1	-1661.766250	-1661.161066	-1657.884425	-71.38	-1657.306428	
4_ac1	-1661.766183	-1661.160950	-1657.884423	-71.04	-1657.306249	
			diaza2cn			
diaza2ci						-1596.676531
7	-1600.943505	-1600.393752	-1597.233106	16.78	-1596.676962	
6	-1600.943019	-1600.392983	-1597.231793	16.90	-1596.675319	
12	-1600.941839	-1600.392107	-1597.227821	10.92	-1596.673929	
4	-1600.941870	-1600.391860	-1597.227872	11.51	-1596.673480	
11	-1600.945102	-1600.395076	-1597.225443	10.50	-1596.671417	
8	-1600.945447	-1600.394979	-1597.225799	11.34	-1596.671012	
10	-1600.945330	-1600.394774	-1597.225661	11.67	-1596.670658	
9	-1600.945404	-1600.395177	-1597.225864	13.43	-1596.670522	
diaza2ci_ac						-1749.321484
12_ac2	-1753.939652	-1753.335058	-1749.891164	*		
7_ac1	-1753.941715	-1753.337110	-1749.895734	-81.04	-1749.321998	
4_ac1**	-1753.943862	-1753.339171	-1749.901724	-63.76	-1749.321319	
6_ac2	-1753.941461	-1753.336828	-1749.895471	-79.96	-1749.321292	
7_ac2	-1753.940520	-1753.335886	-1749.894078	-80.21	-1749.319993	
6_ac1	-1753.940298	-1753.335812	-1749.893874	-78.83	-1749.319412	
12_ac1	-1753.938627	-1753.334255	-1749.889320	-83.89	-1749.316900	
4_ac2	-1753.938925	-1753.334327	-1749.890192	-80.08	-1749.316096	
diaza2tr						-1596.692162
1	-1600.966204	-1600.415847	-1597.246991	10.88	-1596.692490	
2	-1600.966115	-1600.415570	-1597.246901	10.84	-1596.692228	
8	-1600.970127	-1600.419255	-1597.245697	11.25	-1596.690538	
4	-1600.969969	-1600.419133	-1597.245337	11.30	-1596.690198	
diaza2tr_ac						-1749.338218
1_ac1	-1753.965349	-1753.360296	-1749.911629	-84.39	-1749.338719	
2_ac1	-1753.965287	-1753.360099	-1749.911621	-84.35	-1749.338560	
4_ac1	-1753.967892	-1753.362743	-1749.910628	-82.89	-1749.337049	
8_ac1	-1753.967977	-1753.362836	-1749.910648	-82.76	-1749.337028	
1_ac2	-1753.964239	-1753.359155	-1749.909777	-82.97	-1749.336293	
2_ac2	-1753.964166	-1753.359090	-1749.909957	-82.22	-1749.336196	
4_ac2	-1753.966777	-1753.361368	-1749.909059	-81.96	-1749.334869	
8_ac2	-1753.966850	-1753.361456	-1749.908857	-82.05	-1749.334714	
			diaza2ome			
diaza2ome_ci						-1618.854935
10_1	-1623.219989	-1622.634476	-1619.440553	*		
7_2	-1623.218776	-1622.633783	-1619.446630	16.15	-1618.855486	
7_1	-1623.218781	-1622.633783	-1619.446121	15.40	-1618.855259	
6_2	-1623.218661	-1622.633560	-1619.445840	14.81	-1618.855098	
6_1	-1623.218347	-1622.633346	-1619.445917	16.32	-1618.854701	
4_2	-1623.217775	-1622.632594	-1619.443493	11.97	-1618.853755	
4_1	-1623.217768	-1622.632708	-1619.443964	13.72	-1618.853678	
12_2	-1623.217508	-1622.632487	-1619.442853	11.42	-1618.853482	
12_1	-1623.217651	-1622.632544	-1619.443244	15.02	-1618.852416	
8_1	-1623.220211	-1622.634857	-1619.441236	12.64	-1618.851069	

Appendix

Conformer	E_{tot} B98/6-31G(d)	H_{298} B98/6-31G(d)	E_{tot} MP2(FC)/6-31+G(2d,p)	G_{solv}, kJ/mol	H_{298} MP2-5 with solv	$<H_{298}>$ MP2-5 with solv
8_2	-1623.220326	-1622.634761	-1619.440915	11.46	-1618.850984	
10_2	-1623.219993	-1622.634521	-1619.440460	11.21	-1618.850717	
11_2	-1623.219841	-1622.634279	-1619.440531	11.38	-1618.850635	
9_1	-1623.219816	-1622.634489	-1619.440621	12.76	-1618.850434	
11_1	-1623.219460	-1622.634019	-1619.439884	11.88	-1618.849918	
9_2	-1623.219463	-1622.634154	-1619.440340	14.02	-1618.849692	
diaza2ome_ci_ac						-1771.501289
7_1_ac2	-1776.224056	-1775.584240	-1772.113812	-73.43	-1771.501964	
6_2_ac2	-1776.223817	-1775.583998	-1772.113735	-73.43	-1771.501884	
7_2_ac2	-1776.222414	-1775.582659	-1772.112641	-75.52	-1771.501650	
6_1_ac2	-1776.222209	-1775.582524	-1772.112469	-74.56	-1771.501182	
12_2_ac2	-1776.221865	-1775.582070	-1772.110647	-76.73	-1771.500078	
7_1_ac1	-1776.222690	-1775.582887	-1772.111994	-72.97	-1771.499984	
6_2_ac1	-1776.222483	-1775.582579	-1772.111984	-72.80	-1771.499808	
7_2_ac1	-1776.220997	-1775.581292	-1772.110779	-74.98	-1771.499632	
4_2_ac2	-1776.221988	-1775.582092	-1772.111038	-74.68	-1771.499588	
12_1_ac2	-1776.220009	-1775.580385	-1772.108843	-78.91	-1771.499274	
11_2_ac1	-1776.225726	-1775.586027	-1772.110659	-74.01	-1771.499150	
6_1_ac1	-1776.220805	-1775.581056	-1772.110611	-74.06	-1771.499068	
4_1_ac2	-1776.219925	-1775.580373	-1772.108799	-76.69	-1771.498457	
9_1_ac2	-1776.225592	-1775.585506	-1772.110634	-73.22	-1771.498436	
8_2_ac2	-1776.224717	-1775.584887	-1772.109870	-74.52	-1771.498422	
12_2_ac1	-1776.220987	-1775.581331	-1772.108938	-75.73	-1771.498126	
4_2_ac1	-1776.221149	-1775.581136	-1772.109515	-74.64	-1771.497932	
12_1_ac1	-1776.219522	-1775.579930	-1772.107500	-77.11	-1771.497278	
4_1_ac1	-1776.219602	-1775.579729	-1772.107925	-75.65	-1771.496865	
diaza2ome_tr						-1618.872669
2	-1623.242667	-1622.657151	-1619.463047	11.38	-1618.873196	
1	-1623.242764	-1622.656992	-1619.463134	11.46	-1618.872996	
1	-1623.242429	-1622.656825	-1619.462612	12.09	-1618.872403	
2	-1623.242362	-1622.656589	-1619.462488	11.38	-1618.872381	
8	-1623.245855	-1622.660184	-1619.461244	11.25	-1618.871285	
4	-1623.245720	-1622.659765	-1619.461113	11.25	-1618.870872	
8	-1623.245513	-1622.659546	-1619.460639	11.34	-1618.870353	
4	-1623.245407	-1622.659358	-1619.460595	11.34	-1618.870227	
diaza2ome_tr_ac						-1771.520233
1_1_ac2	-1776.249635	-1775.609309	-1772.133535	-72.93	-1771.520985	
2_2_ac2	-1776.249602	-1775.609200	-1772.133494	-72.97	-1771.520884	
1_2_ac2	-1776.249220	-1775.608734	-1772.132863	-73.09	-1771.520217	
2_1_ac2	-1776.249129	-1775.608637	-1772.132729	-73.14	-1771.520093	
8_1_ac2	-1776.252470	-1775.612079	-1772.132731	-71.67	-1771.519638	
4_2_ac2	-1776.252399	-1775.611956	-1772.132723	-71.71	-1771.519595	
8_2_ac2	-1776.252002	-1775.611588	-1772.132032	-72.01	-1771.519043	
2_2_ac1	-1776.248417	-1775.608026	-1772.131779	-72.17	-1771.518878	
4_1_ac1	-1776.251937	-1775.611588	-1772.132013	-71.42	-1771.518867	
1_1_ac1	-1776.248465	-1775.608150	-1772.131629	-71.80	-1771.518659	
1_2_ac1	-1776.248047	-1775.607553	-1772.131162	-72.17	-1771.518157	
2_1_ac1	-1776.247965	-1775.607482	-1772.131045	-72.38	-1771.518132	
4_2_ac1	-1776.251213	-1775.610582	-1772.131070	-70.96	-1771.517467	
8_1_ac1	-1776.251271	-1775.610734	-1772.130730	-70.92	-1771.517345	
8_2_ac1	-1776.250820	-1775.610212	-1772.130364	-71.04	-1771.516815	
4_1_ac1	-1776.250757	-1775.610135	-1772.130360	-70.71	-1771.516670	
cat11un						-743.697086
1	-745.789341	-745.476740	-744.007317	-7.32	-743.697505	
3	-745.786132	-745.473639	-744.004801	-6.61	-743.694826	
5	-745.786239	-745.473876	-744.005104	-6.78	-743.695323	

Appendix

Confor mer	E_{tot} B98/6-31G(d)	H_{298} B98/6-31G(d)	E_{tot} MP2(FC)/6-31+G(2d,p)	G_{solv}, kJ/mol	H_{298} MP2-5 with solv	$<H_{298}>$ MP2-5 with solv
7	-745.784320	-745.472212	-744.004574	-5.52	-743.694569	
cat11un_ac						-896.342735
1_ac1	-898.783948	-898.416667	-896.667002	-108.07	-896.340884	
3_ac1	-898.780328	-898.413066	-896.663496	-107.49	-896.337173	
5_ac1	-898.782909	-898.415444	-896.665798	-107.78	-896.339384	
1_ac2	-898.785210	-898.417833	-896.668872	-109.54	-896.343215	
3_ac2	-898.781440	-898.414080	-896.665088	-109.24	-896.339337	
5_ac2	-898.784199	-898.416921	-896.667691	-109.29	-896.342037	
diaza4						
diaza4tr						-1083.201763
1	-1086.167286	-1085.757767	-1083.608857	-4.69	-1083.201122	
3	-1086.169118	-1085.759289	-1083.611085	-3.26	-1083.202499	
5	-1086.165676	-1085.756122	-1083.606658	-4.27	-1083.198730	
7	-1086.167533	-1085.757840	-1083.608986	-2.97	-1083.200424	
9	-1086.165697	-1085.756307	-1083.608195	-2.93	-1083.199921	
11	-1086.163902	-1085.754685	-1083.605952	-4.85	-1083.198583	
13	-1086.165521	-1085.756162	-1083.607797	-3.81	-1083.199888	
15	-1086.161798	-1085.752904	-1083.605293	-4.14	-1083.197977	
16	-1086.163169	-1085.754080	-1083.606815	-3.05	-1083.198890	
diaza4tr_ac						-1235.844670
1_ac1	-1239.160699	-1238.696631	-1236.267209	-101.04	-1235.841627	
3_ac1	-1239.161682	-1238.697692	-1236.268697	-99.91	-1235.842762	
5_ac1	-1239.159800	-1238.695566	-1236.266017	-100.75	-1235.840157	
7_ac1	-1239.160696	-1238.696587	-1236.267525	-99.58	-1235.841343	
1_ac2	-1239.161830	-1238.697925	-1236.268971	-103.01	-1235.844301	
3_ac2	-1239.162820	-1238.698909	-1236.270428	-101.92	-1235.845337	
5_ac2	-1239.160967	-1238.696903	-1236.267827	-102.63	-1235.842855	
7_ac2	-1239.161880	-1238.697946	-1236.269305	-101.46	-1235.844015	
diaza4ci						-1083.184514
3	-1086.143874	-1085.735266	-1083.592692	-3.26	-1083.185327	
1	-1086.143648	-1085.734583	-1083.592500	-3.89	-1083.184917	
5	-1086.143012	-1085.734190	-1083.590107	-5.86	-1083.183516	
7	-1086.142952	-1085.734068	-1083.590081	-6.90	-1083.183826	
10	-1086.142228	-1085.733622	-1083.590472	-3.64	-1083.183252	
13	-1086.142078	-1085.733042	-1083.590420	-2.76	-1083.182436	
22	-1086.141395	-1085.732493	-1083.587711	-6.07	-1083.181119	
17	-1086.141230	-1085.732363	-1083.587592	-5.98	-1083.181004	
12	-1086.140526	-1085.731774	-1083.589705	-2.93	-1083.182068	
9	-1086.140196	-1085.731512	-1083.589638	-3.68	-1083.182356	
15	-1086.139978	-1085.731465	-1083.589321	-3.14	-1083.182003	
21	-1086.139832	-1085.730762	-1083.587435	-5.73	-1083.180548	
19	-1086.139739	-1085.730814	-1083.587495	-5.36	-1083.180611	
26	-1086.139692	-1085.731080	-1083.587200	-5.94	-1083.180851	
25	-1086.139565	-1085.730713	-1083.587004	-6.57	-1083.180654	
27	-1086.139546	-1085.730610	-1083.588358	-3.05	-1083.180586	
29	-1086.138217	-1085.729901	-1083.589158	-2.43	-1083.181766	
32	-1086.137515	-1085.729105	-1083.586518	-5.02	-1083.180020	
34	-1086.137450	-1085.728956	-1083.587681	-1.84	-1083.179888	
diaza4ci_ac						-1235.825975
1_ac2	-1239.135850	-1238.672353	-1236.249995	-106.11	-1235.826911	
3_ac2	-1239.135831	-1238.672421	-1236.249621	-105.06	-1235.826227	
7_ac2	-1239.136091	-1238.672728	-1236.248119	-108.53	-1235.826095	
15_ac2	-1239.135103	-1238.671812	-1236.248842	-105.81	-1235.825853	
5_ac2	-1239.135702	-1238.672307	-1236.247714	-107.74	-1235.825353	
13_ac2	-1239.134662	-1238.671307	-1236.248157	-104.64	-1235.824658	
19_ac2	-1239.135013	-1238.671779	-1236.246727	-107.95	-1235.824608	

Appendix

Confor mer	E_{tot} B98/6-31G(d)	H_{298} B98/6-31G(d)	E_{tot} MP2(FC)/6-31+G(2d,p)	G_{solv}, kJ/mol	H_{298} MP2-5 with solv	$<H_{298}>$ MP2-5 with solv
1_ac1	-1239.134506	-1238.670992	-1236.248159	-104.27	-1235.824357	
9_ac2	-1239.132590	-1238.669443	-1236.246984	-106.36	-1235.824346	
21_ac2	-1239.134925	-1238.671495	-1236.246532	-107.99	-1235.824233	
3_ac1	-1239.134518	-1238.671304	-1236.247751	-103.22	-1235.823851	
7_ac1	-1239.134941	-1238.671726	-1236.246382	-106.73	-1235.823819	
15_ac1	-1239.133725	-1238.670536	-1236.247112	-104.06	-1235.823557	
5_ac1	-1239.134522	-1238.671428	-1236.245934	-105.98	-1235.823205	
13_ac1	-1239.133292	-1238.670000	-1236.246476	-102.68	-1235.822291	
9_ac1	-1239.131473	-1238.668102	-1236.245544	-105.02	-1235.822172	
19_ac1	-1239.133836	-1238.670708	-1236.244963	-105.90	-1235.822169	
21_ac1	-1239.133691	-1238.670323	-1236.244711	-106.11	-1235.821756	
12_ac1	-1239.130717	-1238.667476	-1236.244250	-103.39	-1235.820387	
25_ac1	-1239.131142	-1238.667870	-1236.242560	-106.94	-1235.820020	

* Convergence problems for the PCM single point calculations

Appendix

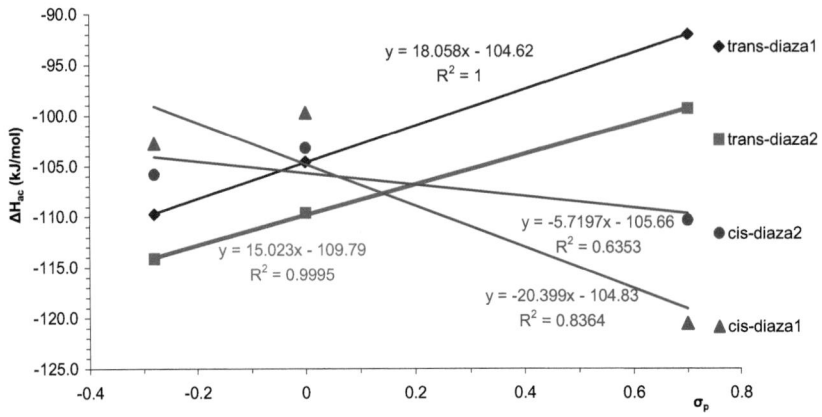

Figure A3.1 Correlation of acetylation enthalpies for *trans* and *cis* isomers (calculated at "MP2-5" = MP2(FC)/6-31+G(2d,p)//B98/6-31G(d) level) with σ-constants of *para*-substituents;
$\sigma_p(OMe) = -0.28$; $\sigma_p(H) = 0$; $\sigma_p(CN) = +0.70$.

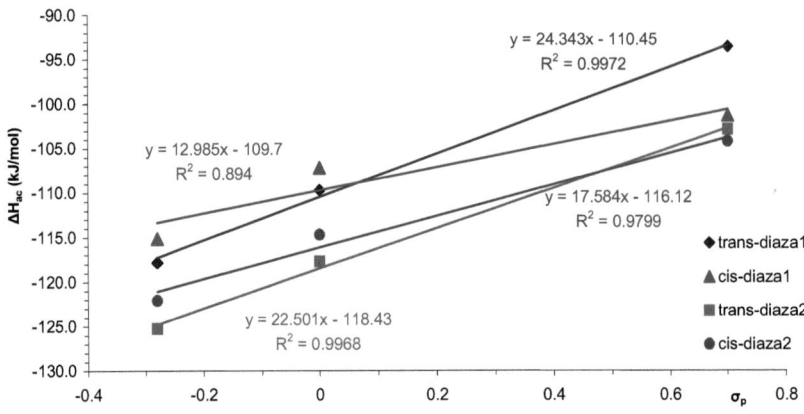

Figure A3.2 Correlation of acetylation enthalpies for *trans* and *cis* isomers (calculated at "B98" = B98/6-31G(d)//B98/6-31G(d) level) with σ-constants of *para*-substituents;
$\sigma_p(OMe) = -0.28$; $\sigma_p(H) = 0$; $\sigma_p(CN) = +0.70$.

Appendix

Table A3.2. NPA atom charges for the best conformations of *cis*- and *trans*- *p*-CN-**diaza2**, as calculated at B98/6-31G(d) level.

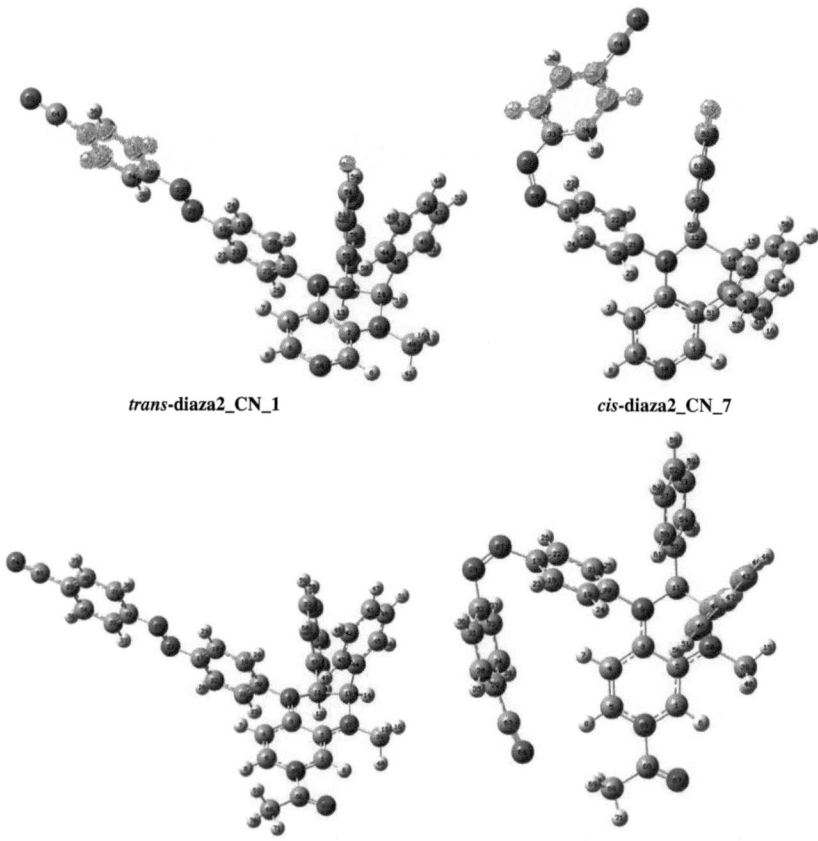

Atom	*cis*-diaza2_CN_7	*cis*-diaza2_CN_4_ac1	*trans*-diaza2_CN_1	*trans*-diaza2_CN_1_ac1
ortho-H(Py)	0.22386	**0.27148**[a]	0.22419	0.26285
C (CN)	0.28089	0.30503	0.27869	0.26866
N (CN)	-0.30963	**-0.34471**[a]	-0.30324	**-0.27177**[a]
Acetyl				
C (66)		0.74807		0.74542
C (67)		-0.51319		-0.51065
O (68)		-0.80390		-0.80136
H (69)		**0.28681**[a]		**0.27320**[a]
H (70)		0.27872		0.27375
H (71)		0.29540		0.29908
q(CH$_3$CO)=		0.29191		0.27944

[a] Charges of the nitrogen atom of CN group and neighbouring hydrogen atoms (of acetyl group and pyridine ring) are marked bold.

A3.2 Relative acetylation enthalpies for paracyclophane derivatives

Table A3.3. Calculated energies of conformers of 3-paracyclophane-4-amino and 3,4,5-trialkylpyridines, and corresponding acetyl intermediates.

Conformer	E_{tot} B98/6-31G(d)	H_{298} B98/6-31G(d)	E_{tot} MP2(FC)/6-31+G(2d,p)	G_{solv}, kJ/mol	H_{298} MP2-5 with solv	$<H_{298}>$ MP2-5 withsolv
para1						-997.111684
3	-999.950530	-999.511753	-997.549332	-3.01	-997.111702	
1	-999.945655	-999.507063	-997.543355	-4.02	-997.106292	
2	-999.939406	-999.501227	-997.538564	-2.01	-997.101150	
para1_ac						-1149.755571
3.ac2	-1152.945152	-1152.451635	-1150.208832	-106.40	-1149.755840	
3.ac1	-1152.944895	-1152.451053	-1150.208192	-103.18	-1149.753648	
1.ac2	-1152.942366	-1152.448863	-1150.204785	-107.45	-1149.752206	
1.ac1	-1152.941373	-1152.447829	-1150.204162	-102.80	-1149.749772	
para2						-1074.278104
4	-1077.347015	-1076.870141	-1074.752619	-6.23	-1074.278120	
6	-1077.342715	-1076.866029	-1074.746456	-6.99	-1074.272431	
1	-1077.339728	-1076.863194	-1074.744949	-5.02	-1074.270328	
3	-1077.329692	-1076.853718	-1074.737255	-3.93	-1074.262779	
para2_ac						-1226.926262
4.ac2	-1230.347147	-1229.816244	-1227.417396	-105.14	-1226.926540	
4.ac1	-1230.346804	-1229.815727	-1227.416873	-101.84	-1226.924585	
6.ac2	-1230.344580	-1229.813377	-1227.413721	-105.27	-1226.922613	
6.ac1	-1230.344154	-1229.812748	-1227.413367	-101.42	-1226.920590	
2.ac1	-1230.334257	-1229.803052	-1227.403495	-104.85	-1226.912226	
2.ac2	-1230.331957	-1229.800557	-1227.400420	-107.49	-1226.909960	
para3						-1113.447048
1	-1116.647618	-1116.141499	-1113.952361	-2.59	-1113.447230	
2	-1116.644983	-1116.139024	-1113.950626	-0.21	-1113.444746	
3	-1116.643679	-1116.137275	-1113.946626	-3.77	-1113.441656	
4	-1116.640671	-1116.134776	-1113.944643	-1.59	-1113.439353	
para3_ac						-1266.095295
1.ac2	-1269.648565	-1269.087791	-1266.617858	-101.04	-1266.095570	
1.ac1	-1269.647448	-1269.086953	-1266.616781	-96.65	-1266.093098	
3.ac2	-1269.646061	-1269.085151	-1266.614104	-100.75	-1266.091568	
2.ac2	-1269.646320	-1269.085355	-1266.614078	-100.37	-1266.091343	
3.ac1	-1269.645202	-1269.084814	-1266.613683	-96.02	-1266.089868	
2.ac1	-1269.645181	-1269.084299	-1266.612920	-95.48	-1266.088404	
para4						-1281.725678
21	-1285.281471	-1284.745912	-1282.260154	-4.06	-1281.726141	
9	-1285.281982	-1284.745600	-1282.261284	-3.05	-1281.726065	
7	-1285.284861	-1284.748063	-1282.261928	-2.18	-1281.725959	
3	-1285.283215	-1284.746804	-1282.259371	-6.53	-1281.725447	
5	-1285.281273	-1284.745171	-1282.258752	-3.77	-1281.724084	
31	-1285.278318	-1284.743691	-1282.256444	-5.15	-1281.723776	
4	-1285.279574	-1284.743039	-1282.259638	-1.42	-1281.723644	
6	-1285.281750	-1284.745424	-1282.259941	0.08	-1281.723583	
28	-1285.282558	-1284.745726	-1282.258257	-3.93	-1281.722923	
12	-1285.279446	-1284.743044	-1282.258692	-1.46	-1281.722849	
15	-1285.278823	-1284.742347	-1282.258811	-0.88	-1281.722669	
14	-1285.279122	-1284.743074	-1282.255104	-6.44	-1281.721510	
13	-1285.277225	-1284.740629	-1282.257653	0.38	-1281.720913	
27	-1285.275894	-1284.739619	-1282.254845	-3.05	-1281.719733	
33	-1285.277196	-1284.740894	-1282.254187	-3.93	-1281.719383	
22	-1285.276901	-1284.740452	-1282.254087	-3.85	-1281.719104	
29	-1285.279875	-1284.743182	-1282.254450	-3.35	-1281.719032	
23	-1285.274377	-1284.738607	-1282.252439	-1.76	-1281.717339	

Appendix

Conformer	E_{tot} B98/6-31G(d)	H_{298} B98/6-31G(d)	E_{tot} MP2(FC)/6-31+G(2d,p)	G_{solv}, kJ/mol	H_{298} MP2-5 with solv	$<H_{298}>$ MP2-5 withsolv
para4_ac						-1434.379104
29_ac1	-1438.297571	-1437.705821	-1434.938889	-85.77	-1434.379807	
3_ac1	-1438.296478	-1437.705109	-1434.936891	-87.45	-1434.378828	
33_ac2	-1438.292381	-1437.700646	-1434.936752	-88.70	-1434.378802	
33_ac1	-1438.296438	-1437.704875	-1434.937393	-85.60	-1434.378435	
3_ac2	-1438.290944	-1437.699489	-1434.935021	-90.42	-1434.378003	
29_ac2	-1438.291622	-1437.700685	-1434.934609	-88.07	-1434.377217	
5_ac1	-1438.293252	-1437.702156	-1434.933042	-86.73	-1434.374981	
7_ac2	-1438.283480	-1437.692216	-1434.925131	-105.98	-1434.374233	
6_ac1	-1438.283807	-1437.693144	-1434.928622	-94.89	-1434.374102	
5_ac2	-1438.287670	-1437.696588	-1434.930812	-89.87	-1434.373960	
14_ac1	-1438.281252	-1437.690293	-1434.925017	-94.47	-1434.370041	
4_ac1	-1438.280597	-1437.690163	-1434.923718	-91.71	-1434.368216	
37a						-365.001576
cat10	-366.086672	-365.904186	-365.181353	-7.11	-365.001576	
37a_ac						-517.634369
cat10_ac	-519.060128	-518.823483	-517.822649	-126.98	-517.634369	
37b						-482.485398
4	-483.975105	-483.702005	-482.758292	-1.67	-482.485830	
1	-483.973719	-483.700644	-482.756390	-1.92	-482.484048	
2	-483.972675	-483.699673	-482.755393	-2.64	-482.483395	
3	-483.972119	-483.699057	-482.753971	-3.77	-482.482344	
37b_ac						-635.118726
41	-636.951414	-636.624166	-635.402799	-115.06	-635.119375	
12	-636.949865	-636.622659	-635.400863	-115.14	-635.117513	
11	-636.949773	-636.622428	-635.400705	-114.77	-635.117072	
32	-636.949652	-636.622472	-635.399800	-115.27	-635.116523	
31	-636.949816	-636.622611	-635.399914	-114.68	-635.116389	
22	-636.948798	-636.621373	-635.399634	-114.89	-635.115970	
21	-636.948571	-636.621367	-635.399474	-113.68	-635.115569	
37c						-599.969258
001	-601.867333	-601.504627	-600.335730	8.66	-599.969725	
005	-601.868235	-601.505463	-6.00E+002	8.79	-599.969691	
003	-601.866425	-601.503546	-600.335656	8.54	-599.969526	
004	-601.865755	-601.503017	-600.335223	8.95	-599.969075	
006	-601.866606	-601.503817	-600.335327	9.12	-599.969064	
052	-601.866109	-601.503417	-600.334732	7.95	-599.969012	
047	-601.864819	-601.501971	-600.334716	8.33	-599.968696	
015	-601.864972	-601.502256	-600.334668	9.00	-599.968526	
027	-601.865940	-601.503323	-600.333754	8.12	-599.968046	
025	-601.866884	-601.504187	-600.333874	8.45	-599.967958	
37c_ac						-752.602459
5_ac1	-754.846894	-754.429794	-752.982879	-98.28	-752.603213	
1_ac1	-754.845981	-754.428934	-752.982842	-97.49	-752.602926	
1_ac2	-754.846060	-754.429106	-752.983008	-96.78	-752.602915	
6_ac1	-754.845109	-754.428275	-752.982213	-97.74	-752.602605	
52_ac1	-754.844513	-754.427544	-752.981423	-99.58	-752.602382	
6_ac2	-754.845179	-754.428012	-752.982244	-97.82	-752.602335	
52_ac2	-754.844582	-754.427449	-752.981463	-99.66	-752.602289	
3_ac1	-754.845169	-754.428013	-752.982907	-95.69	-752.602196	
4_ac1	-754.844386	-754.427448	-752.982380	-96.19	-752.602079	
4_ac2	-754.844169	-754.427096	-752.982190	-96.78	-752.601977	
47_ac2	-754.843066	-754.426161	-752.981197	-98.91	-752.601965	
47_ac1	-754.842990	-754.425959	-752.981070	-98.78	-752.601663	
15_ac1	-754.843354	-754.426214	-752.981490	-97.74	-752.601577	
25_ac2	-754.845342	-754.428404	-752.980898	-98.53	-752.601490	

Appendix

Conformer	E_{tot} B98/6-31G(d)	H_{298} B98/6-31G(d)	E_{tot} MP2(FC)/6-31+G(2d,p)	G_{solv}, kJ/mol	H_{298} MP2-5 with solv	$<H_{298}>$ MP2-5 withsolv
25_ac1	-754.845243	-754.428388	-752.980743	-98.62	-752.601449	
27_ac2	-754.844323	-754.427377	-752.980639	-98.07	-752.601047	
27_ac1	-754.844558	-754.427506	-752.980918	-97.24	-752.600902	

Table A3.4 Acetylation enthalpies and structure parameters for 3-paracyclophane-4-amino and 3,4,5-trialkylpyridines, as calculated at MP2/6-31+G(2d,p)//B98/6-31G(d) level with inclusion of solvent effects at PCM/UAHF/RHF/6-31G(d) level.

Catalyst	On the basis of the energetically lowest conformer					Averaged
	q_{NPA} (Ac)[a,b]	r(C-N)[pm][a]	ΔH_{ac} (MP2-5) [kJ/mol]	$\Delta\Delta G_{solv}$ [kJ/mol]	ΔH_{ac} (MP2-5/solv) [kJ/mol]	ΔH_{ac} (MP2-5/solv) [kJ/mol]
py	0.368	153.3	0.0	0.0	0.0	0.0
para1	0.284	147.4	-86.3	30.2	-56.1	-55.4
para2	0.278	147.0	-102.0	34.7	-67.3	-66.7
para3	0.276	146.9	-102.2	35.1	-67.1	-66.9
para4	0.279	148.1	-131.6	50.5	-81.1	-80.5
37a	0.341	151.4	-40.0	13.7	-26.3	-26.3
37b	0.336	151.0	-48.5	20.2	-28.3	-27.7
37c	0.333	150.8	-55.4	27.3	-28.1	-27.4

[a] Charge and distance parameters of the most favorable conformer
[b] In units of elemental charge e
[c] $\Delta\Delta G_{solv} = \Delta H_{rxn}$ (MP2-5/solv) - ΔH_{rxn} (MP2-5)

Appendix

A3.3 Relative isobutyrylation enthalpies for 4-dilakylaminopyridines and chiral 3,4-diaminopyridines

Table A3.5 Calculated total energies at B98/6-31G(d) level, H_{298} - values at B98/6-31G(d) level, single point energies at MP2(FC)/6-31+G(2d,p) level and solvation free energies at PCM/UAHF/RHF/6-31G(d) level.

Conformer	E_{tot} B98/6-31G(d)	H_{298} B98/6-31G(d)	E_{sp} MP2(FC)/6-31+G(2d,p)	G_{solv}, kJ/mol	H_{298} MP2-5 with solv	$<H_{298}>$ MP2-5 with solv
py_ac						-478.447696
PY_ac1	-479.737870	-479.529828	-478.605182	-132.76	-478.447706	
PY_ac2	-479.732554	-479.525252	-478.599072	-130.96	-478.441650	
DMAP_ac						-611.984799
DMAPac1	-613.687635	-613.401137	-612.224310	-123.39	-611.984807	
DMAPac2	-613.682307	-613.395747	-612.218576	-121.92	-611.978453	
PPY_ac						-689.155002
PPY_ac3	-691.089652	-690.765721	-689.432153	-122.97	-689.155058	
PPY_ac1	-691.089645	-690.765625	-689.432128	-122.97	-689.154944	
PPY_ac2	-691.084320	-690.760178	-689.426253	-121.42	-689.148358	
6a_ac						-766.342304
TCAP_a_ac1	-768.501504	-768.139728	-766.656418	-125.90	-766.342593	
TCAP_a_ac2	-768.501454	-768.139758	-766.656655	-125.98	-766.342542	
TCAP_b_ac1	-768.496078	-768.134079	-766.650539	-123.55	-766.335599	
TCAP_b_ac3	-768.500712	-768.138925	-766.655348	-125.85	-766.341497	
TCAP_a_ac3	-768.495310	-768.133390	-766.649650	-123.51	-766.334773	
TCAP_b_ac2	-768.500697	-768.138871	-766.655344	-125.81	-766.341438	
5b_ac						-978.176049
cat1_1_ac23	-981.009357	-980.510057	-978.637587	-101.29	-978.176868	
cat1_1_ac21	-981.009317	-980.509913	-978.637562	-100.96	-978.176612	
cat1_16_ac23	-981.007903	-980.508512	-978.635991	-103.01	-978.175834	
cat1_1_ac11	-981.007913	-980.508491	-978.635988	-98.49	-978.174080	
cat1_16_ac21	-981.007824	-980.508426	-978.635941	-102.93	-978.175746	
cat1_1_ac13	-981.007941	-980.508428	-978.635932	-99.96	-978.174490	
cat1_4_ac21	-981.007019	-980.507696	-978.635290	-101.67	-978.174691	
cat1_4_ac23	-981.007030	-980.507711	-978.635250	-101.38	-978.174544	
cat1_2_ac23	-981.006732	-980.507609	-978.634679	-101.63	-978.174265	
cat1_2_ac21	-981.006656	-980.507419	-978.634670	-101.55	-978.174109	
cat1_7_ac23	-981.006708	-980.507451	-978.634340	-101.96	-978.173919	
cat1_6_ac23	-981.005733	-980.506768	-978.633990	-102.93	-978.174227	
cat1_7_ac21	-981.006631	-980.507345	-978.634309	-101.88	-978.173827	
cat1_6_ac21	-981.005758	-980.506391	-978.633394	-102.76	-978.173166	
cat1_6_ac11	-981.004577	-980.505390	-978.632642	-101.46	-978.172099	
5k_ac						-1360.474646
cat2_2_ac11	-1364.322712	-1363.712149	-1361.05588	-78.87	-1360.475355	
cat2_9_ac11	-1364.32143	-1363.71065	-1361.05547	-79.71	-1360.475051	
cat2_2_ac13	-1364.322379	-1363.71151	-1361.05521	-76.86	-1360.473611	
cat2_15_ac11	-1364.321277	-1363.71049	-1361.05470	-76.53	-1360.473059	
cat2_9_ac13	-1364.32101	-1363.71033	-1361.05444	-78.16	-1360.473534	
cat2_6_ac11	-1364.32142	-1363.71039	-1361.05451	-83.39	-1360.475239	
cat2_4_ac11	-1364.32158	-1363.71061	-1361.05425	-78.95	-1360.473358	
cat2_1_ac13	-1364.32233	-1363.71137	-1361.05412	-77.49	-1360.472675	
cat2_15_ac13	-1364.320884	-1363.710043	-1361.05393	-74.77	-1360.471566	
cat2_6_ac13	-1364.32101	-1363.71017	-1361.05340	-80.63	-1360.473273	
cat2_1_ac11	-1364.32173	-1363.71093	-1361.05286	-76.40	-1360.471159	
cat2_4_ac13	-1364.32317	-1363.71018	-1361.05317	-75.19	-1360.470689	
cat2_6_ac23	-1364.32028	-1363.71012	-1361.05167	-85.44	-1360.474048	
cat2_6_ac21	-1364.32023	-1363.70990	-1361.05153	-85.40	-1360.473721	
cat2_1_ac21	-1364.32150	-1363.71059	-1361.05210	-79.54	-1360.471485	
cat2_1_ac23	-1364.32139	-1363.71051	-1361.05169	-78.53	-1360.470720	
cat2_13_ac21	-1364.32045	-1363.70975	-1361.050912	-81.71	-1360.471333	

Appendix

Conformer	E_{tot} B98/6-31G(d)	H_{298} B98/6-31G(d)	E_{sp} MP2(FC)/6-31+G(2d,p)	G_{solv}, kJ/mol	H_{298} MP2-5 with solv	$<H_{298}>$ MP2-5 with solv
cat2_13_ac23	-1364.32058	-1363.70955	-1361.050999	-84.10	-1360.471996	
38a						-824.346010
cat4_1	-826.767297	-826.347628	-824.763161	-6.61	-824.346010	
38a_ac						-1055.324548
cat4_1_ac21	-1058.368174	-1057.833981	-1055.821373	-98.91	-1055.324852	
cat4_1_ac23	-1058.368168	-1057.833690	-1055.821329	-99.08	-1055.324587	
cat4_1_ac13	-1058.366847	-1057.832646	-1055.819892	-97.07	-1055.322662	
cat4_1_ac11	-1058.366953	-1057.832303	-1055.819716	-97.70	-1055.322277	
38b						-902.672005
cat5_4	-905.362847	-904.883912	-903.152015	1.42	-902.672538	
cat5_3	-905.363084	-904.883612	-903.151703	2.43	-902.671307	
cat5_2	-905.360185	-904.880747	-903.150093	0.42	-902.670495	
cat5_1	-905.360387	-904.880813	-903.150063	0.08	-902.670457	
38b_ac						-1133.652315
cat5_4_ac23	-1136.965091	-1136.371010	-1134.212916	-88.91	-1133.652700	
cat5_4_ac21	-1136.965072	-1136.371034	-1134.212860	-89.12	-1133.652766	
cat5_3_ac21	-1136.964986	-1136.370871	-1134.212323	-87.36	-1133.651482	
cat5_3_ac23	-1136.964942	-1136.370737	-1134.212266	-87.15	-1133.651255	
cat5_4_ac13	-1136.963643	-1136.369442	-1134.211198	-86.86	-1133.650080	
cat5_4_ac11	-1136.963684	-1136.369514	-1134.211194	-85.90	-1133.649740	
cat5_3_ac13	-1136.963612	-1136.369586	-1134.210675	-83.97	-1133.648633	
cat5_3_ac11	-1136.963530	-1136.369428	-1134.210447	-85.65	-1133.648966	
cat5_1_ac21	-1136.961045	-1136.366558	-1134.208083	-91.76	-1133.648544	
5l_ac						-1515.951619
cat6_1_ac13	-1520.277491	-1519.572113	-1516.631825	-67.82	-1515.952280	
cat6_1_ac11	-1520.277070	-1519.571617	-1516.630983	-66.44	-1515.950837	
cat6_2_ac11	-1520.275783	-1519.570166	-1516.629816	-67.70	-1515.949983	
cat6_1_ac21	-1520.276412	-1519.570426	-1516.629160	-70.33	-1515.949961	
cat6_1_ac23	-1520.276271	-1519.570513	-1516.628921	-71.55	-1515.950413	
cat6_2_ac13	-1520.275453	-1519.569660	-1516.628809	-64.31	-1515.947510	
cat6_4_ac11	-1520.276095	-1519.570118	-1516.628632	-62.01	-1515.946272	
cat6_4_ac13	-1520.275523	-1519.569494	-1516.628018	-61.09	-1515.945255	
5l'						-1284.973738
cat6b_1	-1288.674669	-1288.083841	-1285.568506	9.79	-1284.973949	
cat6b_2	-1288.673788	-1288.082655	-1285.567509	12.26	-1284.971706	
cat6b_3	-1288.672912	-1288.082215	-1285.566185	20.59	-1284.967647	
cat6b_4	-1288.673440	-1288.082701	-1285.565790	17.53	-1284.968373	
5l'_ac						-1515.951375
cat6b_1_ac21	-1520.278321	-1519.572611	-1516.632336	-66.65	-1515.952013	
cat6b_1_ac23	-1520.277927	-1519.572302	-1516.632031	-65.19	-1515.951234	
cat6b_2_ac21	-1520.276838	-1519.571262	-1516.630693	-65.52	-1515.950073	
cat6b_1_ac11	-1520.276923	-1519.571530	-1516.629734	-71.34	-1515.951512	
cat6b_2_ac23	-1520.276457	-1519.570743	-1516.630051	-63.68	-1515.948592	
cat6b_1_ac13	-1520.277043	-1519.571348	-1516.629954	-70.50	-1515.951111	
cat6b_2_ac13	-1520.276711	-1519.571229	-1516.629335	-69.50	-1515.950322	
cat6b_2_ac11	-1520.276718	-1519.570961	-1516.629272	-68.53	-1515.949618	

Appendix

3.4 Relative acetylation enthalpies for ferrocenyl catalysts

Table A3.6 Calculated total energies at B3LYP/6-31G(d) level, H_{298} - values at B3LYP/6-31G(d) level, single point energies at B3LYP/6-311+G(d,p)level and H_{298} - values at B3LYP/6-311+G(d,p)//B3LYP /6-31G(d) level, written as "H_{298} SP", Boltzmann factors w_i and averaged enthalpies $<H_{298}>$ at the B3LYP/6-311+G(d,p)//B3LYP/6-31G(d) level of theory.

Conformer	E_{tot} BLYP/6-31G(d)	H_{298} B3LYP/6-31G(d)	E_{sp} B3LYP/6-311+G(d,p)	"H_{298} SP"	w_i	$<H_{298}>$
py	-248.284973	-248.190708	-248.351162	-248.256898	1.00	-248.256898
py_ac	-401.299539	-401.151170	-401.401972	-401.253603	1.00	-401.253603
p-cpstol						-2641.520030
p-cpstol2	-2641.525196	-2641.149546	-2641.895706	-2641.520056	0.995	
p-cpstol	-2641.520133	-2641.143248	-2641.891998	-2641.515113	0.005	
p-cpstol_ac						-2794.542393
p-cpstol2_ac1	-2794.566429	-2794.135549	-2794.973542	-2794.542662	0.701	
p-cpstol2_ac2	-2794.565695	-2794.134775	-2794.972758	-2794.541838	0.293	
p-cpstol_ac1	-2794.559784	-2794.129248	-2794.968451	-2794.537915	0.005	
p-cpstol_ac2	-2794.558414	-2794.127782	-2794.967094	-2794.536462	0.001	
p-cpstol_en						-2641.517623
2	-2641.524208	-2641.147449	-2641.894451	-2641.517692	0.982	
3	-2641.520189	-2641.143548	-2641.890513	-2641.513872	0.017	
4	-2641.514755	-2641.139157	-2641.885966	-2641.510367	0.000	
p-cpstol_en.ac						-2794.540381
3.ac1	-2794.564954	-2794.134073	-2794.971520	-2794.540639	0.744	
1.ac2	-2794.564016	-2794.133143	-2794.970506	-2794.539633	0.256	
m-cpstol						-2641.520858
m-cpstol4	-2641.525265	-2641.150597	-2641.895686	-2641.521018	0.944	
m-cpstol3	-2641.523477	-2641.147903	-2641.893889	-2641.518315	0.054	
m-cpstol1	-2641.520443	-2641.143549	-2641.892257	-2641.515363	0.002	
m-cpstol2	-2641.517870	-2641.141091	-2641.889467	-2641.512688	0.000	
m-cpstol_ac						-2794.540230
m-cpstol3_ac1	-2794.563901	-2794.133447	-2794.970856	-2794.540402	0.935	
m-cpstol2_ac1	-2794.559871	-2794.129466	-2794.968258	-2794.537853	0.063	
m-cpstol4_ac2	-2794.557337	-2794.126744	-2794.964429	-2794.533837	0.001	
m-cpstol3_ac2	-2794.557541	-2794.127066	-2794.963892	-2794.533417	0.001	
m-cpstol4_ac1	-2794.556479	-2794.125930	-2794.963426	-2794.532877	0.000	
m-cpstol1_ac1	-2794.551146	-2794.120624	-2794.959879	-2794.529358	0.000	
m-cpstol1_ac2	-2794.549414	-2794.118876	-2794.957931	-2794.527393	0.000	
o-cpstol						-2641.518490
o-cpstol1	-2641.523992	-2641.147738	-2641.894983	-2641.518729	0.861	
o-cpstol2	-2641.521159	-2641.144880	-2641.893288	-2641.517009	0.139	
o-cpstol_ac						-2794.522348
o-cpstol2_ac1	-2794.545702	-2794.115430	-2794.952621	-2794.522349	1.000	
o-cpstol1_ac1	-2794.535927	-2794.105727	-2794.944405	-2794.514204	0.000	
p-cpp						-1897.800692
p-cpp	-1897.798514	-1897.544643	-1898.054563	-1897.800692	1.000	
p-cpp_ac						-2050.821563
ac2	-2050.835765	-2050.528584	-2051.128996	-2050.821815	0.727	
ac1	-2050.835758	-2050.527669	-2051.128979	-2050.820890	0.273	
m-cpstolDMAP						-2775.438854
1	-2775.483587	-2775.030545	-2775.892110	-2775.439069	0.526	
5	-2775.486754	-2775.032647	-2775.892984	-2775.438877	0.429	

Appendix

Conformer	E_{tot} BLYP/6-31G(d)	H_{298} B3LYP/6-31G(d)	E_{sp} B3LYP/6-311+G(d,p)	"H_{298} SP"	w_i	$<H_{298}>$
2	-2775.483118	-2775.028816	-2775.890428	-2775.436125	0.023	
6	-2775.480849	-2775.028578	-2775.888308	-2775.436038	0.021	
m-cpstolDMAP_ac						-2928.480661
5_ac1	-2928.546729	-2928.037635	-2928.989907	-2928.480814	0.967	
2_ac2	-2928.541046	-2928.032636	-2928.985105	-2928.476696	0.012	
5_ac2	-2928.541247	-2928.032387	-2928.984970	-2928.476109	0.007	
1_ac2	-2928.540749	-2928.031600	-2928.985217	-2928.476069	0.006	
6_ac2	-2928.541251	-2928.031790	-2928.985235	-2928.475774	0.005	
1_ac1	-2928.539285	-2928.030160	-2928.983677	-2928.474552	0.001	
2_ac1	-2928.538001	-2928.029744	-2928.982621	-2928.474364	0.001	
6_ac1	-2928.539066	-2928.029643	-2928.982899	-2928.473476	0.000	
m-cpstolDMAP_en						-2775.439192
5	-2775.486823	-2775.032743	-2775.893581	-2775.439501	0.828	
1	-2775.486135	-2775.031888	-2775.892161	-2775.437914	0.154	
2	-2775.484235	-2775.029989	-2775.890123	-2775.435878	0.018	
7	-2775.475036	-2775.021038	-2775.881117	-2775.427118	0.000	
m-cpstolDMAP_en_ac						-2928.479394
1.ac1	-2928.546546	-2928.037438	-2928.988542	-2928.479434	0.992	
1.ac2	-2928.540985	-2928.031968	-2928.983597	-2928.474580	0.006	
2.ac2	-2928.539926	-2928.030668	-2928.982753	-2928.473496	0.002	
2.ac1	-2928.538092	-2928.029036	-2928.980224	-2928.471168	0.000	
5.ac1	-2928.534007	-2928.025129	-2928.977636	-2928.468758	0.000	
m-cpstolPPY						-2852.843110
2	-2852.910759	-2852.418562	-2853.335570	-2852.843373	0.900	
5	-2852.909142	-2852.417031	-2853.333120	-2852.841010	0.074	
1	-2852.906531	-2852.414569	-2853.332018	-2852.840056	0.027	
m-cpstolPPY_ac						-3005.886863
5.ac2	-3005.973507	-3005.426384	-3006.434206	-3005.887084	0.575	
1.ac2	-3005.972444	-3005.425264	-3006.433911	-3005.886731	0.396	
5.ac1	-3005.971054	-3005.424084	-3006.431236	-3005.884265	0.029	
m-cpstolPPY_en						-2852.843379
4	-2852.911826	-2852.419668	-2853.335948	-2852.843789	0.799	
3	-2852.909311	-2852.417428	-2853.333791	-2852.841907	0.109	
2	-2852.910473	-2852.418399	-2853.333775	-2852.841701	0.087	
5	-2852.906692	-2852.414428	-2853.331182	-2852.838918	0.005	
m-cpstolPPY_en.ac						-3005.888817
2.ac1	-3005.976478	-3005.429528	-3006.435938	-3005.888988	0.945	
3.ac1	-3005.973277	-3005.426382	-3006.433050	-3005.886155	0.047	
2.ac2	-3005.971591	-3005.424681	-3006.431354	-3005.884444	0.008	
5.ac1	-3005.968303	-3005.421422	-3006.428183	-3005.881302	0.000	
4.ac1	-3005.965487	-3005.418748	-3006.426102	-3005.879363	0.000	
4.ac2	-3005.965511	-3005.418806	-3006.426059	-3005.879354	0.000	

Chapter 4. Computational details.
Acetylation enthalpies of 3-(thio)urea-4-aminopyridines.

Table A4.1 Calculated energies of conformers of 3-(thio)urea-4-aminopyridines and corresponding acetylcations, as calculated at MP2/6-31+G(2d,p)//B98/6-31G(d) level with inclusion of solvent effects at PCM/UAHF/RHF/6-31G(d) level.

Conformer	E_{tot} B98/6-31G(d)	H_{298} B98/6-31G(d)	E_{tot} MP2(FC)/6-31+G(2d,p)	G_{solv}, kJ/mol	H_{298} MP2-5 with solv	$<H_{298}>$ MP2-5 with solv
Py						
Py	-248.181767	-248.087627	-247.589439	-9.00	-247.498727	
Py_ac	-401.140004	-400.991691	-400.215516	-142.55	-400.121498	
cat81ur1						-911.978235
2c2	-914.434079	-914.090468	-912.317918	-10.67	-911.978371	
1a2	-914.434029	-914.090433	-912.318133	-10.21	-911.978425	
1a1	-914.426811	-914.083226	-912.312367	-12.72	-911.973626	
2c1	-914.426240	-914.082568	-912.312192	-11.92	-911.973062	
6b1	-914.426045	-914.082395	-912.312277	-12.26	-911.973296	
5a1	-914.424404	-914.080903	-912.307462	-20.38	-911.971722	
5c1	-914.423689	-914.080334	-912.315623	-10.50	-911.976268	
2d1	-914.417843	-914.074596	-912.313649	-9.92	-911.974178	
cat81ur1_ac						-1064.625490
5c1.ac2	-1067.419733	-1067.021912	-1064.979632	-115.10	-1064.625651	
5c1.ac1	-1067.419506	-1067.021484	-1064.978040	-111.80	-1064.622599	
5a1.ac1	-1067.423787	-1067.025409	-1064.972701	-119.83	-1064.619964	
2c2.ac1	-1067.420543	-1067.022367	-1064.971265	-119.29	-1064.618522	
1a2.ac2	-1067.419955	-1067.021814	-1064.970518	-119.20	-1064.617779	
2c1.ac2	-1067.420060	-1067.021911	-1064.970339	-124.89	-1064.619759	
5a1.ac2	-1067.421393	-1067.023001	-1064.970543	-125.06	-1064.619783	
1a1.ac1	-1067.419465	-1067.021227	-1064.969697	-125.10	-1064.619108	
1a1.ac2	-1067.413825	-1067.015669	-1064.964263	-117.49	-1064.610856	
2c1.ac1	-1067.413409	-1067.015332	-1064.963629	-117.11	-1064.610157	
cat81ur1_Nac						
5c1.Nac1	-1067.415831	-1067.017653	-1064.981045	-127.99	-1064.631615	
cat81ur2						-1234.548690
5c1	-1237.349369	-1237.008059	-1234.887294	-8.08	-1234.549060	
2c2	-1237.353109	-1237.012007	-1234.884799	-9.29	-1234.547235	
1a2	-1237.352762	-1237.011295	-1234.884784	-8.54	-1234.546567	
6b1	-1237.344729	-1237.003506	-1234.878701	-11.17	-1234.541733	
1a1	-1237.345239	-1237.003885	-1234.878563	-11.72	-1234.541672	
2c1	-1237.344843	-1237.003367	-1234.878686	-10.79	-1234.541322	
5a1	-1237.342558	-1237.001353	-1234.874410	-18.07	-1234.540089	
cat81ur2_ac						-1387.195529
5c1.ac2	-1390.341289	-1389.945337	-1387.548349	-113.64	-1387.195680	
5c1.ac1	-1390.340696	-1389.944628	-1387.546834	-110.67	-1387.192917	
2c2.ac2	-1390.338523	-1389.942546	-1387.539578	-117.32	-1387.188285	
1a1.ac2	-1390.340294	-1389.944269	-1387.539443	-116.73	-1387.187879	
2c1.ac1	-1390.340212	-1389.944253	-1387.539019	-117.65	-1387.187872	
1a2.ac1	-1390.337928	-1389.941812	-1387.538909	-117.57	-1387.187573	
2c2.ac1	-1390.339476	-1389.943424	-1387.541004	-111.04	-1387.187246	
1a1.ac1	-1390.338690	-1389.942862	-1387.539083	-115.27	-1387.187159	
1a2.ac2	-1390.338817	-1389.942724	-1387.540235	-110.88	-1387.186372	
cat81ur2_Nac						
5c1.Nac1	-1390.331331	-1389.935396	-1387.547812	-125.14	-1387.199541	
cat81ur1f						-1584.721825
5c1	-1588.285390	-1587.924995	-1585.085573	7.57	-1584.722294	
2c2	-1588.294546	-1587.933999	-1585.083081	3.31	-1584.721275	
1a2	-1588.294453	-1587.933819	-1585.083140	3.64	-1584.721120	
2d1	-1588.280242	-1587.919944	-1585.080983	6.40	-1584.718247	
1a1	-1588.287273	-1587.926830	-1585.077348	1.00	-1584.716523	

Appendix

Conformer	E_{tot} B98/6-31G(d)	H_{298} B98/6-31G(d)	E_{tot} MP2(FC)/6-31+G(2d,p)	G_{solv}, kJ/mol	H_{298} MP2-5 with solv	$<H_{298}>$ MP2-5 with solv
2c1	-1588.286641	-1587.926165	-1585.077215	1.80	-1584.716053	
6b1	-1588.286486	-1587.925938	-1585.077096	1.34	-1584.716037	
5a1	-1588.284347	-1587.924094	-1585.071775	-9.25	-1584.715044	
cat81ur1f_ac						-1737.364100
5c1_ac2	-1741.273285	-1740.858485	-1737.741174	-99.96	-1737.364445	
5c1_ac1	-1741.273221	-1740.858543	-1737.740388	-97.95	-1737.363017	
2d1_ac2	-1741.266890	-1740.852307	-1737.735856	-100.83	-1737.359679	
5a1_ac2	-1741.274732	-1740.859560	-1737.727750	-123.39	-1737.359573	
5a1_ac1	-1741.276775	-1740.861341	-1737.729497	-118.37	-1737.359146	
2c2_ac2	-1741.273378	-1740.858265	-1737.727391	-121.38	-1737.358508	
2c1_ac2	-1741.273409	-1740.858246	-1737.727364	-121.34	-1737.358416	
1a2_ac2	-1741.272771	-1740.857736	-1737.726798	-121.04	-1737.357866	
1a1_ac2	-1741.272773	-1740.857741	-1737.726768	-121.00	-1737.357823	
2c2_ac1	-1741.273007	-1740.858154	-1737.727087	-116.23	-1737.356505	
1a2_ac1	-1741.272366	-1740.857597	-1737.726521	-115.35	-1737.355688	
2d1_ac1	-1741.264332	-1740.849653	-1737.730829	-99.29	-1737.353966	
1a1_ac1	-1741.265501	-1740.850553	-1737.719649	-113.51	-1737.347935	
2c1_ac1	-1741.265102	-1740.850127	-1737.719024	-113.30	-1737.347203	
cat81ur3						-951.139531
66	-953.720149	-953.346038	-951.512076	-6.32	-951.140372	
3	-953.728990	-953.354938	-951.510948	-6.65	-951.139429	
13	-953.729002	-953.354944	-951.510263	-8.03	-951.139264	
1	-953.727898	-953.354108	-951.508627	-10.50	-951.138837	
43	-953.720648	-953.346404	-951.511698	-3.51	-951.138792	
8	-953.727915	-953.354067	-951.508479	-10.79	-951.138742	
51	-953.716724	-953.342413	-951.509357	-6.78	-951.137627	
10	-953.725258	-953.351046	-951.507128	-9.41	-951.136501	
5	-953.724136	-953.350028	-951.505124	-12.09	-951.135621	
59	-953.721730	-953.347771	-951.502535	-18.33	-951.135556	
29	-953.722029	-953.348205	-951.504849	-11.34	-951.135344	
79	-953.720851	-953.347290	-951.501199	-19.04	-951.134889	
40	-953.720565	-953.346962	-951.500709	-20.29	-951.134835	
31	-953.721492	-953.347835	-951.504478	-9.79	-951.134549	
23	-953.721228	-953.347218	-951.504555	-10.42	-951.134514	
20	-953.720903	-953.347136	-951.502888	-13.26	-951.134173	
17	-953.720063	-953.346551	-951.502611	-12.93	-951.134023	
25	-953.720250	-953.346534	-951.502851	-12.43	-951.133868	
67	-953.719746	-953.346131	-951.499482	-20.71	-951.133756	
37	-953.719684	-953.345811	-951.503434	-10.33	-951.133497	
57	-953.719787	-953.345898	-951.503285	-10.75	-951.133491	
cat81ur3_ac						-1103.786798
43_ac2	-1106.714182	-1106.285660	-1104.172555	-114.35	-1103.787586	
66_ac2	-1106.714270	-1106.285678	-1104.172447	-113.76	-1103.787185	
43_ac1	-1106.714372	-1106.285911	-1104.172430	-111.34	-1103.786374	
66_ac1	-1106.714477	-1106.285912	-1104.172753	-110.00	-1103.786084	
8_ac1	-1106.721229	-1106.292731	-1104.171018	-114.01	-1103.785945	
31_ac1	-1106.721229	-1106.292733	-1104.171009	-113.97	-1103.785923	
1_ac1	-1106.720621	-1106.292151	-1104.170198	-113.97	-1103.785138	
31_ac2	-1106.717524	-1106.289247	-1104.167130	-118.41	-1103.783952	
8_ac2	-1106.718999	-1106.290627	-1104.166520	-120.21	-1103.783932	
59_ac1	-1106.720160	-1106.291615	-1104.166802	-119.24	-1103.783675	
51_ac1	-1106.710621	-1106.281806	-1104.168901	-114.01	-1103.783512	
1_ac2	-1106.718433	-1106.289943	-1104.165967	-120.37	-1103.783324	
59_ac2	-1106.722393	-1106.293523	-1104.168583	-113.97	-1103.783123	
13_ac2	-1106.718937	-1106.290460	-1104.165824	-119.54	-1103.782877	
51_ac2	-1106.710786	-1106.281843	-1104.169144	-110.88	-1103.782431	

Appendix

Conformer	E_{tot} B98/6-31G(d)	H_{298} B98/6-31G(d)	E_{tot} MP2(FC)/6-31+G(2d,p)	G_{solv}, kJ/mol	H_{298} MP2-5 with solv	$<H_{298}>$ MP2-5 with solv
5_ac1	-1106.717408	-1106.288926	-1104.166701	-114.52	-1103.781836	
13_ac1	-1106.718754	-1106.290350	-1104.165685	-115.65	-1103.781328	
3_ac1	-1106.718190	-1106.289624	-1104.165199	-114.98	-1103.780425	
5_ac2	-1106.715253	-1106.286823	-1104.162302	-121.25	-1103.780054	
29_ac2	-1106.715119	-1106.286368	-1104.163596	-116.61	-1103.779258	
10_ac2	-1106.715212	-1106.286625	-1104.161572	-120.25	-1103.778785	
10_ac1	-1106.714990	-1106.286295	-1104.161433	-116.65	-1103.777167	
29_ac1	-1106.711614	-1106.283072	-1104.158111	-112.51	-1103.772422	
cat81ur3_Nac						
66_Nac	-1106.716904	-1106.288111	-1104.179205	-121.42	-1103.796659	
cat81ur4						-1273.713229
66	-1276.646192	-1276.274479	-1274.083967	-4.18	-1273.713848	
43	-1276.646484	-1276.274391	-1274.082629	-4.64	-1273.712304	
8	-1276.652803	-1276.281307	-1274.079678	-7.61	-1273.711082	
51	-1276.642701	-1276.270715	-1274.081077	-4.73	-1273.710892	
1	-1276.652540	-1276.280945	-1274.079720	-6.95	-1273.710770	
13	-1276.652567	-1276.280910	-1274.079616	-6.15	-1273.710302	
3	-1276.652256	-1276.280459	-1274.079688	-4.98	-1273.709787	
5	-1276.648669	-1276.276966	-1274.075857	-9.00	-1273.707580	
10	-1276.648469	-1276.276796	-1274.075996	-7.45	-1273.707159	
59	-1276.642405	-1276.270880	-1274.070274	-16.40	-1273.704996	
29	-1276.644500	-1276.272793	-1274.073124	-9.20	-1273.704923	
cat81ur4_ac						-1426.358050
66_ac2	-1429.636696	-1429.210560	-1426.742103	-111.92	-1426.358595	
43_ac2	-1429.636549	-1429.210083	-1426.742158	-112.59	-1426.358576	
66_ac1	-1429.636834	-1429.210483	-1426.742368	-107.99	-1426.357148	
51_ac2	-1429.636578	-1429.210123	-1426.741964	-108.57	-1426.356862	
43_ac1	-1429.636578	-1429.210132	-1426.741950	-108.49	-1426.356826	
8_ac1	-1429.641606	-1429.215534	-1426.739869	-112.80	-1426.356761	
1_ac1	-1429.640949	-1429.214915	-1426.738920	-113.09	-1426.355961	
51_ac1	-1429.633008	-1429.206525	-1426.738490	-112.17	-1426.354732	
8_ac2	-1429.640289	-1429.214271	-1426.736434	-116.06	-1426.354622	
1_ac2	-1429.639696	-1429.213804	-1426.735780	-115.81	-1426.353999	
31_ac2	-1429.640429	-1429.214425	-1426.734924	-115.06	-1426.352744	
5_ac1	-1429.637815	-1429.211674	-1426.735612	-113.55	-1426.352722	
59_ac1	-1429.638951	-1429.212943	-1426.734284	-115.48	-1426.352260	
59_ac2	-1429.642318	-1429.216012	-1426.735925	-111.13	-1426.351945	
13_ac1	-1429.637872	-1429.212769	-1426.734260	-110.37	-1426.351197	
5_ac2	-1429.636609	-1429.210509	-1426.732301	-117.11	-1426.350806	
31_ac1	-1429.642224	-1429.211688	-1426.734334	-111.25	-1426.346173	
cat81ur4_Nac						
66_Nac	-1429.631376	-1429.205023	-1426.741361	-120.00	-1426.360712	
cat81ur5						-990.305293
51	-993.016889	-992.613295	-990.709272	-0.33	-990.305805	
1	-993.026732	-992.623532	-990.707005	-3.89	-990.305287	
8	-993.026296	-992.623031	-990.705549	-5.19	-990.304261	
13	-993.026778	-992.623553	-990.706483	-1.92	-990.303991	
3	-993.023813	-992.620151	-990.703642	-3.60	-990.301350	
5	-993.022525	-992.619020	-990.702260	-6.49	-990.301226	
10	-993.023113	-992.619423	-990.703596	-2.97	-990.301037	
66	-993.009985	-992.606578	-990.702752	-3.31	-990.300603	
43	-993.010006	-992.606489	-990.701191	-3.10	-990.298853	
cat81ur5_ac						-1142.950561
51_ac2	-1146.011622	-1145.553539	-1143.369987	-103.55	-1142.951345	
8_ac1	-1146.020847	-1145.563102	-1143.368097	-105.65	-1142.950591	
51_ac1	-1146.011776	-1145.553691	-1143.369785	-100.54	-1142.949994	

Appendix

Conformer	E_{tot} B98/6-31G(d)	H_{298} B98/6-31G(d)	E_{tot} MP2(FC)/6-31+G(2d,p)	G_{solv}, kJ/mol	H_{298} MP2-5 with solv	$<H_{298}>$ MP2-5 with solv
8_ac2	-1146.017913	-1145.560112	-1143.364405	-112.93	-1142.949615	
1_ac2	-1146.018170	-1145.560290	-1143.364116	-112.88	-1142.949232	
13_ac2	-1146.017581	-1145.559823	-1143.363172	-112.38	-1142.948218	
43_ac2	-1146.006452	-1145.548454	-1143.364694	-108.16	-1142.947891	
66_ac2	-1146.006673	-1145.548746	-1143.364619	-107.61	-1142.947680	
66_ac1	-1146.006958	-1145.548815	-1143.365662	-104.18	-1142.947199	
43_ac1	-1146.006572	-1145.548440	-1143.364851	-104.98	-1142.946703	
5_ac1	-1146.017080	-1145.559110	-1143.363921	-106.69	-1142.946587	
13_ac1	-1146.017617	-1145.559675	-1143.363240	-108.03	-1142.946445	
1_ac1	-1146.017604	-1145.559802	-1143.363445	-105.98	-1142.946009	
5_ac2	-1146.014159	-1145.556344	-1143.360056	-114.10	-1142.945698	
3_ac2	-1146.013645	-1145.555567	-1143.358912	-115.39	-1142.944785	
10_ac2	-1146.013846	-1145.555773	-1143.359013	-113.68	-1142.944238	
3_ac1	-1146.013425	-1145.555457	-1143.358779	-110.75	-1142.942994	
10_ac1	-1146.013864	-1145.555670	-1143.358953	-108.74	-1142.942177	
cat81ur6						-1084.668315
63	-1087.652398	-1087.200677	-1085.121493	1.88	-1084.669055	
12	-1087.651791	-1087.200278	-1085.120114	4.52	-1084.666879	
137	-1087.649225	-1087.197328	-1085.119494	3.68	-1084.666195	
143	-1087.651631	-1087.199821	-1085.118574	-0.29	-1084.666876	
60	-1087.649246	-1087.197399	-1085.118472	1.67	-1084.665987	
1	-1087.647362	-1087.196005	-1085.116869	3.05	-1084.664349	
105	-1087.649716	-1087.198264	-1085.116692	2.80	-1084.664173	
81	-1087.648282	-1087.196410	-1085.116456	0.46	-1084.664409	
21	-1087.650401	-1087.199026	-1085.115575	-1.00	-1084.664582	
117	-1087.648878	-1087.197476	-1085.115318	0.59	-1084.663693	
180	-1087.648841	-1087.197243	-1085.115277	-1.72	-1084.664332	
7	-1087.646359	-1087.194718	-1085.115136	-0.25	-1084.663591	
95	-1087.649410	-1087.197794	-1085.114569	0.75	-1084.662666	
16	-1087.648725	-1087.196862	-1085.114783	-0.75	-1084.663207	
31	-1087.649276	-1087.197748	-1085.113878	-3.05	-1084.663514	
42	-1087.647809	-1087.196124	-1085.113806	-0.50	-1084.662312	
20	-1087.646745	-1087.195228	-1085.113271	-2.34	-1084.662646	
56	-1087.648489	-1087.196613	-1085.113473	-3.77	-1084.663031	
25	-1087.647748	-1087.195868	-1085.113211	-3.18	-1084.662542	
13	-1087.646463	-1087.194582	-1085.113174	-2.05	-1084.662074	
68	-1087.647010	-1087.195210	-1085.112812	-3.43	-1084.662320	
9	-1087.648730	-1087.197030	-1085.112542	-0.84	-1084.661161	
19	-1087.647413	-1087.195962	-1085.112071	-2.55	-1084.661592	
113	-1087.647841	-1087.195938	-1085.112286	-1.92	-1084.661116	
216	-1087.647494	-1087.195746	-1085.111837	-4.85	-1084.661938	
92	-1087.648854	-1087.197578	-1085.110395	-6.36	-1084.661542	
346	-1087.648113	-1087.196665	-1085.110361	-5.61	-1084.661049	
55	-1087.647241	-1087.195537	-1085.110475	-6.23	-1084.661146	
64	-1087.647228	-1087.196191	-1085.109215	-5.90	-1084.660425	
210	-1087.648095	-1087.197060	-1085.108979	-6.61	-1084.660462	
196	-1087.648804	-1087.197091	-1085.109270	-8.70	-1084.660872	
167	-1087.647590	-1087.195942	-1085.109187	-3.31	-1084.658798	
303	-1087.647705	-1087.196233	-1085.108817	-7.78	-1084.660309	
317	-1087.649858	-1087.198732	-1085.108116	-12.47	-1084.661739	
2	-1087.647902	-1087.196027	-1085.108658	-6.99	-1084.659445	
30	-1087.648332	-1087.197078	-1085.107656	-8.45	-1084.659621	
313	-1087.647327	-1087.195744	-1085.107574	-12.80	-1084.660867	
235	-1087.646355	-1087.194586	-1085.107444	-7.70	-1084.658607	
40	-1087.646701	-1087.194763	-1085.107460	-5.77	-1084.657721	
90	-1087.647561	-1087.196092	-1085.106581	-14.23	-1084.660530	

Appendix

Conformer	E_{tot} B98/6-31G(d)	H_{298} B98/6-31G(d)	E_{tot} MP2(FC)/6-31+G(2d,p)	G_{solv}, kJ/mol	H_{298} MP2-5 with solv	$<H_{298}>$ MP2-5 with solv
39	-1087.647091	-1087.195525	-1085.105994	-11.21	-1084.658699	
247	-1087.646496	-1087.195150	-1085.105365	-10.38	-1084.657972	
277	-1087.646510	-1087.194762	-1085.105597	-13.72	-1084.659077	
132	-1087.646679	-1087.195178	-1085.104841	-11.80	-1084.657834	
260	-1087.647668	-1087.196330	-1085.104396	-19.58	-1084.660516	
401	-1087.646485	-1087.195123	-1085.104023	-20.92	-1084.660629	
285	-1087.646653	-1087.195135	-1085.104055	-19.37	-1084.659915	
cat81ur6_ac						-1237.324964
143_ac2	-1240.658226	-1240.151779	-1237.794427	-98.37	-1237.325445	
131_ac2	-1240.656011	-1240.149484	-1237.793683	-95.31	-1237.323459	
137_ac2	-1240.655066	-1240.148568	-1237.792044	-98.58	-1237.323091	
1_ac2	-1240.656859	-1240.150413	-1237.792807	-90.88	-1237.320974	
21_ac2	-1240.659746	-1240.153034	-1237.792123	-93.01	-1237.320836	
1_ac1	-1240.656875	-1240.150481	-1237.793081	-88.45	-1237.320375	
95_ac2	-1240.659973	-1240.153007	-1237.792199	-92.22	-1237.320356	
117_ac1	-1240.657664	-1240.151125	-1237.792033	-90.92	-1237.320123	
12_ac1	-1240.659296	-1240.152651	-1237.791337	-90.25	-1237.319065	
105_ac1	-1240.658992	-1240.152272	-1237.790361	-90.63	-1237.318158	
117_ac2	-1240.657066	-1240.150558	-1237.789005	-93.55	-1237.318130	
63_ac2	-1240.656813	-1240.150008	-1237.786877	-98.16	-1237.317459	
143_ac1	-1240.655000	-1240.148430	-1237.784389	-102.55	-1237.316879	
42_ac1	-1240.656702	-1240.149396	-1237.788676	-92.97	-1237.316780	
180_ac1	-1240.647355	-1240.140726	-1237.785837	-97.03	-1237.316163	
95_ac1	-1240.656375	-1240.149514	-1237.787206	-92.26	-1237.315484	
21_ac1	-1240.656166	-1240.149156	-1237.787030	-92.84	-1237.315382	
17_ac2	-1240.657595	-1240.150828	-1237.785318	-94.98	-1237.314726	
31_ac2	-1240.656912	-1240.150251	-1237.784087	-97.07	-1237.314397	
63_ac1	-1240.654464	-1240.147824	-1237.780357	-104.10	-1237.313365	
137_ac1	-1240.651900	-1240.145146	-1237.780923	-102.51	-1237.313212	
131_ac1	-1240.653824	-1240.147541	-1237.781193	-100.54	-1237.313204	
105_ac2	-1240.655717	-1240.149289	-1237.785604	-89.29	-1237.313183	
60_ac1	-1240.653822	-1240.146853	-1237.782181	-99.50	-1237.313108	
12_ac2	-1240.655524	-1240.148790	-1237.786552	-87.07	-1237.312981	
16_ac2	-1240.654685	-1240.147864	-1237.782035	-98.95	-1237.312902	
9_ac2	-1240.654118	-1240.147556	-1237.781266	-99.54	-1237.312616	
17_ac1	-1240.655892	-1240.149212	-1237.783053	-94.27	-1237.312277	
60_ac2	-1240.650932	-1240.144139	-1237.778469	-100.37	-1237.309906	
16_ac1	-1240.653149	-1240.146139	-1237.779850	-96.94	-1237.309764	
9_ac1	-1240.652514	-1240.145592	-1237.779011	-97.74	-1237.309315	
PheOMe						-1217.600993
37c	-1220.818907	-1220.367388	-1218.055831	7.28	-1217.601540	
59	-1220.818626	-1220.367166	-1218.054727	4.73	-1217.601466	
22c	-1220.817798	-1220.366737	-1218.052799	2.68	-1217.600718	
51c	-1220.818592	-1220.367420	-1218.051629	-0.04	-1217.600473	
31c	-1220.818546	-1220.367443	-1218.051845	0.79	-1217.600439	
27c	-1220.817258	-1220.365975	-1218.052742	6.19	-1217.599101	
35c	-1220.817017	-1220.365884	-1218.050281	2.64	-1217.598144	
2c	-1220.816511	-1220.365417	-1218.051169	5.27	-1217.598067	
9	-1220.815538	-1220.364317	-1218.049583	1.09	-1217.597948	
43	-1220.817055	-1220.365854	-1218.050277	4.18	-1217.597483	
10c	-1220.816671	-1220.365736	-1218.048175	-0.29	-1217.597352	
1c	-1220.816717	-1220.365321	-1218.048329	0.29	-1217.596822	
23c	-1220.816396	-1220.365268	-1218.047329	-0.42	-1217.596360	
7c	-1220.816572	-1220.365276	-1218.048514	3.05	-1217.596054	
26c	-1220.815338	-1220.364183	-1218.050307	8.16	-1217.596045	
37	-1220.805182	-1220.353853	-1218.042694	3.31	-1217.590106	

Appendix

Conformer	E_{tot} B98/6-31G(d)	H_{298} B98/6-31G(d)	E_{tot} MP2(FC)/6-31+G(2d,p)	G_{solv}, kJ/mol	H_{298} MP2-5 with solv	$<H_{298}>$ MP2-5 with solv
59	-1220.804902	-1220.353688	-1218.041534	1.09	-1217.589906	
26	-1220.804262	-1220.352990	-1218.040464	2.09	-1217.588395	
9	-1220.807054	-1220.355474	-1218.039800	0.13	-1217.588172	
43	-1220.803357	-1220.351847	-1218.039662	0.33	-1217.588024	
23	-1220.806936	-1220.355485	-1218.038933	-1.13	-1217.587912	
27	-1220.803445	-1220.352060	-1218.039662	1.30	-1217.587783	
22	-1220.802957	-1220.352100	-1218.038180	-0.92	-1217.587674	
31	-1220.803350	-1220.352526	-1218.037316	-2.51	-1217.587449	
51	-1220.803403	-1220.352514	-1218.037027	-3.39	-1217.587429	
1	-1220.805238	-1220.354131	-1218.037532	-2.47	-1217.587366	
10	-1220.805257	-1220.354011	-1218.037228	-3.56	-1217.587336	
35	-1220.805380	-1220.354094	-1218.038614	0.46	-1217.587153	
2	-1220.803970	-1220.352687	-1218.039906	3.93	-1217.587125	
7	-1220.805667	-1220.354119	-1218.039334	1.84	-1217.587085	
33	-1220.805615	-1220.353954	-1218.037533	-2.64	-1217.586876	
PheOMe_ac						-1370.250656
22c_ac2	-1373.819870	-1373.313895	-1370.723518	-88.53	-1370.251264	
51c_ac2	-1373.819816	-1373.313527	-1370.721354	-90.46	-1370.249519	
31c_ac2	-1373.819291	-1373.313231	-1370.720949	-90.17	-1370.249231	
37c_ac1	-1373.812755	-1373.306715	-1370.721301	-86.73	-1370.248296	
37c_ac2	-1373.812494	-1373.306331	-1370.721581	-85.73	-1370.248070	
31c_ac1	-1373.812936	-1373.307616	-1370.715111	-96.48	-1370.246539	
51c_ac1	-1373.813435	-1373.307616	-1370.715299	-96.99	-1370.246420	
59c_ac2	-1373.812549	-1373.306560	-1370.718810	-87.15	-1370.246015	
22c_ac1	-1373.812139	-1373.306133	-1370.714567	-92.63	-1370.243843	
35c_ac1	-1373.812196	-1373.306219	-1370.710882	-97.95	-1370.242211	
2c_ac1	-1373.809303	-1373.303762	-1370.710711	-95.35	-1370.241488	
1c_ac1	-1373.813237	-1373.307102	-1370.708710	-98.99	-1370.240280	
37_ac1	-1373.799696	-1373.293649	-1370.710251	-88.32	-1370.237845	
59_ac1	-1373.799182	-1373.293026	-1370.708526	-91.21	-1370.237111	
51_ac1	-1373.798173	-1373.292415	-1370.701257	-98.74	-1370.233109	
31_ac1	-1373.797747	-1373.292136	-1370.701139	-98.37	-1370.232993	
22_ac1	-1373.798257	-1373.292373	-1370.701331	-95.44	-1370.231797	
1_ac1	-1373.803280	-1373.297260	-1370.701377	-95.65	-1370.231787	
43_ac1	-1373.789510	-1373.283535	-1370.695759	-107.15	-1370.230596	
23_ac1	-1373.797398	-1373.291271	-1370.695833	-107.15	-1370.230519	
10_ac1	-1373.798459	-1373.292551	-1370.696276	-105.27	-1370.230463	
35_ac1	-1373.797998	-1373.292396	-1370.697079	-101.21	-1370.230026	
27_ac1	-1373.789082	-1373.283065	-1370.695584	-106.06	-1370.229965	
7_ac1	-1373.797585	-1373.292158	-1370.696829	-100.54	-1370.229696	
2_ac1	-1373.795533	-1373.290100	-1370.697786	-97.28	-1370.229404	
9_ac1	-1373.796992	-1373.291092	-1370.695497	-99.66	-1370.227557	
1_ac2	-1373.797975	-1373.292207	-1370.695972	-104.52	-1370.230012	
37_ac2	-1373.799485	-1373.293287	-1370.710784	-86.94	-1370.237702	
59_ac2	-1373.798952	-1373.292792	-1370.708311	-89.33	-1370.236174	
51_ac2	-1373.804967	-1373.299183	-1370.707094	-93.43	-1370.236895	
31_ac2	-1373.804481	-1373.298651	-1370.706600	-93.30	-1370.236307	
22_ac2	-1373.801711	-1373.295922	-1370.703764	-89.66	-1370.232126	
PhePh$_2$OH						-1565.165014
19	-1569.423153	-1568.812959	-1565.785580	24.43	-1565.166079	
2	-1569.423897	-1568.813823	-1565.784863	24.06	-1565.165626	
14	-1569.416471	-1568.806523	-1565.781972	20.54	-1565.164200	
20	-1569.418389	-1568.808881	-1565.782299	23.10	-1565.163995	
6	-1569.417069	-1568.807055	-1565.781524	19.79	-1565.163972	
1	-1569.422489	-1568.812587	-1565.782267	22.55	-1565.163775	
9	-1569.419183	-1568.809427	-1565.783086	26.36	-1565.163290	

Appendix

Conformer	E_{tot} B98/6-31G(d)	H_{298} B98/6-31G(d)	E_{tot} MP2(FC)/6-31+G(2d,p)	G_{solv}, kJ/mol	H_{298} MP2-5 with solv	$<H_{298}>$ MP2-5 with solv
58	-1569.415513	-1568.805376	-1565.783279	26.02	-1565.163230	
21	-1569.415423	-1568.805615	-1565.783112	26.48	-1565.163216	
3	-1569.419540	-1568.809048	-1565.782558	23.85	-1565.162982	
97	-1569.415502	-1568.806904	-1565.779452	20.79	-1565.162934	
85	-1569.415858	-1568.805537	-1565.782035	24.31	-1565.162454	
46	-1569.415858	-1568.805542	-1565.782039	24.43	-1565.162415	
8	-1569.415009	-1568.805399	-1565.778183	16.32	-1565.162358	
10	-1569.416677	-1568.806679	-1565.782265	26.19	-1565.162292	
7	-1569.419202	-1568.809151	-1565.781317	23.77	-1565.162214	
31	-1569.417408	-1568.807781	-1565.781665	26.65	-1565.161887	
100	-1569.416529	-1568.806752	-1565.779738	21.63	-1565.161722	
17	-1569.415427	-1568.805379	-1565.781350	25.40	-1565.161628	
43	-1569.415961	-1568.806438	-1565.779546	22.05	-1565.161625	
16	-1569.418117	-1568.808197	-1565.780256	22.93	-1565.161603	
18	-1569.413387	-1568.803047	-1565.781806	26.44	-1565.161394	
37	-1569.415739	-1568.805843	-1565.778898	20.13	-1565.161337	
48	-1569.417116	-1568.807750	-1565.779759	24.73	-1565.160975	
13	-1569.415204	-1568.805095	-1565.780083	23.81	-1565.160906	
77	-1569.414553	-1568.805121	-1565.778024	20.25	-1565.160879	
45	-1569.414227	-1568.804541	-1565.779718	25.19	-1565.160438	
4	-1569.418489	-1568.808299	-1565.778605	21.46	-1565.160240	
34	-1569.416072	-1568.806141	-1565.777485	19.54	-1565.160112	
36	-1569.414370	-1568.804388	-1565.778009	21.67	-1565.159772	
49	-1569.414562	-1568.804852	-1565.777111	20.59	-1565.159561	
15	-1569.414599	-1568.804989	-1565.778290	24.27	-1565.159438	
88	-1569.413605	-1568.803945	-1565.778396	24.64	-1565.159349	
40	-1569.414422	-1568.804928	-1565.777128	22.34	-1565.159124	
22	-1569.414532	-1568.804863	-1565.777494	22.84	-1565.159124	
53	-1569.416095	-1568.805682	-1565.777505	22.09	-1565.158677	
30	-1569.415026	-1568.805431	-1565.776143	22.18	-1565.158102	
57	-1569.414879	-1568.805198	-1565.773722	15.94	-1565.157970	
52	-1569.415081	-1568.805224	-1565.775730	20.96	-1565.157889	
26	-1569.414921	-1568.805025	-1565.775119	19.62	-1565.157749	
86	-1569.414174	-1568.804561	-1565.775785	23.14	-1565.157359	
51	-1569.411256	-1568.801253	-1565.776018	22.97	-1565.157266	
35	-1569.414227	-1568.804355	-1565.774939	20.54	-1565.157242	
32	-1569.414266	-1568.804348	-1565.775168	21.25	-1565.157154	
134	-1569.414613	-1568.805000	-1565.771280	13.85	-1565.156392	
PhePh$_2$OH ac						-1717.808742
58_ac2	-1722.412842	-1721.747866	-1718.448540	-67.82	-1717.809397	
43_ac1	-1722.416753	-1721.752659	-1718.448210	-66.11	-1717.809296	
20_ac1	-1722.418811	-1721.753840	-1718.449764	-64.31	-1717.809286	
9_ac1	-1722.417503	-1721.752879	-1718.449044	-64.94	-1717.809153	
15_ac1	-1722.417503	-1721.752879	-1718.449043	-64.94	-1717.809152	
58_ac1	-1722.413553	-1721.748503	-1718.449947	-61.92	-1717.808482	
9_ac2	-1722.412895	-1721.748697	-1718.444820	-72.26	-1717.808143	
22_ac1	-1722.412477	-1721.748113	-1718.443559	-75.98	-1717.808134	
97_ac2	-1722.414067	-1721.749576	-1718.443841	-74.14	-1717.807589	
43_ac2	-1722.412250	-1721.748169	-1718.443646	-73.30	-1717.807485	
37_ac1	-1722.414644	-1721.750062	-1718.443751	-74.01	-1717.807361	
2_ac1	-1722.414643	-1721.750051	-1718.443759	-73.97	-1717.807342	
3_ac1	-1722.414643	-1721.750050	-1718.443759	-73.97	-1717.807340	
97_ac1	-1722.415918	-1721.751538	-1718.445489	-67.66	-1717.806878	
40_ac1	-1722.412980	-1721.748336	-1718.444030	-72.13	-1717.806860	
19_ac2	-1722.410357	-1721.745861	-1718.441628	-77.70	-1717.806725	
20_ac2	-1722.412476	-1721.748127	-1718.443336	-72.09	-1717.806445	

Appendix

Conformer	E_{tot} B98/6-31G(d)	H_{298} B98/6-31G(d)	E_{tot} MP2(FC)/6-31+G(2d,p)	G_{solv}, kJ/mol	H_{298} MP2-5 with solv	$<H_{298}>$ MP2-5 with solv
3_ac2	-1722.410508	-1721.745920	-1718.440033	-80.79	-1717.806218	
2_ac2	-1722.410509	-1721.745890	-1718.440057	-80.75	-1717.806195	
48_ac2	-1722.412180	-1721.748294	-1718.440870	-76.02	-1717.805940	
36_ac1	-1722.415612	-1721.750496	-1718.446171	-65.14	-1717.805867	
19_ac1	-1722.414595	-1721.749742	-1718.442483	-73.64	-1717.805677	
31_ac2	-1722.411658	-1721.747351	-1718.440413	-75.06	-1717.804695	
77_ac1	-1722.413810	-1721.749176	-1718.442759	-69.62	-1717.804643	
77_ac2	-1722.410627	-1721.746258	-1718.439712	-76.53	-1717.804490	
1_ac2	-1722.410358	-1721.745635	-1718.437952	-81.80	-1717.804384	
49_ac1	-1722.413276	-1721.748874	-1718.442319	-68.87	-1717.804148	
49_ac2	-1722.410288	-1721.745708	-1718.439636	-76.15	-1717.804060	
17_ac2	-1722.409472	-1721.745321	-1718.440275	-72.97	-1717.803917	
15_ac2	-1722.408755	-1721.744440	-1718.440211	-72.93	-1717.803673	
48_ac1	-1722.413612	-1721.749089	-1718.441671	-68.45	-1717.803219	
36_ac2	-1722.409252	-1721.744772	-1718.439783	-73.05	-1717.803128	
53_ac2	-1722.408913	-1721.745778	-1718.437687	-74.64	-1717.802981	
16_ac2	-1722.408927	-1721.744767	-1718.438382	-74.68	-1717.802668	
31_ac1	-1722.413027	-1721.748471	-1718.441145	-67.99	-1717.802485	
cat11ur1						-1067.475665
3	-1070.423467	-1069.982556	-1067.915604	-3.64	-1067.476079	
1	-1070.423386	-1069.982330	-1067.915845	-2.55	-1067.475761	
2	-1070.413566	-1069.973034	-1067.914926	-1.46	-1067.474951	
5	-1070.413944	-1069.973160	-1067.914381	-2.80	-1067.474665	
17	-1070.416647	-1069.976280	-1067.909665	-2.85	-1067.470382	
38	-1070.415144	-1069.974447	-1067.907180	-9.20	-1067.469989	
9	-1070.415997	-1069.975389	-1067.909126	-2.38	-1067.469426	
57	-1070.414694	-1069.974084	-1067.906181	-9.67	-1067.469252	
11	-1070.415461	-1069.974210	-1067.908275	-5.06	-1067.468951	
12	-1070.417298	-1069.976682	-1067.908213	-3.10	-1067.468776	
26	-1070.414924	-1069.974572	-1067.907035	-3.56	-1067.468037	
13	-1070.416485	-1069.975755	-1067.907239	-3.77	-1067.467943	
15	-1070.414110	-1069.973283	-1067.906784	-4.44	-1067.467646	
19	-1070.407328	-1069.966771	-1067.907064	-2.85	-1067.467590	
22	-1070.407855	-1069.967175	-1067.907018	-2.55	-1067.467310	
23	-1070.414363	-1069.973434	-1067.904675	-1.63	-1067.464368	
cat11ur1_ac						-1220.122465
2_ac2	-1223.412017	-1222.916634	-1220.580324	-99.45	-1220.122821	
2_ac1	-1223.411551	-1222.916025	-1220.578576	-97.19	-1220.120070	
5_ac1	-1223.411550	-1222.916101	-1220.579484	-101.55	-1220.122711	
5_ac2	-1223.411260	-1222.916138	-1220.578013	-99.54	-1220.120803	
22_ac1	-1223.409480	-1222.914113	-1220.577280	-99.04	-1220.119633	
19_ac2	-1223.408020	-1222.912673	-1220.575804	-100.75	-1220.118831	
12_ac2	-1223.414189	-1222.918990	-1220.574269	-102.55	-1220.118130	
13_ac2	-1223.412727	-1222.917350	-1220.572754	-104.60	-1220.117217	
1_ac1	-1223.412252	-1222.917001	-1220.570761	-109.66	-1220.117133	
3_ac2	-1223.411813	-1222.916458	-1220.569769	-111.25	-1220.116788	
12_ac1	-1223.414566	-1222.919395	-1220.574891	-96.52	-1220.116484	
22_ac2	-1223.408839	-1222.913493	-1220.574789	-96.11	-1220.116049	
38_ac2	-1223.413704	-1222.918356	-1220.571036	-105.90	-1220.116022	
38_ac1	-1223.416237	-1222.920590	-1220.573403	-100.08	-1220.115875	
13_ac1	-1223.413208	-1222.917813	-1220.573392	-98.53	-1220.115526	
3_ac1	-1223.412436	-1222.916955	-1220.570621	-105.44	-1220.115299	
1_ac2	-1223.412895	-1222.917233	-1220.571444	-103.55	-1220.115223	
57_ac2	-1223.412333	-1222.917090	-1220.569117	-107.65	-1220.114878	
57_ac1	-1223.414728	-1222.919303	-1220.571446	-102.01	-1220.114873	
19_ac1	-1223.407283	-1222.911520	-1220.573152	-98.32	-1220.114838	

Appendix

Conformer	E_{tot} B98/6-31G(d)	H_{298} B98/6-31G(d)	E_{tot} MP2(FC)/6-31+G(2d,p)	G_{solv}, kJ/mol	H_{298} MP2-5 with solv	$<H_{298}>$ MP2-5 with solv
11_ac2	-1223.406891	-1222.911274	-1220.564930	-106.61	-1220.109918	
11_ac1	-1223.407080	-1222.911554	-1220.565216	-101.38	-1220.108304	
cat11ur2						-1106.639006
52	-1109.710389	-1109.239257	-1107.111618	2.22	-1106.639641	
10	-1109.717647	-1109.246583	-1107.107426	-1.80	-1106.637047	
8	-1109.718429	-1109.247514	-1107.107708	-0.59	-1106.637016	
5	-1109.717447	-1109.246311	-1107.107703	-0.88	-1106.636901	
3	-1109.718298	-1109.247222	-1107.108049	0.92	-1106.636623	
7	-1109.717219	-1109.246398	-1107.105779	-3.85	-1106.636424	
1	-1109.717086	-1109.246146	-1107.105985	-3.10	-1106.636224	
85	-1109.709819	-1109.238618	-1107.102247	-3.01	-1106.632194	
42	-1109.711503	-1109.240266	-1107.100650	-7.07	-1106.632107	
59	-1109.709999	-1109.238952	-1107.101729	-3.22	-1106.631909	
101	-1109.711023	-1109.240059	-1107.099840	-7.49	-1106.631729	
31	-1109.710334	-1109.239324	-1107.099084	-8.28	-1106.631229	
57	-1109.711907	-1109.241168	-1107.102041	0.21	-1106.631222	
74	-1109.709949	-1109.239033	-1107.098232	-9.16	-1106.630806	
29	-1109.711185	-1109.240329	-1107.101468	0.38	-1106.630469	
18	-1109.710633	-1109.239361	-1107.100562	-3.10	-1106.630469	
32	-1109.710804	-1109.240026	-1107.100085	-3.01	-1106.630454	
35	-1109.711626	-1109.240802	-1107.100538	-1.38	-1106.630239	
15	-1109.710131	-1109.239423	-1107.099789	-2.55	-1106.630053	
14	-1109.709384	-1109.238209	-1107.098476	-6.15	-1106.629644	
45	-1109.710895	-1109.239909	-1107.099730	-1.76	-1106.629413	
26	-1109.711765	-1109.240611	-1107.099831	0.08	-1106.628645	
23	-1109.707964	-1109.237036	-1107.096897	-5.52	-1106.628072	
16	-1109.709719	-1109.239109	-1107.096836	-3.51	-1106.627564	
48	-1109.707274	-1109.236131	-1107.097913	-1.97	-1106.627519	
28	-1109.707629	-1109.236407	-1107.097437	-2.68	-1106.627234	
97	-1109.707638	-1109.236456	-1107.096719	0.25	-1106.625441	
71	-1109.709240	-1109.237952	-1107.097069	2.43	-1106.624856	
38	-1109.708067	-1109.237225	-1107.094809	-1.92	-1106.624700	
cat11ur2_ac						-1259.285088
52_ac2	-1262.705793	-1262.180543	-1259.773476	-98.53	-1259.285755	
52_ac1	-1262.706096	-1262.180578	-1259.773460	-95.35	-1259.284260	
10_ac1	-1262.712980	-1262.187371	-1259.770484	-100.50	-1259.283154	
7_ac1	-1262.712980	-1262.187369	-1259.770481	-100.50	-1259.283149	
5_ac1	-1262.713320	-1262.187607	-1259.771098	-98.91	-1259.283058	
5_ac2	-1262.709628	-1262.184080	-1259.767444	-103.30	-1259.281242	
10_ac2	-1262.710557	-1262.185047	-1259.765669	-106.78	-1259.280828	
7_ac2	-1262.710557	-1262.185048	-1259.765666	-106.78	-1259.280826	
1_ac2	-1262.711127	-1262.185170	-1259.766543	-105.39	-1259.280729	
1_ac1	-1262.712327	-1262.186681	-1259.767835	-98.70	-1259.279782	
8_ac2	-1262.710391	-1262.184538	-1259.764927	-106.36	-1259.279584	
42_ac2	-1262.712256	-1262.186257	-1259.767037	-101.17	-1259.279571	
42_ac1	-1262.714522	-1262.188303	-1259.769087	-95.19	-1259.279123	
8_ac1	-1262.710282	-1262.184508	-1259.764851	-102.17	-1259.277993	
85_ac2	-1262.705228	-1262.179640	-1259.763164	-106.02	-1259.277959	
59_ac2	-1262.705223	-1262.179431	-1259.762961	-104.93	-1259.277136	
85_ac1	-1262.705692	-1262.179705	-1259.765714	-96.61	-1259.276523	
59_ac1	-1262.705192	-1262.179645	-1259.762836	-99.96	-1259.275360	
3_ac1	-1262.710695	-1262.184676	-1259.765592			
3_ac2	-1262.710927	-1262.184897	-1259.765808			

Appendix

Conformer	E_{tot} B98/6-31G(d)	H_{298} B98/6-31G(d)	E_{tot} MP2(FC)/6-31+G(2d,p)	G_{solv}, kJ/mol	H_{298} MP2-5 with solv	$<H_{298}>$ MP2-5 with solv
cat11ur3						-1740.218663
3	-1744.284109	-1743.826345	-1740.680601	10.38	-1740.218884	
1	-1744.283927	-1743.825878	-1740.680928	11.38	-1740.218544	
2	-1744.275711	-1743.818183	-1740.681784	15.10	-1740.218503	
22	-1744.269732	-1743.812360	-1740.677057	15.23	-1740.213884	
19	-1744.270097	-1743.813337	-1740.674869	14.31	-1740.212659	
cat11ur3_ac						-1892.865628
2_ac2	-1897.268588	-1896.756942	-1893.345766	-82.93	-1892.865705	
5_ac1	-1897.265973	-1896.75427	-1893.341298	-84.35	-1892.861722	
22_ac2	-1897.265161	-1896.752881	-1893.340591	-82.47	-1892.859721	
2_ac1	-1897.266552	-1896.754381	-1893.340332	-82.09	-1892.859427	
1_ac2	-1897.265901	-1896.753773	-1893.328081	-104.31	-1892.855681	
3_ac2	-1897.265527	-1896.753306	-1893.327164	-106.36	-1892.855452	
19_ac1	-1897.261455	-1896.749254	-1893.333955	-87.91	-1892.855235	
19_ac2	-1897.260641	-1896.749487	-1893.333560	-85.40	-1892.854931	
3_ac1	-1897.265260	-1896.753040	-1893.327202	-100.79	-1892.853372	
1_ac1	-1897.265518	-1896.753272	-1893.327986	-98.49	-1892.853254	
22_ac1	-1897.262560	-1896.750403	-1893.334073	-81.50	-1892.852959	
cat11ur4						-2062.790621
2	-2067.200675	-2066.744915	-2063.253991	19.62	-2062.790756	
1	-2067.201781	-2066.746290	-2063.247853	15.82	-2062.786338	
3	-2067.202084	-2066.746712	-2063.247805	13.10	-2062.787446	
22	-2067.188412	-2066.732989	-2063.242722	18.33	-2062.780319	
19	-2067.188454	-2066.733163	-2063.240309	17.82	-2062.778229	
12	-2067.190116	-2066.735290	-2063.236557	14.27	-2062.776297	
17	-2067.191850	-2066.736639	-2063.239438	13.85	-2062.778953	
cat11ur4_ac						-2215.436046
5_ac1	-2220.189289	-2219.679729	-2215.915665	-79.29	-2215.436304	
2_ac2	-2220.190093	-2219.680114	-2215.915665	-78.03	-2215.435406	
5_ac2	-2220.187042	-2219.676993	-2215.910507	-77.15	-2215.429844	
2_ac1	-2220.187278	-2219.677516	-2215.909551	-77.40	-2215.429270	
22_ac2	-2220.183935	-2219.673952	-2215.904140	-79.20	-2215.424323	
19_ac1	-2220.182435	-2219.67242	-2215.902864	-81.42	-2215.423860	
3_ac2	-2220.181559	-2219.671945	-2215.894151	-103.09	-2215.423803	
1_ac2	-2220.182036	-2219.672315	-2215.894908	-98.99	-2215.422892	
22_ac1	-2220.183542	-2219.673645	-2215.903479	-74.77	-2215.422059	
19_ac2	-2220.182249	-2219.672149	-2215.902581	-76.61	-2215.421660	
1_ac1	-2220.182068	-2219.672216	-2215.895351	-92.17	-2215.420606	
Precursors						
precat81						-513.380991
1	-514.826985	-514.599147	-513.601666	-19.50	-513.381253	
2	-514.824791	-514.597088	-513.600142	-20.17	-513.380120	
precat11						-668.880735
3	-670.822318	-670.497593	-669.200633	-14.18	-668.881310	
2	-670.821772	-670.496960	-669.199833	-14.73	-668.880631	
5	-670.819726	-670.494870	-669.199371	-14.43	-668.880012	
1	-670.817937	-670.493030	-669.198858	-14.69	-668.879544	
4	-670.817554	-670.492712	-669.198311	-14.85	-668.879126	
Isocyanates						
PhNCO						-398.567395
phnco	-399.571317	-399.459352	-398.678660	-1.84	-398.567395	
BnNCO						-437.726064
1	-438.861524	-438.719228	-437.866591	-4.23	-437.725905	
2	-438.861088	-438.719114	-437.865409	-5.40	-437.725491	
4	-438.861152	-438.719040	-437.866837	-4.35	-437.726383	

Appendix

Conformer	E_{tot} B98/6-31G(d)	H_{298} B98/6-31G(d)	E_{tot} MP2(FC)/6-31+G(2d,p)	G_{solv}, kJ/mol	H_{298} MP2-5 with solv	$<H_{298}>$ MP2-5 with solv
PhNCS						-721.145101
phncs	-722.503732	-722.393824	-721.254403	-1.59	-721.145101	
BnNCS						-760.302848
1	-761.7932982	-761.653183	-760.4413391	-4.56	-760.302961	
2	-761.7930038	-761.653085	-760.4400484	-5.98	-760.302408	
4	-761.7932963	-761.653176	-760.4413452	-4.60	-760.302978	
3,5-(CF$_3$)$_2$PhNCO						-1071.307075
cf3phnco	-1073.428426	-1073.299579	-1071.439794	10.17	-1071.307075	
PhCH$_3$CHNCO						-476.893811
ur_3	-478.162177	-477.990559	-477.065281	-0.50	-476.893854	
ur_1	-478.162184	-477.990577	-477.065258	-0.42	-476.893810	
ur_2	-478.162111	-477.990515	-477.064804	-1.46	-476.893766	
3,5-(CF$_3$)$_2$PhNCS						-1393.883496
cf3phncs	-1396.359623	-1396.232920	-1394.014757	11.97	-1393.883496	

Table A4.2 Acetylation enthalpies for 3-(thio)urea-4-aminopyridines, as calculated at MP2/6-31+G(2d,p)//B98/6-31G(d) level with inclusion of solvent effects in chloroform at PCM/UAHF/RHF/6-31G(d) level.

Catalyst	Stability [a] [kJ/mol]	ΔH_{ac} (MP2-5/solv) [kJ/mol] best conf	ΔH_{ac} (MP2-5/solv) [kJ/mol] averaged
py	-	0.0	0.0
cat81ur1	78.4	-64.2	-64.3
cat81ur2	59.3	-62.6	-63.2
cat81ur1f	88.6	-50.9	-51.2
cat81ur3	85.3	-64.2	-64.3
cat81ur4	77.2	-57.7	-57.9
cat81ur5	80.1	-59.8	-59.1
cat81ur6		-88.3	-88.9
PheOMe		-70.8	-70.6
PhePh$_2$OH		-53.9	-55.0
cat11ur1	72.3	-62.9	-63.1
cat11ur2	84.6	-61.3	-61.2
cat11ur3	81.0	-63.1	-63.5
cat11ur4	69.3	-59.8	-59.5

[a] Enthalpies for the reaction of (thio)ureas formation schown in Scheme 4.2.

Appendix

Table A4.3. Comparison of non-stacked and stacked conformers energies.

				free catalyst			acylated catalyst		
	X	R^1, R^2	R^3	non-stacked	stacked	$\Delta\Delta E^a$ kJ/mol	non-stacked	stacked	$\Delta\Delta E^a$ kJ/mol
cat8lur1	O	H	Ph	1a2	5c1	20.6	5a1.ac1	5c1.ac2	28.8
cat8lur2	S	H	Ph	2c2	5c1	16.4	1a1.ac2	5c1.ac2	20.8
cat8lur1f	O	H	3,5-$(CF_3)_2C_6H_3$	2c2	5c1	30.6	5a1.ac1	5c1.ac2	39.8
cat8lur3	O	H	CH_2Ph	13	66	28.0	59.ac2	66.ac1	31.7
cat8lur4	S	H	CH_2Ph	8	66	29.2	59.ac2	66.ac1	31.3
cat8lur5	O	H	(S)-PhMeCH	13	51	33.3	8.ac1	51.ac2	29.2
cat1lur1	O	$(-CH_2-)_4$	Ph	3	2	24.2	38.ac1	2.ac2	29.3
cat1lur2	O	$(-CH_2-)_4$	CH_2Ph	8	52	31.4	42.ac1	52.ac2	34.4
cat1lur3	O	$(-CH_2-)_4$	3,5-$(CF_3)_2C_6H_3$	3	2	25.2	1.ac2	2.ac2	39.4
cat1lur4	S	$(-CH_2-)_4$	3,5-$(CF_3)_2C_6H_3$	3	2	21.0	1.ac1	5.ac1	34.4

[a] $\Delta\Delta E = [E_{stacked}(MP2) - E_{non-stacked}(MP2)] - [E_{stacked}(B98) - E_{non-stacked}(B98)]$

Appendix

Chapter 5. Computational details.

Stationary points (reactant, product and transition state geometries) were optimized and characterized by frequency analysis at the B3LYP/6-31G(d) level of theory. The conformational space of transition states **ts1, ts2, ts2a** and **ts3** for catalyst **PPY**, as well as TS **65** and TS **67** for catalyst **59a**, has initially been studied with the OPLS-AA force field searched using the Monte Carlo conformational search. The conformational space of transition state TS **67** for catalyst **59b** and **59c** has also initially been studied with the OPLS-AA force field. The energetically most favorable conformers identified in this way have subsequently been reoptimized at the B3LYP/6-31G(d) level of theory. In order to save computational cost, the structures of transition states TS **67** for catalysts **59d-g** were initially built based on the best conformers of TS **67** for catalyst **59a** and then reoptimized at the B3LYP/6-31G(d) level of theory. Single point calculations have been performed at the B3LYP/6-311+G(d, p) level of theory, as well as at MP2(FC)/6-311+G(d,p) and MP2(FC)/6-31+G(2d,p) levels, with Gaussian 03.[35] Dispersion corrections to DFT (termed DFT-D) proposed by S. Grimme[76] were used to calculate the accurate dispersion interaction by the ORCA 2.6.4 program package.[81] Thermochemical corrections to free energies (G_{298}) and enthalpies at 298.15 K (H_{298}), as well as at 195.15 K, have been calculated at the same level as that used for geometry optimization.

Appendix

A5.1 Catalytic system with PPY: Theoretical study of the catalytic cycle

Table A5.1. Calculated energies of conformers at B3LYP/6-311+G(d,p)//B3LYP/6-31G(d) level. Δ<H₁₉₅> are relative enthalpies of conformers. Thermal corrections are calculated at 195 K.

Conformer	E_{tot} B3LYP/6-31G(d)	H_{195} B3LYP/6-31G(d)	E_{sp} B3LYP/6-311+G(d,p)	<H₁₉₅> B3LYP/6-311+G(d,p)// B3LYP/6-31G(d)	Δ<H₁₉₅>, kJ/mol	G_{195} B3LYP/6-31G(d)	<G₁₉₅> B3LYP/6-311+G(d,p)// B3LYP/6-31G(d)
PPY							
ppy	-459.684287	-459.479522	-459.804214	-459.599449		-459.504836	-459.624763
Alcohol							
60_1	-539.728903	-539.514254	-539.877522	-539.662873	3.95	-539.540592	-539.704416
60_2	-539.729933	-539.515372	-539.878222	-539.663661	1.88	-539.541740	-539.705250
60_3	-539.730441	-539.515885	-539.878932	-539.664376	0.00	-539.542353	-539.706113
60_4	-539.728042	-539.513604	-539.877255	-539.662817	4.09	-539.539999	-539.704453
60_5	-539.726038	-539.511512	-539.875044	-539.660518	10.13	-539.537832	-539.702031
60_6	-539.726776	-539.512339	-539.875370	-539.660933	9.04	-539.538686	-539.702494
Anhydride							
61_1	-538.985402	-538.764844	-539.145797	-538.925239	0.00	-538.795313	-538.973339
61_2	-538.985018	-538.764432	-539.145246	-538.924660	1.52	-538.795084	-538.972941
61_3	-538.984625	-538.764027	-539.144695	-538.924097	3.00	-538.795313	-538.970977
61_4	-538.984149	-538.763479	-539.144666	-538.923996	3.26	-538.794665	-538.972884
61_5	-538.983751	-538.763060	-539.144074	-538.923383	4.87	-538.794556	-538.971155
61_6	-538.983219	-538.762470	-539.143629	-538.922880	6.19	-538.793787	-538.972171
61_7	-538.984043	-538.763099	-539.143768	-538.922824	6.34	-538.793752	-538.967261
61_8	-538.983872	-538.763103	-539.143592	-538.922823	6.34	-538.793360	-538.973339
61_9	-538.982916	-538.762111	-539.143492	-538.922687	6.70	-538.793643	-538.971826
Reactant complex							-538.792854	-538.972405
13	-1538.425645	-1537.781720	-1538.846119	-1538.202203		-1537.842122	-1538.264297
23	-1538.425636	-1537.781719	-1538.846106	-1538.202180	0.06	-1537.841978	-1538.262439
235	-1538.424856	-1537.781720	-1538.846119	-1538.202203	0.00	-1537.842122	-1538.262605
134	-1538.424149	-1537.780902	-1538.845392	-1538.201438	2.01	-1537.842023	-1538.262559
232	-1538.424472	-1537.780405	-1538.845012	-1538.201268	2.45	-1537.841920	-1538.262783
33	-1538.422021	-1537.780522	-1538.844844	-1538.200894	3.44	-1537.841092	-1538.261464
132	-1538.424458	-1537.778349	-1538.844835	-1538.201164	2.73	-1537.841482	-1538.264297
2	-1538.423466	-1537.780524	-1538.844832	-1538.200899	3.42	-1537.841055	-1538.261430
1	-1538.423449	-1537.779514	-1538.843691	-1538.199740	6.47	-1537.839891	-1538.260117
135	-1538.424835	-1537.779549	-1538.843682	-1538.199783	6.35	-1537.840255	-1538.260489
		-1537.780089	-1538.844181	-1538.199435	7.27	-1537.839411	-1538.258757

256

Appendix

Conformer	E_{tot} B3LYP/6-31G(d)	H_{195} B3LYP/6-31G(d)	E_{sp} B3LYP/6-311+G(d,p)	<H_{195}> B3LYP/6-311+G(d,p)// B3LYP/6-31G(d)	Δ<H_{195}>, kJ/mol	G_{195} B3LYP/6-31G(d)	<G_{195}> B3LYP/6-311+G(d,p)// B3LYP/6-31G(d)
PPY-alcohol complex							
13	-999.430826	-999.009492	-999.696570	-999.275188	0.03	-999.051233	-999.316967
23	-999.430874	-999.009442	-999.696560	-999.275176	0.00	-999.051233	-999.316967
23	-999.430874	-999.009492	-999.696570	-999.275188	0.00	-999.051051	-999.316747
33	-999.430874	-999.004900	-999.696569	-999.270596	12.06	-999.051046	-999.316742
ppyts1							
		-1537.758945	-1538.823475	-1538.180005		-1537.814767	-1538.236604
8	-1538.402436	-1537.758945	-1538.822901	-1538.179410	1.56	-1537.813978	-1538.234443
32	-1538.402282	-1537.758813	-1538.823475	-1538.180005	0.00	-1537.814209	-1538.235401
11	-1538.401923	-1537.758431	-1538.822131	-1538.178640	3.59	-1537.814433	-1538.234642
2	-1538.401839	-1537.758504	-1538.822955	-1538.179620	1.01	-1537.814410	-1538.235526
7	-1538.401719	-1537.758301	-1538.822784	-1538.179366	1.68	-1537.814022	-1538.235087
10	-1538.401077	-1537.757806	-1538.822708	-1538.179437	1.49	-1537.814157	-1538.235788
12	-1538.400964	-1537.757686	-1538.822655	-1538.179377	1.65	-1537.814451	-1538.236142
3	-1538.400757	-1537.757530	-1538.823026	-1538.179799	0.54	-1537.813801	-1538.236070
6	-1538.400391	-1537.757296	-1538.822116	-1538.179021	2.58	-1537.814767	-1538.236492
1	-1538.399867	-1537.756632	-1538.822115	-1538.178880	2.96	-1537.813808	-1538.236056
5	-1538.399828	-1537.756632	-1538.821961	-1538.178765	3.25	-1537.814471	-1538.236604
4	-1538.399091	-1537.755895	-1538.820647	-1538.177451	6.71	-1537.813117	-1538.234673
9	-1538.399040	-1537.755732	-1538.820985	-1538.177677	6.11	-1537.813249	-1538.235194
16	-1538.398439	-1537.755014	-1538.820323	-1538.176898	8.16	-1537.811676	-1538.233560
pyac (intermediate)							
		-1537.764200	-1538.831355	-1538.186977		-1537.821321	-1538.244308
234	-1538.408578	-1537.764200	-1538.831355	-1538.186977	0.00	-1537.821321	-1538.244098
232	-1538.408170	-1537.763830	-1538.830601	-1538.186261	1.88	-1537.820793	-1538.243224
233	-1538.408066	-1537.763509	-1538.831341	-1538.186784	0.51	-1537.820814	-1538.244089
23	-1538.407590	-1537.763011	-1538.830466	-1538.185887	2.86	-1537.820505	-1538.243381
334	-1538.407449	-1537.762839	-1538.830790	-1538.186179	2.09	-1537.819573	-1538.242913
1	-1538.406247	-1537.761927	-1538.829991	-1538.185671	3.43	-1537.818328	-1538.242072
335	-1538.406194	-1537.761759	-1538.830125	-1538.185690	3.38	-1537.819494	-1538.243425
336	-1538.406146	-1537.761824	-1538.829849	-1538.185528	3.80	-1537.820604	-1538.244308
4	-1538.406105	-1537.761749	-1538.829472	-1538.185116	4.89	-1537.818329	-1538.241696
12	-1538.405981	-1537.761609	-1538.829076	-1538.184704	5.97	-1537.818934	-1538.242029
3	-1538.405921	-1537.761503	-1538.829547	-1538.185128	4.85	-1537.817685	-1538.241310
2	-1538.405829	-1537.761401	-1538.829159	-1538.184731	5.90	-1537.818230	-1538.241560

Appendix

Conformer	E_{tot} B3LYP/6-31G(d)	H_{195} B3LYP/6-31G(d)	E_{sp} B3LYP/6-311+G(d,p)	$\langle H_{195}\rangle$ B3LYP/6-311+G(d,p)// B3LYP/6-31G(d)	$\Delta\langle H_{195}\rangle$, kJ/mol	G_{195} B3LYP/6-31G(d)	$\langle G_{195}\rangle$ B3LYP/6-311+G(d,p)// B3LYP/6-31G(d)
ppyts3							
1	-1538.401443	-1537.762222	-1538.824112	-1538.184891		-1537.817509	-1538.240112
2	-1538.401442	-1537.762160	-1538.824080	-1538.184797	0.25	-1537.816992	-1538.239629
3	-1538.400980	-1537.762222	-1538.824112	-1538.184891	0.00	-1537.817163	-1538.239832
4	-1538.401385	-1537.761718	-1538.823863	-1538.184601	0.76	-1537.817229	-1538.240112
5	-1538.401370	-1537.762135	-1538.824075	-1538.184825	0.17	-1537.817272	-1538.239962
6	-1538.400958	-1537.762087	-1538.824025	-1538.184742	0.39	-1537.817229	-1538.239884
7	-1538.401226	-1537.761688	-1538.823837	-1538.184567	0.85	-1537.816817	-1538.239696
8	-1538.401266	-1537.762169	-1538.822320	-1538.183262	4.28	-1537.817509	-1538.238602
9	-1538.401218	-1537.761965	-1538.822600	-1538.183299	4.18	-1537.816534	-1538.237868
10	-1538.400820	-1537.762045	-1538.822547	-1538.183373	3.99	-1537.816263	-1538.237591
11	-1538.400972	-1537.761403	-1538.822428	-1538.183011	4.94	-1537.815165	-1538.236773
12	-1538.401393	-1537.761533	-1538.822448	-1538.183009	4.94	-1537.815170	-1538.236646
44	-1538.398097	-1537.762201	-1538.822541	-1538.183349	4.05	-1537.816060	-1538.237208
84	-1538.398202	-1537.759134	-1538.819984	-1538.181021	10.16	-1537.814897	-1538.236784
		-1537.759121	-1538.820059	-1538.180978	10.28	-1537.814378	-1538.236235
ppyts2							
6	-1538.392513	-1537.751287	-1538.810758	-1538.169532	0.00	-1537.805961	-1538.224206
		-1537.751287	-1538.810758	-1538.169532	0.00	-1537.805961	-1538.224206
63	-1538.391968	-1537.750137	-1538.809406	-1538.167575	5.14	-1537.804522	-1538.221960
62	-1538.391694	-1537.749600	-1538.809343	-1538.167249	6.00	-1537.803758	-1538.221407
2	-1538.388341	-1537.746801	-1538.806251	-1538.164710	12.66	-1537.801309	-1538.219218
7	-1538.387212	-1537.745682	-1538.805195	-1538.163664	15.41	-1537.801006	-1538.218988
ppyts2a		-1537.750715	-1538.808332	-1538.167109		-1537.805343	-1538.221774
31	-1538.391266	-1537.750715	-1538.807660	-1538.167109	0.00	-1537.805004	-1538.221398
21	-1538.391283	-1537.750665	-1538.807671	-1538.167053	0.15	-1537.804780	-1538.221168
11	-1538.391849	-1537.750419	-1538.808280	-1538.166850	0.68	-1537.805343	-1538.221774
4	-1538.391895	-1537.750399	-1538.808332	-1538.166836	0.72	-1537.804623	-1538.221060
2	-1538.390707	-1537.750054	-1538.806894	-1538.166241	2.28	-1537.803952	-1538.220139
3	-1538.390714	-1537.750048	-1538.806902	-1538.166237	2.29	-1537.804050	-1538.220239
1	-1538.389971	-1537.748491	-1538.806758	-1538.165278	4.81	-1537.803112	-1538.219899
41	-1538.389997	-1537.748451	-1538.806801	-1538.165256	4.87	-1537.802651	-1538.219456
Product complex		-1537.809239	-1538.873790	-1538.229349		-1537.871751	-1538.292858
1	-1538.452983	-1537.808702	-1538.873365	-1538.229084	0.70	-1537.869808	-1538.290190
2	-1538.453003	-1537.808695	-1538.873410	-1538.229102	0.65	-1537.869744	-1538.290151
12	-1538.453455	-1537.809239	-1538.873139	-1538.228923	1.12	-1537.870527	-1538.290211

Appendix

Conformer	E_{tot} B3LYP/6-31G(d)	H_{195} B3LYP/6-31G(d)	E_{sp} B3LYP/6-311+G(d,p)	<H_{195}> B3LYP/6-311+G(d,p)// B3LYP/6-31G(d)	Δ<H_{195}>, kJ/mol	G_{195} B3LYP/6-31G(d)	<G_{195}> B3LYP/6-311+G(d,p)// B3LYP/6-31G(d)
13	-1538.452683	-1537.808242	-1538.873790	-1538.229349	0.00	-1537.871751	-1538.292858
22	-1538.452807	-1537.808557	-1538.873261	-1538.229010	0.89	-1537.870584	-1538.291037
23	-1538.452683	-1537.808264	-1538.873768	-1538.229349	0.00	-1537.870892	-1538.291977
15	-1538.452525	-1537.808236	-1538.872644	-1538.228354	2.61	-1537.869402	-1538.289520

Table A5.2. Calculated energies of conformers at MP2/**//B3LYP/6-31G(d) level. Δ<H_{195}> are relative enthalpies of conformers at MP2/6-311+G(d,p)//B3LYP/6-31G(d) level. Thermal corrections are calculated at 195 K.

Confo rmer	E_{tot} B3LYP/6-31G(d)	H_{195} B3LYP/6-31G(d)	E_{sp} MP2/6-311+G(d,p)	<H_{195}> MP2/6-311+G(d,p)// B3LYP/6-31G(d)	Δ<H_{195}>, kJ/mol	E_{sp} MP2/6-311+G(2d,p)	<H_{195}> MP2/6-31+G(2d,p)// B3LYP/6-31G(d)
PPY							
ppy	-459.684287	-459.479522	-458.425670	-458.220905		-458.383981	-458.179217
Alcohol							
60_1	-539.728903	-539.514254	-538.059059	-538.065289	5.55	-538.236463	-538.021814
60_2	-539.729933	-539.515372	-538.060919	-538.067146	0.67	-538.237940	-538.023379
60_3	-539.730441	-539.515885	-538.061183	-538.067401	0.00	-538.238128	-538.023572
60_4	-539.728042	-539.513604	-538.059674	-538.065916	3.90	-538.236418	-538.021980
60_5	-539.726038	-539.511512	-538.057104	-538.063317	10.72	-538.234279	-538.019753
60_6	-539.726776	-539.512339	-538.058306	-538.064542	7.51	-538.234954	-538.020517
Anhydride							
	-538.764844	-538.764844	-537.6324719	-537.4119139		-537.5681654	-537.3476074
61_1	-538.985402	-538.764844	-537.632472	-537.411914	0.00	-537.568165	-537.347607
61_2	-538.985018	-538.764432	-537.631765	-537.411179	1.93	-537.567548	-537.346962
61_3	-538.984625	-538.764027	-537.631038	-537.410440	3.87	-537.566879	-537.346281
61_4	-538.984149	-538.763479	-537.631810	-537.411140	2.03	-537.567527	-537.346857
61_5	-538.983751	-538.763060	-537.631081	-537.410390	4.00	-537.566868	-537.346177
61_6	-538.983219	-538.762470	-537.630639	-537.409890	5.31	-537.567164	-537.346415
61_7	-538.984043	-538.763099	-537.630459	-537.409515	6.30	-537.567082	-537.346138
61_8	-538.983872	-538.763103	-537.629918	-537.409149	7.26	-537.566532	-537.345763
61_9	-538.982916	-538.762111	-537.631130	-537.410325	4.17	-537.566888	-537.346083
Reactant complex							
132	-1538.424458	-1537.780524	-1534.382283	-1533.738349	0.00	-1534.230998	-1533.587065
232	-1538.424472	-1537.780522	-1534.382199	-1533.738249	0.26	-1534.230914	-1533.586964
13	-1538.425645	-1537.781719	-1534.380273	-1533.736348	5.25	-1534.229654	-1533.585728
23	-1538.425636	-1537.781720	-1534.380481	-1533.736565	4.69	-1534.229606	-1533.585690
33	-1538.422021	-1537.778349	-1534.370375	-1533.726704	30.58		

259

Appendix

Confo rmer	E_{tot} B3LYP/6-31G(d)	H_{195} B3LYP/6-31G(d)	E_{ap} MP2/6-311+G(d,p)	$<H_{195}>$ MP2/6-311+G(d,p)// B3LYP/6-31G(d)	$\Delta <H_{195}>$, kJ/mol	E_{ap} MP2/6-31+G(2d,p)	$<H_{195}>$ MP2/6-31+G(2d,p)// B3LYP/6-31G(d)
134	-1538.424149	-1537.780405	-1534.379232	-1533.735488	7.51		
235	-1538.424856	-1537.780902	-1534.378612	-1533.734658	9.69		
2	-1538.423466	-1537.779514	-1534.376493	-1533.732542	15.25		
1	-1538.423449	-1537.779549	-1534.376537	-1533.732638	14.99		
135	-1538.424835	-1537.780089	-1534.376803	-1533.732057	16.52		
ppyts1							
	-1537.758945	-1534.368235	-1533.724744		-1534.220358	-1533.576867	
8	-1538.402436	-1537.758945	-1534.368235	-1533.724744	0.00	-1534.220358	-1533.576867
32	-1538.402282	-1537.758813	-1534.366900	-1533.723430	3.45	-1534.219618	-1533.576149
11	-1538.401923	-1537.758431	-1534.368145	-1533.724653	0.24	-1534.219874	-1533.576383
2	-1538.401839	-1537.758504	-1534.364735	-1533.721400	8.78	-1534.217374	-1533.574039
7	-1538.401719	-1537.758301	-1534.362936	-1533.719518	13.72	-1534.215788	-1533.572369
10	-1538.401077	-1537.757806	-1534.367667	-1533.724397	0.91	-1534.219893	-1533.576623
12	-1538.400964	-1537.757686	-1534.367351	-1533.724073	1.76	-1534.219525	-1533.576247
3	-1538.400757	-1537.757573	-1534.366470	-1533.723243	3.94	-1534.219072	-1533.575845
6	-1538.400391	-1537.757296	-1534.358347	-1533.715253	24.92	-1534.211781	-1533.568686
1	-1538.399867	-1537.756632	-1534.357226	-1533.713991	28.23	-1534.211190	-1533.567955
5	-1538.399828	-1537.756632	-1534.356701	-1533.713505	29.51		
4	-1538.399091	-1537.755895	-1534.358838	-1533.715342	24.68		
9	-1538.399040	-1537.755732	-1534.356742	-1533.713435	29.69		
16	-1538.398439	-1537.755014	-1534.352439	-1533.709014	41.30		
pyac							
	-1537.745027	-1534.371118	-1533.726778		-1534.223054	-1533.578714	
232	-1538.408170	-1537.763830	-1534.371118	-1533.726778	0.00	-1534.223054	-1533.578714
3	-1538.405921	-1537.761503	-1534.369730	-1533.725312	3.85	-1534.221822	-1533.577404
234	-1538.408578	-1537.764200	-1534.369666	-1533.725288	3.91	-1534.221858	-1533.577480
1	-1538.406247	-1537.761927	-1534.369194	-1533.724874	5.00	-1534.221345	-1533.577025
233	-1538.408066	-1537.763509	-1534.365542	-1533.720985	15.21	-1534.218642	-1533.574085
336	-1538.406146	-1537.761824	-1534.360801	-1533.716479	27.04	-1534.214626	-1533.570305
23	-1538.407590	-1537.763011	-1534.366058	-1533.721479	13.91		
334	-1538.407449	-1537.762839	-1534.368132	-1533.723522	8.55		
335	-1538.406194	-1537.761759	-1534.361314	-1533.716879	25.99		
4	-1538.406105	-1537.761749	-1534.367142	-1533.722786	10.48		
12	-1538.405981	-1537.761609	-1534.368711	-1533.724339	6.40		
2	-1538.405829	-1537.761401	-1534.366821	-1533.722393	11.51		
ppyts3							
	-1538.401443	-1537.762160	-1534.366188	-1533.726906	34.59	-1534.228749	-1533.589388
1	-1538.401443	-1537.762160	-1534.366188	-1533.726906	34.59	-1534.228749	-1533.589388

Appendix

Confo-rmer	E_{tot} B3LYP/6-31G(d)	H_{195} B3LYP/6-31G(d)	E_{sp} MP2/6-311+G(d,p)	<H_{195}> MP2/6-311+G(d,p)//B3LYP/6-31G(d)	Δ <H_{195}>, kJ/mol	E_{sp} MP2/6-31+G(2d,p)	<H_{195}> MP2/6-31+G(2d,p)//B3LYP/6-31G(d)
2	-1538.401442	-1537.762222	-1534.366165	-1533.726944	34.49	-1534.217662	-1533.578380
3	-1538.400980	-1537.761718	-1534.366323	-1533.727061	34.18	-1534.217656	-1533.578436
4	-1538.401385	-1537.762135	-1534.366110	-1533.726861	34.71	-1534.217831	-1533.578569
5	-1538.401370	-1537.762087	-1534.366101	-1533.726818	34.82	-1534.217594	-1533.578345
6	-1538.400958	-1537.761688	-1534.366244	-1533.726974	34.41	-1534.217577	-1533.578294
7	-1538.401226	-1537.762169	-1534.379138	-1533.740080	0.00	-1534.217752	-1533.578482
8	-1538.401266	-1537.761965	-1534.378737	-1533.739436	1.69	-1534.228445	-1533.589388
9	-1538.401218	-1537.762045	-1534.378825	-1533.739652	1.12	-1534.228057	-1533.588756
10	-1538.400820	-1537.761403	-1534.379266	-1533.739850	0.61	-1534.228184	-1533.589011
11	-1538.400972	-1537.761533	-1534.379417	-1533.739977	0.27	-1534.228554	-1533.589137
12	-1538.401393	-1537.762201	-1534.379195	-1533.740003	0.20	-1534.217662	-1533.578392
44	-1538.398097	-1537.759134	-1534.359476	-1533.720513	51.37	-1534.211597	-1533.572634
84	-1538.398202	-1537.759121	-1534.359244	-1533.720163	52.29	-1534.211399	-1533.572318
ppyts2		-1537.751287	-1534.354815	-1533.712721		-1534.205887	-1533.563793
6	-1538.392513	-1537.751287	-1534.350811	-1533.709584	8.24	-1534.201711	-1533.560485
63	-1538.391968	-1537.750137	-1534.351797	-1533.709966	7.23	-1534.203534	-1533.561703
62	-1538.391694	-1537.749600	-1534.354815	-1533.712721	0.00	-1534.205887	-1533.563793
2	-1538.388341	-1537.746801	-1534.346644	-1533.705104	20.00	-1534.197724	-1533.556184
7	-1538.387212	-1537.745682	-1534.346754	-1533.705224	19.68	-1534.197785	-1533.556255
ppyts2a		-1537.750715	-1534.350840	-1533.710274		-1534.201916	-1533.561361
31	-1538.391266	-1537.750715	-1534.350825	-1533.710274	0.00	-1534.201913	-1533.561361
21	-1538.391283	-1537.750665	-1534.350840	-1533.710222	0.14	-1534.201916	-1533.561298
11	-1538.391849	-1537.750419	-1534.346851	-1533.705421	12.74	-1534.198954	-1533.557523
4	-1538.391895	-1537.750399	-1534.346854	-1533.705358	12.91	-1534.198942	-1533.557446
2	-1538.390707	-1537.750054	-1534.350309	-1533.709656	1.62		
3	-1538.390714	-1537.750048	-1534.350380	-1533.709714	1.47		
1	-1538.389971	-1537.748491	-1534.344726	-1533.703246	18.45		
41	-1538.389997	-1537.748451	-1534.344762	-1533.703216	18.53		
Product complex			-1534.400615	-1533.756399		-1534.251260	-1533.607044
12	-1538.453455	-1537.809239	-1534.400615	-1533.756399	0.00	-1534.251260	-1533.607044
13	-1538.452683	-1537.808242	-1534.397095	-1533.752654	9.83	-1534.247913	-1533.603473
1	-1538.452983	-1537.808702	-1534.399209	-1533.754928	3.86		
2	-1538.453003	-1537.808695	-1534.399233	-1533.754925	3.87		
22	-1538.452807	-1537.808557	-1534.398675	-1533.754425	5.18		

Appendix

Table A5.3. Calculated energies of conformers at B3LYP-D/6-311+G(d,p)//B3LYP/6-31G(d) level. Thermal corrections are calculated at 195 K.

Compound	Conformer	E_{tot} B3LYP/6-31G(d)	H_{195} B3LYP/6-31G(d)	E_{tot} B3LYP-D/6-311+G(d,p)	Dispersion correction, B3LYP-D/6-311+G(d,p)	$<H_{195}>$ B3LYP-D/6-311+G(d,p)//B3LYP/6-31G(d)	G_{195} B3LYP/6-31G(d)	$<G_{195}>$ B3LYP-D/6-311+G(d,p)//B3LYP/6-31G(d)
PPY	PPY	-459.684287	-459.479522	-459.804214	-0.026174	-459.625624	-459.504836	-459.650938
alc	63	-539.730441	-539.515885	-539.878932	-0.029987	-539.694363	-539.542353	-539.736100
anh	1	-538.985402	-538.764844	-539.145797	-0.027680	-538.952919	-538.795084	-539.000621
react	13	-1538.425645	-1537.781719	-1538.846106	-0.103513	-1538.305693	-1537.841978	-1538.365952
	232	-1538.424472	-1537.780522	-1538.844844	-0.105560	-1538.306454	-1537.841092	-1538.367024
	132	-1538.424458	-1537.780524	-1538.844832	-0.105616	-1538.306515	-1537.841055	-1538.367046
pyac	234	-1538.408578	-1537.764200	-1538.831355	-0.112127	-1538.299104	-1537.821321	-1538.356225
	232	-1538.408170	-1537.763830	-1538.830601	-0.113826	-1538.300087	-1537.820793	-1538.357050
prod	13	-1538.452683	-1537.808242	-1538.873790	-0.096493	-1538.325842	-1537.871751	-1538.389351
ppyts1	32	-1538.402282	-1537.758813	-1538.823475	-0.113301	-1538.293306	-1537.814209	-1538.348702
	8	-1538.402436	-1537.758945	-1538.822901	-0.114581	-1538.293991	-1537.813978	-1538.349024
ppyts2	62	-1538.391694	-1537.749600	-1538.809343	-0.116927	-1538.284175	-1537.803758	-1538.338333
	6	-1538.392513	-1537.751287	-1538.810758	-0.113312	-1538.282844	-1537.805961	-1538.337518
ppyts2a	21	-1538.391283	-1537.750665	-1538.807671	-0.115553	-1538.282606	-1537.804780	-1538.336721
	31	-1538.391266	-1537.750715	-1538.807660	-0.115597	-1538.282706	-1537.805004	-1538.336995
ppyts3	1	-1538.401443	-1537.762160	-1538.824080	-0.114485	-1538.299282	-1537.816992	-1538.354114
	7	-1538.401226	-1537.762169	-1538.822320	-0.122458	-1538.305721	-1537.817509	-1538.361061

Table A5.4. Calculated energies of the best conformers of transition states with alcohol hydrogen H, substituted by deuterium D, at B3LYP /6-311+G(d,p)//B3LYP/6-31G(d) level. Thermal corrections are calculated at 195 K and 298 K.

Compound	Conformer	E_{tot} B3LYP/6-31G(d)	H_{298} B3LYP/6-31G(d)	H_{195} B3LYP/6-31G(d)	G_{298} B3LYP/6-31G(d)	G_{195} B3LYP/6-31G(d)
ppyts3	4	-1538.401385	-1537.745123	-1537.764124	-1537.852451	-1537.819307
ppyts1	32	-1538.402282	-1537.743326	-1537.762333	-1537.851094	-1537.817799
ppyts2	6	-1538.392513	-1537.735408	-1537.754277	-1537.841853	-1537.808989
ppyts2a	21	-1538.391283	-1537.734312	-1537.753187	-1537.839922	-1537.807348
alc	3	-539.730441	-539.512816	-539.519115	-539.561153	-539.545766

A5.2 Catalytic system with Spivey's catalyst: prediction of the selectivity

A5.2.1 Number of conformations needed for the selectivity prediction.

Table A5.5. Numbers of conformers used for Boltzmann-averaging to obtain energy differences $\Delta H(S-R)$ and $\Delta G(S-R)$ of the diastereomers of TS **67** for catalysts **59a-g** from Table 5.7. The same numbers of TSs *(R)*- and *(S)*-**67** conformers were used for averaging.

catalyst	B3LYP/6-311+G(d,p)//B3LYP/6-31G(d)				B3LYP-D/6-311+G(d,p)//B3LYP/6-31G(d)				B3LYP/6-31G(d) level		
	ΔH_{298}	ΔG_{298}	ΔH_{195}	ΔG_{195}	ΔH_{298}	ΔG_{298}	ΔH_{195}	ΔG_{195}	ΔH_{298}	ΔH_{195}	ΔG_{195}
59a	16	16	6	6	9	9	6	6	9	6	6
59b	6	6	5	5	5	5	5	5	6	5	5
59c	8	8	5	5	8	8	5	5	8	5	5
59e	7	7	7	7	5	5	5	5	7	7	7
59d	7	7	7	7	7	7	7	7	7	7	7
59f	5	5	5	5	5	5	5	5	5	5	5
59g	5	5	5	5	4	4	4	4	5	5	5

Table A5.6. Dependence of the calculated enthalpy differences $\Delta H(S-R)$ (in kJ/mol) on the number of conformations, used for Boltzmann-averaging.

Number of conformers	$\Delta H_{298}(S-R)$[a], kJ/mol				
	59d	**59a**	**59b**	**59c**	**59e**
7 conformers	9.29	5.99	6.12	5.75	6.01
6 conformers	9.28	5.92	6.13	5.66	5.96
5 conformers	9.12	5.95	6.15	5.54	5.94
4 conformers	9.15	5.94	6.10	5.59	5.93
3 conformers	9.18	5.95	6.07	5.61	5.92
2 conformers	9.50	5.97	5.91	5.90	5.96
1 conformer	9.82	6.04	5.85	6.13	5.73

[a] B3LYP/6-311+G(d,p)//B3LYP/6-31G(d) level

A graphical representation of the data from Table A5.6 is shown in Figure A5.1.

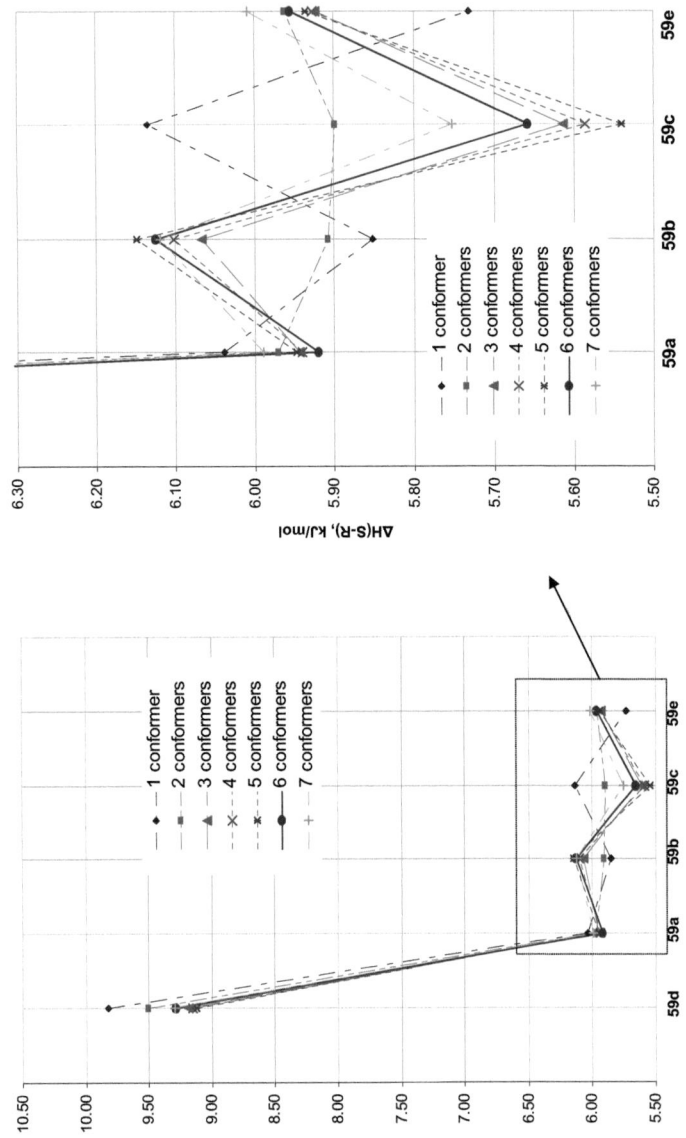

Figure A5.1. Dependence of the enthalpy difference $\Delta H^\ddagger(S-R)$ between TS (*R*)- and (*S*)-**67** for catalysts **59a-d** on the number of conformations, used for the Boltzmann averaging.

A5.2.2. Types of conformations needed for the selectivity prediction.

Analysis of the conformational space for TS **67** with catalysts **59a** and **59d** shows that for TSs (*R*)-**67** taking into account type (*R*)-**I** conformations, and for (*S*)-**67** – both types (*S*)-**II** and (*S*)-**III** is necessary to find the most stable conformations of TSs **67**. This method was successively used to calculate enthalpy differences $\Delta H_{298}(S\text{-}R)$ for other catalysts **59e**, **59f** and **59g**. Comparison of the relative energies of TS **67** conformers with different catalysts was carried out (Figure A5.2).

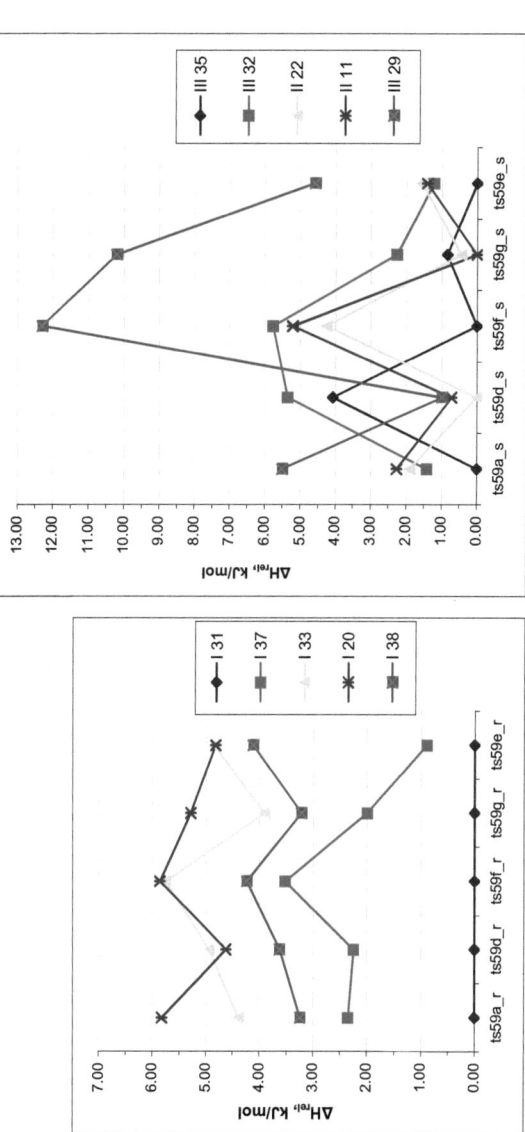

Figure A5.2. Relative energies of (*R*)-TS **67** (left) and (*S*)-TS **67** (right), as calculated at B3LYP/6-31G(d) level, for catalysts **59a**, **59d-g**. The types of TS **67** conformers together with the names used for TS **67** with **59a** (according pattern "spi2atsXX", see Table A5.7 in SI), are also listed.

Table A5.7. Relative enthalpies of TS **67** conformers with catalyst **59a** at B3LYP/6-311+G(d,p)//B3LYP/6-31G(d) level.

(S)-**67** type	Conformation	ΔH_{298}, kJ/mol	(R)-**67** type	Conformation	ΔH_{298}, kJ/mol
III	spi2ats35	6.04	I	spi2ats31	0.00
III	spi2ats32	7.20	I	spi2ats37	1.44
II	spi2ats22	10.34	I	spi2ats33	3.23
III	spi2ats29	10.80	I	spi2ats38	3.28
II	spi2ats11	11.01	I	spi2ats20	4.00
III	spi2ats28	12.63	I	spi2ats4	4.70
III	spi2ats34	14.27	I	spi2ats26	11.28
II	spi2ats19	15.80	I	spi2ats27	12.25
II	spi2ats7	16.31	IV	spi2ats25	15.22
II	spi2ats13	17.38	III	spi2ats21	15.36
II	spi2ats6	17.75	IV	spi2ats30	15.87
I	spi2ats3	25.15	III	spi2ats24	16.85
III	spi2ats12	26.28	II	spi2ats23	18.23
I	spi2ats17	26.60	III	spi2ats10	20.74
I	spi2ats8	26.75	III	spi2ats9	21.17
IV	spi2ats1	32.68	II	spi2ats14	21.25
			IV	spi2ats2	22.07
			II	spi2ats18	30.80
			II	spi2ats16	31.03
			II	spi2ats5	32.12
			III	spi2ats15	36.12

A5.2.3. Energies at B3LYP/6-311+G(d,p)//B3LYP/6-31G(d) level and thermal corrections at 298 K.

Table A5.8. Total Energies, Enthalpies and Free Energies (in Hartree) for TS **67** with Catalyst **59c**.

59c	conformer	B3LYP/ 6-31G(d) E(total, E_h)	B3LYP/ 6-31G(d) H_{298}	B3LYP/ 311+G(d,p)// B3LYP/6-31G(d) E(total, E_h)	B3LYP/ 311+G(d,p)// B3LYP/6-31G(d) "H_{298}"	B3LYP/ 6-31G(d) G_{298}	B3LYP/ 311+G(d,p)// B3LYP/6-31G(d) "G_{298}"
59c	spipyr1	-2154.144501	-2153.265439	-2154.70852	-2153.829458	-2153.399735	-2153.963754
	spipyr1_002	-2154.144418	-2153.265388	-2154.708707	-2153.829677	-2153.399573	-2153.963862
	spipyr1_001	-2154.142919	-2153.263907	-2154.707312	-2153.828300	-2153.398012	-2153.962405
(R)-67	spipyr1_004	-2154.142842	-2153.263855	-2154.707137	-2153.828150	-2153.398493	-2153.962788
	spipyr1_003	-2154.142821	-2153.263846	-2154.707101	-2153.828126	-2153.398454	-2153.962734
	spipyr1_007	-2154.141846	-2153.262782	-2154.704800	-2153.825736	-2153.395792	-2153.958746
	spipytsr4_1	-2154.140627	-2153.261536	-2154.703487	-2153.824396	-2153.394392	-2153.957252
	spipytsr4_2	-2154.138947	-2153.259612	-2154.701439	-2153.822104	-2153.391030	-2153.953522
	spipys3	-2154.142837	-2153.263746	-2154.706432	-2153.827340	-2153.396505	-2153.960099
	spipys3_003	-2154.142808	-2153.263626	-2154.706390	-2153.827208	-2153.396468	-2153.960050
	spipys5	-2154.142508	-2153.263597	-2154.705203	-2153.826292	-2153.396601	-2153.959296
	spipys4	-2154.141178	-2153.261986	-2154.703323	-2153.824131	-2153.393490	-2153.955635
(S)-67	spipys1	-2154.140006	-2153.260935	-2154.702061	-2153.822990	-2153.393611	-2153.955666
	ts59c_s_6	-2154.140352	-2153.261245	-2154.703932	-2153.824825	-2153.394110	-2153.957690
	ts59c_s_7	-2154.142882	-2153.263762	-2154.706446	-2153.827326	-2153.396107	-2153.959671
	ts59c_s_8	-2154.141323	-2153.262112	-2154.704914	-2153.825703	-2153.394354	-2153.957945
	ts59c_s_9	-2154.141111	-2153.261751	-2154.704920	-2153.825560	-2153.394361	-2153.958170

Table A5.9. Total Energies, Enthalpies and Free Energies (in Hartree) for TS **67** with Catalyst **59d** and Substrate **60**.

59d		type	B3LYP/6-31G(d) E(total, E_h)	B3LYP/6-31G(d) H_{298}	B3LYP/6-311+G(d,p)// B3LYP/6-31G(d) E(total, E_h)	B3LYP/6-311+G(d,p)// B3LYP/6-31G(d) "H_{298}"	B3LYP/6-31G(d) G_{298}	B3LYP/6-311+G(d,p)// B3LYP/6-31G(d) "G_{298}"
	1	I	-2233.980833	-2233.020723	-2234.566646	-2233.606535	-2233.166649	-2233.752461
	2	I	-2233.979980	-2233.019945	-2234.566097	-2233.606062	-2233.165804	-2233.751921
	5	I	-2233.979074	-2233.018961	-2234.565678	-2233.605565	-2233.168977	-2233.755581
	3	I	-2233.979453	-2233.019411	-2234.565378	-2233.605336	-2233.164133	-2233.750058
	4	I	-2233.978952	-2233.018915	-2234.565251	-2233.605214	-2233.165101	-2233.751400
(R)-67	10	II	-2233.974920	-2233.015169	-2234.559692	-2233.599941	-2233.162659	-2233.747431
	6	IV	-2233.974703	-2233.014538	-2234.559549	-2233.599384	-2233.159913	-2233.744759
	7	IV	-2233.973934					
	8	III	-2233.974913					
	9	III	-2233.974796					
	6	II	-2233.978138	-2233.017829	-2234.562985	-2233.602676	-2233.162576	-2233.747423
	7	II	-2233.977866	-2233.018002	-2234.562527	-2233.602663	-2233.165841	-2233.750502
	1	III	-2233.976589	-2233.016456	-2234.562152	-2233.602019	-2233.163390	-2233.748953
	4	III	-2233.976564	-2233.016624	-2234.561889	-2233.601950	-2233.161703	-2233.747029
2new		III	-2233.976097	-2233.015888	-2234.561418	-2233.601210	-2233.160562	-2233.745884
(S)-67	8	I	-2233.971590	-2233.011644	-2234.557217	-2233.597271	-2233.160434	-2233.746061
	3	III	-2233.977762	-2233.017399	-2234.563156	-2233.602793	-2233.161725	-2233.747119
	5	III	-2233.972573					
	9	I	-2233.970849					
	10	IV	-2233.965891					

Appendix

Table A5.10. Total Energies, Enthalpies and Free Energies (in Hartree) for TS **67** with Catalyst **59e** and Substrate **60**.

59e		type	B3LYP/6-31G(d) E(total, E_h)	B3LYP/6-31G(d) H_{298}	B3LYP/6-311+G(d,p)// B3LYP/6-31G(d) E(total, E_h)	B3LYP/6-311+G(d,p)// B3LYP/6-31G(d) "H_{298}"	B3LYP/6-31G(d) G_{298}	B3LYP/6-311+G(d,p)// B3LYP/6-31G(d) "G_{298}"
	1	I	-2312.600237	-2311.578883	-2313.209406	-2312.188052	-2311.729122	-2312.338291
	2	I	-2312.599394	-2311.578545	-2313.208682	-2312.187833	-2311.731429	-2312.340717
	3	I	-2312.598521	-2311.577316	-2313.208041	-2312.186836	-2311.727840	-2312.337360
(R)-67	4	I	-2312.598107	-2311.577033	-2313.207991	-2312.186917	-2311.730040	-2312.339924
	5	I	-2312.598489	-2311.577046	-2313.208338	-2312.186895	-2311.728638	-2312.338487
	6	IV	-2312.597954	-2311.576672	-2313.207597	-2312.186315	-2311.728781	-2312.338424
	11	I	-2312.597867	-2311.576537	-2313.206775	-2312.185445	-2311.726478	-2312.335386
	1	III	-2312.598146	-2311.576613	-2313.207402	-2312.185869	-2311.726861	-2312.336117
	2	III	-2312.597513	-2311.576145	-2313.206737	-2312.185369	-2311.726548	-2312.335772
	3	II	-2312.597485	-2311.576015	-2313.205249	-2312.183779	-2311.727517	-2312.335281
(S)-67	4	III	-2312.596380	-2311.574866	-2313.205505	-2312.183991	-2311.724642	-2312.333766
	5	II	-2312.597060	-2311.576077	-2313.204954	-2312.183970	-2311.726615	-2312.334508
	6	III	-2312.595403	-2311.574238	-2313.204878	-2312.183713	-2311.724827	-2312.334302
	7	III	-2312.594810	-2311.572881	-2313.204088	-2312.182158	-2311.723291	-2312.332568
	11	III	-2312.595804	-2311.574434	-2313.204861	-2312.183491	-2311.724432	-2312.333489

269

Appendix

Table A5.11. Total Energies, Enthalpies and Free Energies (in Hartree) for TS **67** with Catalysts **59f** and **59g** and Substrate **60**.

TS 67			B3LYP/6-31G(d) E(total, E_h)	B3LYP/6-31G(d) H_{298}	B3LYP/6-311+G(d,p)//B3LYP/6-31G(d) E(total, E_h)	B3LYP/6-311+G(d,p)//B3LYP/6-31G(d) "H_{298}"	B3LYP/6-31G(d) G_{298}	B3LYP/6-311+G(d,p)//B3LYP/6-31G(d) "G_{298}"
catalyst		type						
59f								
(R)-67	1	I	-2269.867489	-2268.931108	-2270.467980	-2269.531599	-2269.074836	-2269.675327
	2	I	-2269.866516	-2268.930254	-2270.467433	-2269.531172	-2269.074722	-2269.675640
	3	I	-2269.866413	-2268.929984	-2270.467041	-2269.530612	-2269.072499	-2269.673127
	4	I	-2269.865989	-2268.929406	-2270.467083	-2269.530500	-2269.072675	-2269.673769
	5	I	-2269.866155	-2268.929366	-2270.467408	-2269.530619	-2269.073539	-2269.674792
	11	I	-2269.868052	-2268.931596	-2270.468628	-2269.532172	-2269.075381	-2269.675957
	1	III	-2269.866921	-2268.930046	-2270.466782	-2269.529907	-2269.073560	-2269.673421
(S)-67	2	II	-2269.865241	-2268.928428	-2270.464625	-2269.527813	-2269.070996	-2269.670381
	3	II	-2269.864351	-2268.928063	-2270.463898	-2269.527610	-2269.071328	-2269.670875
	11	III	-2269.865350	-2268.928465	-2270.465742	-2269.528857	-2269.071213	-2269.671605
	4	III	-2269.864575	-2268.927850	-2270.465021	-2269.528296	-2269.070463	-2269.670909
59g								
(R)-67	1	I	-2492.381671	-2491.471945	-2493.060709	-2492.150983	-2491.617741	-2492.296779
	2	I	-2492.380779	-2491.471188	-2493.060160	-2492.150569	-2491.618289	-2492.297670
	3	I	-2492.380288	-2491.470717	-2493.059555	-2492.149984	-2491.616893	-2492.296160
	4	I	-2492.380007	-2491.470459	-2493.059652	-2492.150103	-2491.617595	-2492.297239
	5	I	-2492.379714	-2491.469932	-2493.059453	-2492.149671	-2491.617550	-2492.297289
(S)-67	1	III	-2492.380479	-2491.470547	-2493.059013	-2492.149080	-2491.616260	-2492.294793
	2	II	-2492.380310	-2491.470693	-2493.057713	-2492.148096	-2491.616774	-2492.294177
	3	II	-2492.380280	-2491.470866	-2493.057306	-2492.147892	-2491.616768	-2492.293794
	4	III	-2492.379735	-2491.470006	-2493.058388	-2492.148659	-2491.616091	-2492.294744
	7	III	-2492.379086	-2491.469376	-2493.057203	-2492.147493	-2491.615266	-2492.293383

A5.2.4. Dispersion corrections at B3LYP-D/6-311+G(d,p) level and thermal corrections at 195 K at B3LYP/6-31G(d) level.

Table A5.12. Total Energies, Enthalpies and Free Energies (in Hartree) at 195 K for TSs (*R*) and (*S*)-67 with Catalyst **59a** and Substrate **60**.

59a	conformer	type	B3LYP/6-31G(d) E(total, E_h)	B3LYP-D/6-311+G(d,p) dispersion correction	B3LYP/6-31G(d) H_{195}	B3LYP/6-31G(d) G_{195}	B3LYP-D/6-311+G(d,p) E(total, E_h)
	spi2ats31	(R)-I	-2155.344683	-0.178708	-2154.470410	-2154.538174	-2156.090440
	spi2ats37	(R)-I	-2155.343760	-0.179576	-2154.469251	-2154.538603	-2156.090730
	spi2ats38	(R)-I	-2155.343428	-0.179360	-2154.468901	-2154.537573	-2156.089824
	spi2ats33	(R)-I	-2155.343057	-0.178962	-2154.468487	-2154.537831	-2156.089508
(*R*)-67	spi2ats20	(R)-I	-2155.342699	-0.177319	-2154.467923	-2154.537805	-2156.087760
	spi2ats4	(R)-I	-2155.342674	-0.178188			-2156.088301
	spi2ats26	(R)-I	-2155.341445	-0.179567			-2156.087402
	spi2ats27	(R)-I	-2155.341746	-0.181903	-2154.466681	-2154.534775	-2156.089470
	spi2ats25	(R)-IV	-2155.340416	-0.182498			-2156.088958
	spi2ats35	(S)-III	-2155.342717	-0.178305	-2154.468297	-2154.535439	-2156.087901
	spi2ats32	(S)-III	-2155.342022	-0.179260	-2154.467457	-2154.535966	-2156.088258
	spi2ats22	(S)-II	-2155.342041	-0.181225	-2154.467278	-2154.535512	-2156.089227
	spi2ats29	(S)-III	-2155.340842	-0.181047	-2154.465882	-2154.533829	-2156.089050
(*S*)-67	spi2ats11	(S)-II	-2155.341761	-0.180736	-2154.467151	-2154.535597	-2156.088343
	spi2ats28	(S)-III	-2155.340193	-0.180328			-2156.087454
	spi2ats34	(S)-III	-2155.339563	-0.177285			-2156.083987
	spi2ats19	(S)-II	-2155.340634	-0.184258	-2154.465749	-2154.533812	-2156.090308
	spi2ats7	(S)-II	-2155.340504	-0.184686	-2154.465951	-2154.534081	-2156.090220

Appendix

Table A5.13. Total Energies, Enthalpies and Free Energies (in Hartree) at 195 K for TSs (*R*) and (*S*)-**67** with Catalyst **59b** and Substrate **60**.

59b	conformer	type	B3LYP/6-31G(d) E(total, E_h)	B3LYP-D/6-311+G(d,p) dispersion correction	B3LYP/6-31G(d) H_{195}	B3LYP/6-31G(d) G_{195}	B3LYP-D/6-311+G(d,p) E(total, E_h)
	spi1atsr1	I	-2076.716129	-0.164262	-2075.900126	-2075.966204	-2077.427197
	spi1atsr7	I	-2076.716058	-0.164115	-2075.900059	-2075.966026	-2077.426987
	spi1atsr2	I	-2076.713069	-0.163592	-2075.897054	-2075.964205	-2077.424041
(*R*)-67	spi1atsr8	I	-2076.713050	-0.164079			-2077.424454
	spi1atsr4	I	-2076.712340	-0.167035			-2077.425558
	spi1atsr4_1	IV	-2076.712296	-0.169992	-2075.896223	-2075.961661	-2077.427771
	spi1atsr4_2	IV	-2076.710690	-0.172930	-2075.894627	-2075.959291	-2077.428507
	spi1a3s_001	III	-2076.714456	-0.165292	-2075.898314	-2075.964118	-2077.425983
	spi1a3s_006	III	-2076.714447	-0.165288	-2075.898357	-2075.964169	-2077.425974
(*S*)-67	spi1atss2	II	-2076.713839	-0.167305	-2075.897983	-2075.963629	-2077.426561
	spi1atss3	II	-2076.712396	-0.170671	-2075.881467	-2075.946985	-2077.427918
	spi1a3s_003	III	-2076.712305	-0.166723	-2075.896266	-2075.963629	-2077.425629
	spi1a3s_008	I	-2076.705444				-2077.425387

Appendix

Table A5.14. Total Energies, Enthalpies and Free Energies (in Hartree) at 195 K for TSs (*R*) and (*S*)-67 with Catalyst 59c and Substrate 60.

59c	conformer	type	B3LYP/6-31G(d) E(total, E_h)	B3LYP-D/6-311+G(d,p) dispersion correction	B3LYP/6-31G(d) H_{195}	B3LYP/6-31G(d) G_{195}	B3LYP-D/6-311+G(d,p) E(total, E_h)
	spipyr1	I	-2154.144501	-0.171688	-2153.291091	-2153.359153	-2154.880209
	spipyr1_002	I	-2154.144418	-0.171639	-2153.291035	-2153.358843	-2154.880346
	spipytsr4_2	IV	-2154.138947	-0.179740	-2153.285221	-2153.350948	-2154.881179
(*R*)-67	spipyr1_004	I	-2154.142842	-0.171138			-2154.878275
	spipyr1_003	I	-2154.142821	-0.171175			-2154.878276
	spipyr1_007	I	-2154.141846	-0.173649			-2154.878449
	spipytsr4_1	IV	-2154.140627	-0.177097	-2153.287172	-2153.353821	-2154.880584
	spipyr1_001	I	-2154.142919	-0.170879	-2153.289584	-2153.357015	-2154.878191
	spipys3	III	-2154.142837	-0.172839	-2153.289373	-2153.355964	-2154.879270
	spipys3_003	III	-2154.142808	-0.172762	-2153.289251	-2153.355899	-2154.879152
	spipys5	II	-2154.142508	-0.174798	-2153.289263	-2153.355983	-2154.880001
	spipys4	II	-2154.141178	-0.177957	-2153.287630	-2153.353386	-2154.881280
(*S*)-67	spipys1	II	-2154.140006	-0.176555	-2153.286596	-2153.353111	-2154.878616
	ts59c_s_6	III	-2154.140352	-0.173586	-2153.286906	-2153.353539	-2154.877518
	ts59c_s_7	III	-2154.142882	-0.172970	-2153.289385	-2153.355709	-2154.879415
	ts59c_s_8	III	-2154.141323	-0.174464	-2153.287753	-2153.353995	-2154.879377
	ts59c_s_9	III	-2154.141111	-0.171808	-2153.287377	-2153.353871	-2154.876728

Appendix

Table A5.15. Total Energies, Enthalpies and Free Energies (in Hartree) at 195 K for TSs (*R*) and (*S*)-67 with Catalyst **59d** and Substrate **60**.

59d	conformer	type	B3LYP/6-31G(d) E(total, E_h)	B3LYP-D/6-311+G(d,p) dispersion correction	B3LYP/6-31G(d) H_{195}	B3LYP/6-31G(d) G_{195}	B3LYP-D/6-311+G(d,p) E(total, E_h)
	1	I	-2233.980833	-0.189729	-2233.049089	-2233.122111	-2234.756374
	2	I	-2233.979980	-0.190220	-2233.048315	-2233.121290	-2234.756317
	5	I	-2233.979074	-0.187556	-2233.047341	-2233.123029	-2234.753234
(*R*)-67	3	I	-2233.979453	-0.190365	-2233.047770	-2233.120010	-2234.755743
	4	I	-2233.978952	-0.189916	-2233.047301	-2233.120477	-2234.755167
	10	II	-2233.974920	-0.190708	-2233.043579	-2233.117590	-2234.750400
	6	IV	-2233.974703	-0.195197	-2233.042884	-2233.115562	-2234.754746
	3	III	-2233.977762	-0.191199	-2233.045742	-2233.117736	-2234.754355
	7	II	-2233.977866	-0.191535	-2233.046396	-2233.120648	-2234.754062
	6	II	-2233.978138	-0.192045	-2233.046176	-2233.118443	-2234.755030
(*S*)-67	1	III	-2233.976589	-0.189451	-2233.044811	-2233.118503	-2234.751602
	4	III	-2233.976564	-0.190963	-2233.045004	-2233.117461	-2234.752853
	2new	III	-2233.976097	-0.190753	-2233.044223	-2233.116451	-2234.752171
	8	I	-2233.971590	-0.186857	-2233.040018	-2233.114908	-2234.744074

Table A5.16. Total Energies, Enthalpies and Free Energies (in Hartree) at 195 K for TSs (*R*) and (*S*)-67 with Catalyst 59e and Substrate 60.

59e	conformer	type	B3LYP/6-31G(d) E(total, E_h)	B3LYP-D/6-311+G(d,p) dispersion correction	B3LYP/6-31G(d) H_{195}	B3LYP/6-31G(d) G_{195}	B3LYP-D/6-311+G(d,p) E(total, E_h)
(*R*)-67	1	I	-2312.600237	-0.199915	-2311.608133	-2311.683283	-2313.409321
	2	I	-2312.599394	-0.201324	-2311.607860	-2311.684689	-2313.410006
	3	I	-2312.598521	-0.199989	-2311.606576	-2311.681905	-2313.408031
	4	I	-2312.598107	*	-2311.606349	-2311.683258	
	5	I	-2312.598489	-0.197929	-2311.606309	-2311.682334	-2313.406267
	6	IV	-2312.597954	-0.198792	-2311.605961	-2311.682304	-2313.406390
	11	I	-2312.597867		-2311.605779	-2311.680741	
(*S*)-67	1	III	-2312.598146	-0.198110	-2311.605848	-2311.681017	-2313.405513
	2	III	-2312.597513	-0.199347	-2311.605391	-2311.680652	-2313.406084
	3	II	-2312.597485	-0.201404	-2311.605263	-2311.681243	-2313.406652
	4	III	-2312.596380	-0.201558	-2311.604103	-2311.678962	-2313.407063
	5	II	-2312.597060	*	-2311.605386	-2311.680684	
	6	III	-2312.595403	-0.199571	-2311.603518	-2311.678874	-2313.404449
	7	III	-2312.594810	-0.198199	-2311.602101	-2311.677388	-2313.402287
	11	III	-2312.595804		-2311.603666	-2311.678674	

Appendix

Table A5.17. Total Energies, Enthalpies and Free Energies (in Hartree) at 195 K for TSs (*R*) and (*S*)-**67** with Catalysts **59f** and **59g**.

TS 67 catalyst	conformer	type	B3LYP/6-31G(d) E(total, E_h)	B3LYP-D/6-311+G(d,p) dispersion correction	B3LYP/6-31G(d) H_{195}	B3LYP/6-31G(d) G_{195}	B3LYP-D/6-311+G(d,p) E(total, E_h)
59f							
(*R*)-67	1	I	-2269.867489	-0.185380	-2268.959037	-2269.030975	-2270.653360
	2	I	-2269.866516	-0.185580	-2268.958194	-2269.030603	-2270.653014
	3	I	-2269.866413	-0.185545	-2268.957901	-2269.029050	-2270.652586
	4	I	-2269.865989	*	-2268.957334	-2269.028968	*
	5	I	-2269.866155	-0.183696	-2268.957280	-2269.029517	-2270.651104
	11	I	-2269.868052	-0.185339	-2268.959519	-2269.031495	-2270.653968
(*S*)-67	1	III	-2269.866921	-0.185869	-2268.957934	-2269.029777	-2270.652651
	2	II	-2269.865241	-0.187720	-2268.956318	-2269.027525	-2270.652345
	3	II	-2269.864351	-0.187700	-2268.956009	-2269.027626	-2270.651597
	11	III	-2269.865350	-0.184317	-2268.956340	-2269.027676	-2270.650059
	4	III	-2269.864575	-0.185433	-2268.955733	-2269.026974	-2270.650453
59g							
(*R*)-67	1	I	-2492.381671	-0.185495	-2491.500352	-2491.573259	-2493.246204
	2	I	-2492.380779	-0.185874	-2491.499602	-2491.573358	-2493.246034
	3	I	-2492.380288	-0.185937	-2491.499137	-2491.572282	-2493.245492
	4	I	-2492.380007	*	-2491.498898	-2491.572657	*
	5	I	-2492.379714	-0.183669	-2491.498352	-2491.572441	-2493.243121
(*S*)-67	1	III	-2492.380479	-0.186639	-2491.498937	-2491.571804	-2493.245652
	2	II	-2492.380310	-0.188869	-2491.499112	-2491.572198	-2493.246582
	3	II	-2492.380280	-0.188639	-2491.499291	-2491.572255	-2493.245946
	4	III	-2492.379735	-0.187302	-2491.498404	-2491.571508	-2493.245690
	7	III	-2492.379086	-0.189241	-2491.497794	-2491.570755	-2493.246444

A5.3. Prochiral probe approach.

Table A5.18. Calculated energies of conformers of MOSC-adducts of **5b**. Averaged enthalpies $<H_{298}>$ were calculated at MP2(FC)/6-31+G(2d,p)//B98/6-31G(d) level of theory with inclusion of solvent effects in chloroform at PCM/UAHF/RHF/6-31G(d) level.

Conformer	E_{tot} B98/6-31G(d)	H_{298} B98/6-31G(d)	E_{tot} MP2(FC)/6-31+G(2d,p)	"H_{298} MP2-5"	G_{solv}, kJ/mol	$<H_{298}>$ MP2-5 with solv
MOSC	-721.802536		-720.3270031			-720.373983
mosc_1	-721.9716283	-721.802591	-720.4962865	-720.3272492	-123.05	-720.374117
mosc_2	-721.9673426	-721.798547	-720.4944162	-720.3256206	-126.48	-720.373795
si-attack		**-1470.916634**		**-1467.637862**		**-1467.666415**
N_3ax N_4eq						
1	-1471.475458	-1470.917615	-1468.196958	-1467.639115	-73.18	-1467.666987
6	-1471.475010	-1470.917501	-1468.195750	-1467.638241	-74.35	-1467.666559
19	-1471.473966	-1470.916477	-1468.195487	-1467.637998	-75.14	-1467.666619
15	-1471.472676	-1470.915430	-1468.194442	-1467.637196	-73.68	-1467.665259
111	-1471.473741	-1470.916465	-1468.194444	-1467.637167	-76.15	-1467.666171
4	-1471.472205	-1470.914069	-1468.194541	-1467.636404		
80	-1471.473504	-1470.916180	-1468.193631	-1467.636307		
88	-1471.471240	-1470.913922	-1468.193575	-1467.636258		
21	-1471.471874	-1470.914339	-1468.193356	-1467.635821		
65	-1471.471445	-1470.913937	-1468.192930	-1467.635422		
18	-1471.470384	-1470.912719	-1468.192834	-1467.635168		
50	-1471.470539	-1470.913466	-1468.191414	-1467.634341		
73	-1471.470260	-1470.912452	-1468.191566	-1467.633759		
90	-1471.468538	-1470.911390	-1468.190263	-1467.633115		
81	-1471.467891	-1470.911073	-1468.188976	-1467.632157		
117	-1471.467806	-1470.911061	-1468.188806	-1467.632061		
5	-1471.467299	-1470.909941	-1468.189342	-1467.631983		
N_3eq N_4ax						
6	-1471.473464	-1470.915918	-1468.195101	-1467.637555	-73.60	-1467.665586
24	-1471.472778	-1470.915290	-1468.194385	-1467.636896	-74.64	-1467.665326
10	-1471.472124	-1470.914649	-1468.194290	-1467.636816	-72.72	-1467.664513
1	-1471.473179	-1470.915894	-1468.193982	-1467.636697		
111	-1471.472202	-1470.914997	-1468.193845	-1467.636640		
75	-1471.471299	-1470.914014	-1468.193339	-1467.636055		
104	-1471.471999	-1470.915021	-1468.192820	-1467.635841		
2	-1471.472133	-1470.914449	-1468.193462	-1467.635778		
9	-1471.472872	-1470.915298	-1468.193125	-1467.635552		
28	-1471.470590	-1470.913270	-1468.192765	-1467.635445		
71	-1471.471486	-1470.914111	-1468.192711	-1467.635336		
64	-1471.471219	-1470.913884	-1468.192525	-1467.635190		
62	-1471.471271	-1470.913930	-1468.191319	-1467.633978		
re-attack		**-1470.917131**		**-1467.638573**		**-1467.667111**
N_3ax N_4eq						
4	-1471.475220	-1470.917544	-1468.197278	-1467.639602	-73.68	-1467.667665
1	-1471.475527	-1470.918132	-1468.196113	-1467.638717	-75.27	-1467.667386
22	-1471.473954	-1470.916603	-1468.195951	-1467.638600	-75.19	-1467.667237
10	-1471.472200	-1470.914895	-1468.194804	-1467.637499	-73.47	-1467.665483
19	-1471.474207	-1470.916467	-1468.194677	-1467.636936	-76.82	-1467.666195
3	-1471.472382	-1470.914488	-1468.193628	-1467.635734		
12	-1471.470888	-1470.913380	-1468.192558	-1467.635050		
6	-1471.470333	-1470.913053	-1468.191885	-1467.634605		
20	-1471.469527	-1470.912141	-1468.191085	-1467.633699		
25	-1471.469077	-1470.911651	-1468.190318	-1467.632892		
N_3eq N_4ax						
5	-1471.473284	-1470.915767	-1468.195508	-1467.637990	-74.18	-1467.666245
77	-1471.472209	-1470.914659	-1468.194351	-1467.636800	-74.18	-1467.665055

Appendix

Conformer	E_{tot} B98/6-31G(d)	H_{298} B98/6-31G(d)	E_{tot} MP2(FC)/6-31+G(2d,p)	"H_{298} MP2-5"	G_{solv}, kJ/mol	$<H_{298}>$ MP2-5 with solv
2	-1471.473512	-1470.916080	-1468.194134	-1467.636702	-75.60	-1467.665498
15	-1471.472121	-1470.914236	-1468.194348	-1467.636464		
32	-1471.472594	-1470.914890	-1468.194105	-1467.636401		
6	-1471.472185	-1470.914568	-1468.193663	-1467.636046		
129	-1471.471468	-1470.913584	-1468.193555	-1467.635671		
74	-1471.472346	-1470.914962	-1468.192922	-1467.635538		
18	-1471.472574	-1470.915093	-1468.193017	-1467.635537		
11	-1471.470399	-1470.912973	-1468.192905	-1467.635479		
92	-1471.471452	-1470.914103	-1468.192595	-1467.635246		
83	-1471.471146	-1470.913877	-1468.192358	-1467.635088		
20	-1471.470060	-1470.912948	-1468.191803	-1467.634691		
50	-1471.469884	-1470.912528	-1468.191840	-1467.634484		
86	-1471.471063	-1470.913974	-1468.191231	-1467.634142		

Table A5.19. Calculated energies of conformers of MOSC-adducts of **5l**. Averaged enthalpies $<H_{298}>$ were calculated at MP2(FC)/6-31+G(2d,p)//B98/6-31G(d) level of theory with inclusion of solvent effects in chloroform at PCM/UAHF/RHF/6-31G(d) level.

Conformer	E_{tot} B98/6-31G(d)	H_{298} B98/6-31G(d)	E_{tot} MP2(FC)/6-31+G(2d,p)	"H_{298} MP2-5"	G_{solv}, kJ/mol	$<H_{298}>$ MP2-5 with solv
si-attack		**-2009.976727**		**-2005.427713**		-2005.444708
cat6_1mosc_a3	-2010.739459	-2009.976436	-2006.190742	-2005.427719	-45.73	-2005.445137
cat6_1mosc_a1	-2010.737998	-2009.974494	-2006.189427	-2005.425924	-44.52	-2005.442880
cat6_2mosc_a3	-2010.738485	-2009.974889	-2006.189757	-2005.426161	-43.64	-2005.442782
cat6_3mosc_a3	-2010.740839	-2009.977367	-2006.191644	-2005.428173	-36.53	-2005.442085
cat6_3mosc_a1	-2010.739811	-2009.976471	-2006.188350	-2005.425010		
cat6_1mosc_a2	-2010.738540	-2009.974852	-2006.186633	-2005.422945		
cat6_2mosc_a1	-2010.737362	-2009.973726	-2006.183741	-2005.420105		
re-attack		**-2009.976321**		**-2005.426051**		-2005.442188
cat6_1mosc_b3	-2010.739950	-2009.976345	-2006.188553	-2005.424947	-46.53	-2005.442668
cat6_2mosc_b2	-2010.739609	-2009.975906	-2006.188746	-2005.425043	-45.27	-2005.442286
cat6_3mosc_b4	-2010.740263	-2009.977138	-2006.189781	-2005.426656	-38.20	-2005.441205
cat6_3mosc_b2	-2010.739382	-2009.975609	-2006.190313	-2005.426541	-37.40	-2005.440787
cat6_4mosc_b3	-2010.740118	-2009.976491	-2006.188897	-2005.425270	-39.25	-2005.440218
cat6_1mosc_b2	-2010.738922	-2009.975620	-2006.187531	-2005.424228		
cat6_4mosc_b2	-2010.739156	-2009.975714	-2006.187105	-2005.423663		
cat6_1mosc_b1	-2010.738816	-2009.975123	-2006.187292	-2005.423600		
cat6_4mosc_b1	-2010.738946	-2009.975599	-2006.186939	-2005.423592		
cat6_3mosc_b1	-2010.739465	-2009.975972	-2006.186530	-2005.423036		

Table A5.20. Calculated energies of conformers of MOSC-adducts of **59a**. Averaged enthalpies <H_{298}> were calculated at MP2(FC)/6-31+G(2d,p)//B98/6-31G(d) level of theory with inclusion of solvent effects in chloroform at PCM/UAHF/RHF/6-31G(d) level.

Conformer	E_{tot} B98/6-31G(d)	H_{298} B98/6-31G(d)	E_{tot} MP2(FC)/6-31+G(2d,p)	"H_{298} MP2-5"	G_{solv}, kJ/mol	<H_{298}> MP2-5 with solv
si-attack		**-1797.623764**		**-1793.591995**		**-1793.607925**
348	-1798.249486	-1797.624509	-1794.217766	-1793.592789	-38.99	-1793.607642
324	-1798.248044	-1797.623118	-1794.217525	-1793.592599	-40.88	-1793.608168
272	-1798.248075	-1797.622995	-1794.217252	-1793.592172	-42.34	-1793.608299
202	-1798.247764	-1797.622829	-1794.217091	-1793.592156	-41.00	-1793.607774
001	-1798.249816	-1797.624807	-1794.216431	-1793.591422	-43.64	-1793.608043
149	-1798.246989	-1797.622103	-1794.216055	-1793.591169	-41.92	-1793.607137
207	-1798.247348	-1797.622564	-1794.215947	-1793.591163	-40.42	-1793.606557
321	-1798.245662	-1797.621169	-1794.215232	-1793.590740		
010	-1798.248411	-1797.623565	-1794.215578	-1793.590732		
017	-1798.247861	-1797.622740	-1794.215414	-1793.590293		
234	-1798.247258	-1797.622381	-1794.215024	-1793.590147		
034	-1798.248044	-1797.623140	-1794.215042	-1793.590138		
544	-1798.246435	-1797.621386	-1794.215040	-1793.589992		
016	-1798.247880	-1797.622858	-1794.215008	-1793.589986		
045	-1798.247323	-1797.622650	-1794.213949	-1793.589276		
013	-1798.247082	-1797.622158	-1794.213687	-1793.588764		
009	-1798.247531	-1797.622472	-1794.213647	-1793.588588		
014	-1798.245992	-1797.621262	-1794.212751	-1793.588020		
011	-1798.246349	-1797.621096	-1794.212726	-1793.587473		
re-attack		**-1797.623670**		**-1793.592401**		**-1793.608664**
358	-1798.249822	-1797.624956	-1794.218403	-1793.593537	-41.17	-1793.609218
194	-1798.248135	-1797.623426	-1794.216952	-1793.592243	-43.60	-1793.608848
298	-1798.248103	-1797.623227	-1794.217291	-1793.592415	-41.88	-1793.608367
108	-1798.246678	-1797.622099	-1794.215778	-1793.591200	-43.05	-1793.607598
203	-1798.247486	-1797.622797	-1794.215951	-1793.591262	-42.51	-1793.607453
350	-1798.248874	-1797.623834	-1794.216305	-1793.591264	-42.09	-1793.607296
037	-1798.248167	-1797.623219	-1794.217337	-1793.592389		
007	-1798.247579	-1797.622760	-1794.216029	-1793.591210		
048	-1798.247569	-1797.622559	-1794.215875	-1793.590865		
170	-1798.247517	-1797.622302	-1794.215985	-1793.590770		
069	-1798.247499	-1797.622464	-1794.215489	-1793.590453		
084	-1798.247752	-1797.622952	-1794.215240	-1793.590440		
306	-1798.246472	-1797.621755	-1794.215134	-1793.590417		
115	-1798.247754	-1797.623420	-1794.214751	-1793.590416		
359	-1798.247206	-1797.622504	-1794.215054	-1793.590352		
362	-1798.246755	-1797.621761	-1794.214857	-1793.589863		
406	-1798.247221	-1797.622567	-1794.213974	-1793.589319		
379	-1798.246716	-1797.621675	-1794.214047	-1793.589006		
506	-1798.246642	-1797.621936	-1794.213394	-1793.588689		
114	-1798.246359	-1797.621790	-1794.212517	-1793.587948		

Table A5.21. Calculated energies of conformers of MOSC-adducts of **5g**. Averaged enthalpies <H_{298}> were calculated at MP2(FC)/6-31+G(2d,p)//B98/6-31G(d) level of theory with inclusion of solvent effects in chloroform at PCM/UAHF/RHF/6-31G(d) level.

Conformer	E_{tot} B98/6-31G(d)	H_{298} B98/6-31G(d)	E_{tot} MP2(FC)/6-31+G(2d,p)	"H_{298} MP2-5"	G_{solv}, kJ/mol	<H_{298}> MP2-5 with solv
si-attack		**-1932.654751**		**-1928.279939**		**-1928.294158**
a1.001	-1933.383850	-1932.654873	-1929.009177	-1928.280201	-37.32	-1928.294416
b4.006	-1933.383547	-1932.654586	-1929.007881	-1928.278920	-36.94	-1928.292991
re-attack		**-1932.653656**		**-1928.279544**		**-1928.294105**
a2.005	-1933.382643	-1932.653879	-1929.008349	-1928.279585	-37.70	-1928.293944
b1.004	-1933.381012	-1932.651973	-1929.008539	-1928.279500	-38.66	-1928.294225

A5.4 Conformational analysis of transition states.

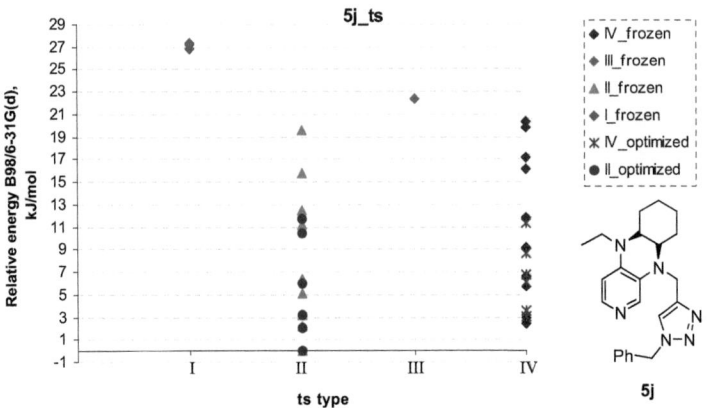

Figure A5.3. Relative energies (in kJ mol^{-1}) of conformers of TSs with catalyst **5j**, as calculated at B98/6-31G(d).

Table A5.22. Relative enthalpies of TSs conformations of different types (Scheme 5.8) as calculated at the B98/6-31G(d) (referred as "B98") and MP2(FC)/6-31+G(2d,p)//B98/6-31G(d) (referred as "MP2-5") levels of theory.

Conformer	ts type	Relative enthalpy H$_{298}$, kJ mol^{-1}	
		B98	MP2-5
5a_ts			
frozen ts			
1_ts1	IV	1.84	2.00
2_ts4	IV	3.27	3.59
3_ts4	IV	6.22	8.13
1_ts2	III	11.07	17.82
2_ts3	III	11.15	
2_ts2	II	0.44	0.53
3_ts2	II	5.31	6.37
1_ts4	II	0.00	0.00
1_ts3	I	11.85	18.99
3_ts1	I	15.11	
2_ts1	I	12.85	
optimized ts			
2_ts4opt	IV	3.57	3.54
2_ts3opt	III	11.45	18.26
1_ts4opt	II	0.27	-0.08
2_ts2opt	II	0.84	0.63

Conformer	ts type	Relative enthalpy H$_{298}$, kJ mol^{-1}	
		B98	MP2-5
5b_ts			
frozen ts			
1_ts4	IV	2.42	2.65
11_ts4	IV	4.46	4.57
2_ts2	IV	10.75	11.20
4_ts4	IV	6.80	7.49
6_ts4	IV	9.40	9.13
13_ts4	IV	11.80	10.98
1_ts3	III	9.37	16.70
11_ts3	III	13.30	20.99
4_ts2	II	5.09	5.41
6_ts2	II	7.20	7.29
13_ts2	II	12.74	12.33
1_ts2	II	0.00	0.00
11_ts2	II	2.57	2.63
2_ts4	II	13.56	14.58
1_ts1	I	11.83	19.64
11_ts1	I	14.96	22.74
optimized ts			
1_ts4opt	IV	2.40	2.21
1_ts3opt	III	9.76	16.97
1_ts2opt	II	0.02	-0.30
11_ts2opt	II	2.71	2.53

Table A5.22 (continued)

Conformer	ts type	Relative enthalpy H_{298}, kJ mol^{-1}	
		B98	MP2-5
5k_ts			
frozen ts			
2_ts4	IV	5.20	0.00
15_ts4	IV	7.05	1.12
13_ts2	IV	0.00	7.28
4_ts2	IV	1.93	9.22
6_ts1	IV	0.02	7.72
6_ts2	III	10.25	17.94
13_ts1	III	8.73	23.80
13_ts4	II	0.31	8.16
6_ts4	II	2.91	4.87
4_ts4	II	3.30	4.32
2_ts2	II	10.40	10.85
15_ts2	II	13.17	12.65
13_ts3	I	8.91	22.79
6_ts3	I	9.74	21.12
optimized ts			
15_ts4opt	IV	6.86	1.20
13_ts2	IV	0.54	7.58
6_ts1	IV	0.49	7.80
4_ts2	IV	1.97	8.58
2_ts4opt	IV	5.20	0.00
6_ts4opt	II	3.70	5.40
13_ts3	I	9.09	22.92
5l_ts			
frozen ts			
2_ts2	IV	0.00	5.52
1_ts4	IV	6.84	13.33
1_ts3	III	11.66	22.77
2_ts1	III	14.77	32.17
1_ts2	II	2.57	0.00
2_ts4	II	5.74	12.36
1_ts1	I	10.36	25.07
2_ts3	I	12.87	28.66
optimized ts			
2_ts2opt	IV	0.50	5.72
1_ts2opt	II	2.99	0.55
5l-Me_ts			
optimized ts			
1_ts4	IV	5.81	10.15
2_ts2	IV	1.56	6.45
2_ts4	II	4.85	10.49
1_ts2	II	0.00	0.00

Conformer	ts type	Relative enthalpy H_{298}, kJ mol^{-1}	
		B98	MP2-5
5j_ts			
frozen ts			
102_ts2	IV	2.49	
33_ts2	IV	2.99	
158_ts2	IV	5.78	
85_ts2	IV	6.62	
106_ts2	IV	9.14	
155_ts2	IV	11.82	
1_ts2	IV	16.15	
46_ts2	IV	17.23	
182_ts2	IV	19.82	
22_ts2	IV	19.83	
7_ts2	IV	20.35	
187_ts2	III	22.38	
102_ts1	II	2.18	
158_ts1	II	3.15	
33_ts1	II	0.00	
187_ts1	II	5.15	
1_ts1	II	11.26	
207_ts1	II	12.45	
85_ts1	II	6.38	
7_ts1	II	19.57	
155_ts1	II	15.77	
182_ts1	I	27.28	
22_ts1	I	27.35	
106_ts1	I	26.86	
optimized ts			
102_ts2	IV	3.10	2.07
33_ts2	IV	3.52	2.34
85_ts2	IV	6.79	6.28
158_ts2	IV	6.10	7.50
106_ts2	IV	8.68	22.54
155_ts2	IV	11.31	26.72
33_ts1	II	-0.01	0.00
158_ts1	II	3.16	2.57
102_ts1	II	1.99	10.64
85_ts1	II	6.03	17.98
1_ts1	II	10.44	14.55
207_ts1	II	11.69	25.99
5l-Ph_ts			
optimized ts			
1_ts4	IV	0.90	9.41
2_ts2	IV	0.50	12.09
2_ts4	II	1.27	17.23
1_ts2	II	0.00	0.00

Table A5.23. Calculated energies of conformers of 3,4-diaminopyridines and corresponding acetyl intermediates and transition states. Averaged enthalpies $<H_{298}>$ were calculated at MP2(FC)/6-31+G(2d,p)//B98/6-31G(d) level of theory with inclusion of solvent effects in chloroform at PCM/UAHF/RHF/6-31G(d) level.

Conformer	E_{tot} B98/6-31G(d)	H_{298} B98/6-31G(d)	E_{tot} MP2(FC)/6-31+G(2d,p)	G_{solv}, kJ/mol	H_{298} MP2-5 with solv	$<H_{298}>$ MP2-5 with solv
Py						
Py	-248.181767	-248.087627	-247.589439	-9.00	-247.498727	
Py_ac	-401.140004	-400.991691	-400.215516	-142.55	-400.121498	
5l-Me						-1441.641383
1	-1445.880901	-1445.172693	-1442.357138	19.37	-1441.641552	
2	-1445.880375	-1445.171564	-1442.356300	22.34	-1441.638979	
3	-1445.878714	-1445.170331	-1442.355406	32.72	-1441.634561	
4	-1445.879184	-1445.170629	-1442.353600	25.73	-1441.635244	
5l-Me_ac						-1594.293769
1_ac1	-1598.889925	-1598.126495	-1595.033168	-63.97	-1594.294104	
1_ac2	-1598.891052	-1598.127564	-1595.034445	-60.54	-1594.294016	
2_ac1	-1598.889015	-1598.125839	-1595.032196	-62.17	-1594.292701	
2_ac2	-1598.889646	-1598.126274	-1595.032726	-59.87	-1594.292159	
3_ac1	-1598.890016	-1598.126825	-1595.033191	-54.73	-1594.290844	
4_ac2	-1598.889550	-1598.126017	-1595.031496	-55.06	-1594.288935	
3_ac2	-1598.885235	-1598.121800	-1595.027104	-49.54	-1594.282537	
5l-Ph						-2206.255203
3	-2212.521031	-2211.587771	-2207.214573	67.86	-2206.255465	
1	-2212.524033	-2211.590934	-2207.203380	42.55	-2206.254074	
2	-2212.521634	-2211.588352	-2207.196322	38.58	-2206.248347	
4	-2212.520811	-2211.587683	-2207.195645	45.81	-2206.245067	
5l-Ph_ac						-2358.904362
1_ac1	-2365.531860	-2364.543720	-2359.878651	-37.03	-2358.904614	
1_ac2	-2365.532457	-2364.544401	-2359.879792	-32.47	-2358.904103	
3_ac2	-2365.526447	-2364.539858	-2359.883327	-12.22	-2358.901391	
3_ac1	-2365.530534	-2364.542603	-2359.876213	-22.38	-2358.896808	
4_ac2	-2365.530981	-2364.543013	-2359.874073	-26.28	-2358.896113	
2_ac1	-2365.528125	-2364.539985	-2359.869431	-38.66	-2358.896016	
Transition states						
5l-Me_ts						-2055.216096
1_ts2	-2061.042343	-2060.082516	-2056.197370	55.06	-2055.216571	
1_ts4	-2061.041406	-2060.080302	-2056.194782	49.33	-2055.214890	
2_ts2	-2061.043440	-2060.081920	-2056.196608	53.68	-2055.214642	
2_ts4	-2061.042318	-2060.080667	-2056.195198	53.35	-2055.213229	
5l-Ph_ts						-2819.828123
1_ts2	-2827.684496	-2826.498719	-2821.045830	82.84	-2819.828499	
1_ts4	-2827.684285	-2826.498376	-2821.042378	77.91	-2819.826796	
2_ts2	-2827.684617	-2826.498528	-2821.041536	78.49	-2819.825551	
2_ts4	-2827.684279	-2826.498235	-2821.039535	76.61	-2819.824312	

9. Kinetics of reactions in homogeneous solution: derivation of the kinetic law.

I. Introduction

The rate of chemical reactions in homogeneous solution depends on a multitude of factors and the following script is intended to provide some orientation in practical rate studies. In order to facilitate the discussion, several important terms should first be defined.

Stochiometry

The stochiometry of a chemical reaction defines the molar quantities of reactants and products of the overall reaction. For the example of the reaction of alcohol **1** with acetic anhydride (**2**) in the presence of triethylamine (NEt$_3$, **3**) and a catalytic base such as 4-dimethylaminopyridine (**4**) the following stochiometric equation applies:

1,4-dioxane (**5**) is used here as an internal standard for ^{1}H NMR measurements. Components not seeing any turnover during the reaction (solvents, catalysts, reference standards) are not included in the stochiometric equation, but are listed on top/below the reaction arrow.

Elementary steps

Each elementary step of a reaction can contain a variable number of individual molecules. The reaction is **unimolecular** when involving transformation of only one molecule, **bimolecular** when involving two molecules, and **trimolecular** with three molecules. Steps involving more than two molecules are statistically quite rare. This implies that the majority of elementary steps in the gas phase and in homogeneous solution are uni- and bimolecular. The following sequence of bimolecular steps has been suggested for reaction (A):

Reaction Schemes

(A1): Acetic anhydride (2) + DMAP (4) → acetate + N-acetylpyridinium

(A2): N-acetylpyridinium + 1-ethynylcyclohexanol (1) → DMAP·H⁺ (6) + acetylated alkynyl cyclohexane

(A3): DMAP·H⁺ + NEt₃ → HNEt₃⁺ + DMAP (4)

Kinetic rate law

A kinetic rate law defines the dependence of the rate of reaction r on the concentration of all components of the reaction mixture, including most commonly those of the reactants and catalysts, but sometimes those of the products and other reagents as well. This usually takes on the form of equation 1:

$$r = k\,[A]^a[B]^b[C]^c \ldots \quad (1)$$

with k – rate constant of the reaction, $[A]$ – the concentration of reaction component A, and a the order of the reaction with respect to reactant A. With $a = 1$ the reaction is called first-order with respect to that reactant, with $a = 2$ the reaction is called second-order with respect to that reactant etc. Values for a often involve integral numbers, but may, in some cases, also become non-integral. The sum of all of the exponents $a, b, c \ldots$ is called the (overall) order of the reaction. The rate of the reaction r may be determined either by measuring the decrease of reactant concentrations (e.g. reactant A):

$$r = -\frac{d[A]}{dt} \quad (2)$$

or by measuring the increase of product concentrations (e.g. product P):

$$r = \frac{d[P]}{dt} \qquad (3)$$

In case the reaction involves the formation of long-lived intermediates these two types of measurements need not give the same results!

II. Experimental procedures

The first step in rate studies is the accurate measurement of the concentration of reactants and products as a function of time. ^{1}H NMR spectroscopy provides a convenient way of determining these concentrations for reactions, which are not too fast. The following two examples will be used to demonstrate the details involved in those types of measurements.

1. Studied reactions:

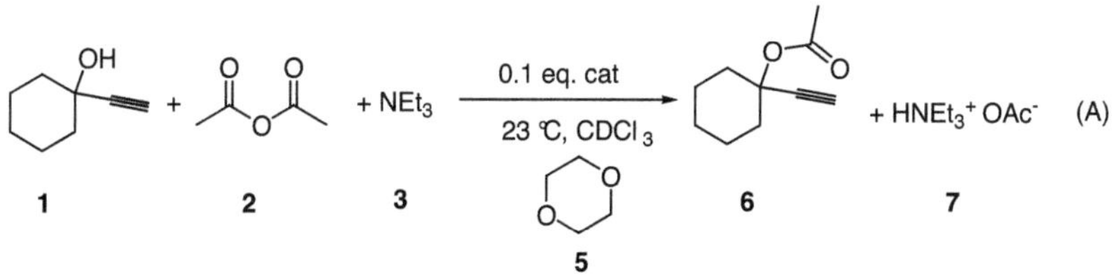

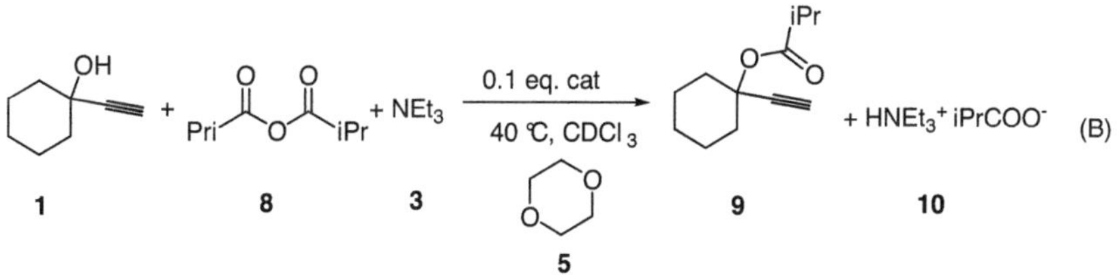

[Reaction A: 1 + 2 + NEt₃ (3) → 6 + HNEt₃⁺ OAc⁻ (A), 0.1 eq. cat, 23 °C, CDCl₃, with 1,4-dioxane (5); compound 7]

[Reaction B: 1 + 8 + NEt₃ (3) → 9 + HNEt₃⁺ iPrCOO⁻ (B), 0.1 eq. cat, 40 °C, CDCl₃, with 1,4-dioxane (5); compound 10]

Conditions: 0.2 M alcohol **1**, 0.4 M anhydride **2** or **8**, 0.6 M triethylamine (**3**), 0.1 M 1,4-dioxane (**5**), 0.02 M catalyst.

2. Sample preparation and kinetic measurements.

CDCl₃ and Et₃N were freshly distilled under N₂ from CaH₂ before use. Acetic and isobutyric anhydrides were distilled under reduced pressure from P₄O₁₀ on anhydrous K₂CO₃, filtered and fractionally distilled under reduced pressure. Both anhydrides were kept over 4 Å molecular sieves. All kinetic measurements were recorded at a constant temperature of 23 °C for reaction A and 40 °C for reaction B on a Varian Mercury 200 NMR spectrometer. The following solutions were prepared in $CDCl_3$ in three dry, calibrated 5 mL flasks:

A: 1.2 M in Ac_2O or isobutyric anhydride, and 0.3 M in anhydrous dioxane;
B: 0.6 M in ethynylcyclohexanol ROH and 1.8 M in Et_3N;
C: 0.06 M in catalyst.

In an NMR tube, 200 μL each of the above mentioned standardized solutions were added under nitrogen using an Eppendorf pipette. The NMR tube was then flame-sealed under N_2. The reaction solution was mixed and immediately inserted into the NMR spectrometer. The reaction was monitored by recording ^1H NMR spectra within a defined time interval until full conversion.

III. Integration process

^1H NMR spectra were analysed with program VNMR 4.3 Rev. G0194. Peak integrals were automatically integrated by using the subprogram listed below.

```
intmod = 'partial'
$i = 1
REPEAT
ds($i)
cz
$height=0
$c=0
peak:$height,cr
integ(cr-2,cr+2):$c
write ('file','daten',$c)
$i = $i + 1
UNTIL $i > arraydim
```

Reaction A. Signals of ester **6** (1.91 ppm, CH_3 group, corresponds to 3H) and dioxane (3.57 ppm, corresponds to 8H) (see Figure 9.1) were integrated automatically during the course of the reaction. Taking into account that the final concentration of ester $[ester]_\infty = [ROH]_0 = 2 \cdot [dioxane]$, one can calculate the conversion according to equation 6.

$$\text{Conversion} = \frac{[ester]}{[ester]_\infty} \cdot 100\% \qquad (4)$$

$$\text{Conversion} = \frac{[ester]}{2[dioxane]} \cdot 100\% \qquad (5)$$

$$\text{Conversion} = \frac{1}{2}\left[\frac{I_{Ester}/3}{I_{Dioxane}/8}\right] \cdot 100\% = \left[\frac{4 I_{Ester}}{3 I_{Dioxane}}\right] \cdot 100\% \qquad (6)$$

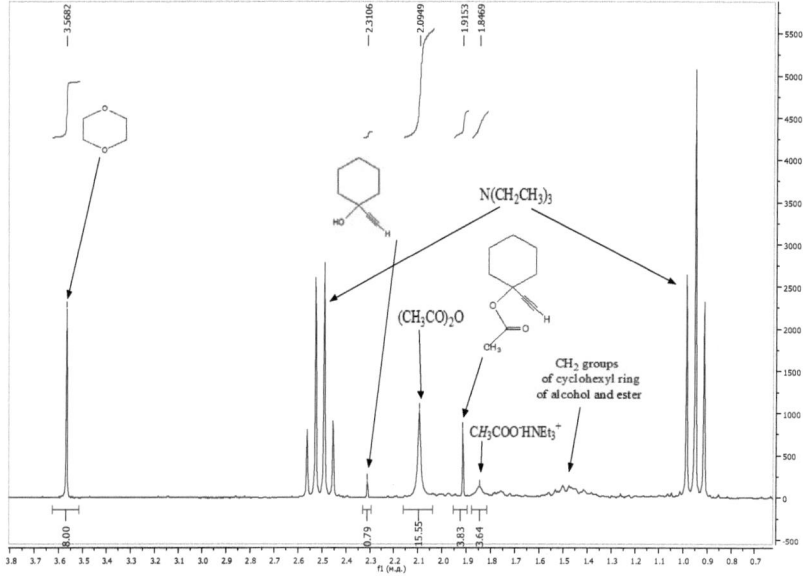

Figure 9.1. ^{1}H NMR (200 MHz) spectrum of reaction mixture for the reaction (A).

Reaction B. Signals of isobutyric anhydride (**8**) (a half of duplet at 1.14 ppm was taken for integration, 2CH$_3$ groups corresponds to 6H) and dioxane (3.57 ppm, corresponds to 8H) were integrated automatically during the course of the reaction. Taking into account that $\frac{[(R'CO)_2O]_0}{[ROH]_0} = 2$ and $\frac{[(R'CO)_2O]_0}{[dioxane]} = 4$, one can calculate the conversion according to equation 12.

$$\text{Conversion} = \frac{[ROH]_0 - [ROH]}{[ROH]_0} \cdot 100\% \qquad (7)$$

$$Conversion = \frac{[(R'CO)_2O]_b - [(R'CO)_2O]}{[ROH]_b} \cdot 100\% \qquad (8)$$

$$Conversion = \frac{[(R'CO)_2O]_b - [(R'CO)_2O]}{\frac{1}{2}[(R'CO)_2O]_b} \cdot 100\% \qquad (9)$$

$$Conversion = \left[2 - 2\frac{[(R'CO)_2O]/[dioxane]}{[(R'CO)_2O]_b/[dioxane]}\right] \cdot 100\% \qquad (10)$$

$$Conversion = \left[2 - 2\frac{I^{half-duplet}_{Anhydride}/6}{I_{Dioxane}/8}\Bigg/4\right] \cdot 100\% \qquad (11)$$

$$Conversion = \left[1 - \frac{I^{half-duplet}_{Anhydride}}{3 I_{Dioxane}}\right] \cdot 200\% \qquad (12)$$

Integration boundaries depend on the peak width. The boundaries should be changed in such a way that the conversion as calculated by equations 6 or 12 reaches 100% at the time when the peak of alcohol **1** at 2.31 ppm is vanished.

IV. Fitting

The obtained plot of conversion y versus time t can be fitted to integrated rate laws of "typical" cases.

1. Zero-order reaction

The rate of many catalyzed processes depends only on the concentration of the catalyst, but not on those of the reactants or products. For reaction (A) one would then expect that the reaction rate is independent of the concentration of the reactants (e.g. alcohol **1**). The appropriate rate law in this case is given by equation 13:

$$-\frac{d[ROH]}{dt} = k_0 \qquad (13)$$

Chapter 9. Kinetics

The integrated form of equation 13 predicts a linear dependence of the concentration [ROH] on the reaction time t (equation 7):

$$[ROH] = [ROH]_0 - k_0 t \qquad (14),$$

which only makes sense as long as there is any alcohol in reaction mixture (until $t = [ROH]_0/k_0$). Conversion y, which is given by equation 7, can be plotted using equation 15:

$$y = (k_0/[ROH]_0) \cdot t \qquad (15)$$

For the particular example of reaction (A) the analysis shows only a poor fit to a zero-order rate law (Figure 9.2). This is not only clear from visual inspection of Figure 9.2, but also from the correlation coefficient R^2.

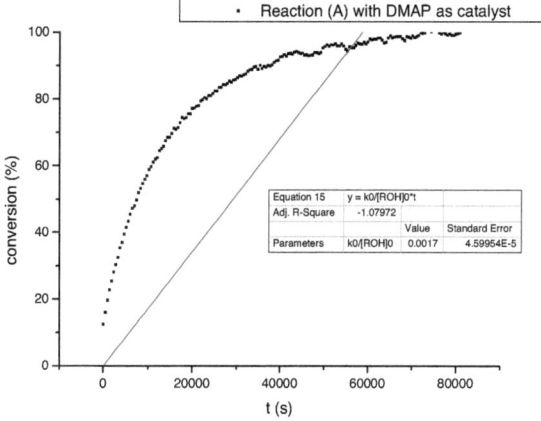

Figure 9.2. Fitting of the experimental data for the reaction (A) with zero-order equation 15.

2. First-order reaction

The rate law for a first-order reaction is given by equation 16:

$$-\frac{d[ROH]}{dt} = k_1 [ROH] \qquad (16)$$

289

The integrated form of equation 16 gives the exponential dependence of the concentration [ROH] on the reaction time t (equation 17):

$$[ROH] = [ROH]_0 \cdot e^{-k_1 t} \qquad (17)$$

The dependence of the conversion y, which is given by equation 7, on the time t is given by equation 18:

$$y = \left(1 - e^{-k_1 t}\right) \cdot 100\% \qquad (18)$$

The function that can be used for fitting of the experimental data is given by equation 19:

$$y = y_0 \cdot \left(1 - e^{-k_1 (t - t_0)}\right) \qquad (19)$$

where k_1 is a rate-constant of the first-order reaction; t_0 has a meaning of time axis offset. With this parameter in the fitting process it's not necessary to measure the starting point of the reaction exactly, which may always be complicated by the time it takes to prepare the sample, calibrate the spectrometer and get the first data point. The variable y_0 allows for rescaling of the conversion axis. Ideally, the value of this variable is equal to 100%, but weighting and mixing of reagents could introduce some error, leading to values of y_0 slightly different from 100 %.

Equation 19 could be replaced by similar equation 20.

$$y = y_0 - y' \cdot e^{-k_1 t} \qquad (20)$$

These two equations give identical result in the fitting process, because some parameters are interdependent (equation 21).

$$y' = y_0 \cdot e^{k_1 t_0} \qquad (21)$$

3. Second-order reaction

An obvious rate-law for reactions involving two reactants is that of a second-order reaction. For reaction A this could apply under the condition that the concentration of the catalyst is assumed to be constant during the reaction and that triethylamine doesn't participate in the rate determining step. The following equation for the second-order reaction can then be written (equation 22).

$$ROH + (R'CO)_2O \xrightarrow{k_2} R'COOR \qquad (22)$$

The rate law for this reaction is given by equation 23.

$$-\frac{d[ROH]}{dt} = k_2[ROH][(R'CO)_2O] \qquad (23)$$

The alcohol concentration can be expressed from conversion y and the initial alcohol concentration $[ROH]_0$ by equation 24.

$$[ROH] = [ROH]_0(1-y) \qquad (24)$$

If the ratio of the initial concentrations of alcohol and anhydride is assumed as n (equation 25), then the anhydride concentration can be expressed by equation 26.

$$\frac{[(R'CO)_2O]_0}{[ROH]_0} = n \quad (n>1) \qquad (25)$$

$$[(R'CO)_2O] = [ROH]_0(n-y) \qquad (26)$$

With taking into account equations 24 and 26, the rate law can be written as equation 27.

$$-[ROH]_0 \frac{d(1-y)}{dt} = k_2[ROH]_0^2(1-y)(n-y) \qquad (27)$$

Rearranging of variables gives equation 28.

$$\frac{d(1-y)}{(1-y)(n-y)} = -k \cdot dt \qquad (28),$$

where $k = k_2[ROH]_0$ (29)

Integration of equation 28 after some transformations gives equation 35.

$$\int_0^y \frac{d(1-y)}{(1-y)(n-y)} = -\int_{t_0}^t k\,dt \qquad (30)$$

$$\frac{1}{(1-y)(n-y)} = \left(\frac{1}{1-y} - \frac{1}{n-y}\right)/(n-1) \qquad (31)$$

$$\ln\left(\frac{1-y}{1}\right) - \ln\left(\frac{n-y}{n}\right) = -(n-1)\cdot k \cdot (t-t_0) \qquad (32)$$

$$\frac{1}{[Ac_2O]_0 - [ROH]_0} \ln\left(\frac{[Ac_2O]\cdot[ROH]_0}{[Ac_2O]_0 \cdot [ROH]}\right) = k\cdot t \qquad (33)$$

$$\frac{n-y}{n\cdot(1-y)} = e^{k(n-1)(t-t_0)} \qquad (34)$$

$$y = 1 - \frac{n-1}{ne^{k(n-1)(t-t_0)} - 1} \qquad (35)$$

Equation 35 expresses conversion for the ideal second-order reaction. While fitting the real kinetics it is again necessary to take into account errors of preparing and mixing of solutions for the reaction. This can be achieved by introducing one more variable y_0, which again acts as conversion axes rescaling parameter, as in the case of first-order reaction fitting (see above). The final equation for fitting is given by equation 36.

$$y = y_0\left(1 - \frac{n-1}{ne^{k(n-1)(t-t_0)} - 1}\right) \qquad (36)$$

For the situation that $n = 2$ (as is the case in reaction A), it becomes slightly simpler (equation 37).

Chapter 9. Kinetics

$$y = y_0\left(1 - \frac{1}{2e^{k(t-t_0)} - 1}\right) \qquad (37)$$

4. Calculation of the half-lives

In case of complex rate laws involving multiple variables the direct comparison of two different reactions is not so trivial. It is therefore practical to derive from the rate-law a single performance number such as the reaction half-life $\tau_{1/2}$. Two slightly different definitions can be given for the half-life:

1) *Kinetic half-life* τ^{kin} is the time taken for a given concentration to decrease to half of its initial value.

2) *Synthetic half-life* τ^{syn} is the time interval between the start of reaction and the time, when conversion is equal to 50%.

The difference between these two definitions is not immediately obvious, and for the case of an ideal first-order reaction these two definitions actually are identical. But while fitting the experimental data they could give different results. This can most readily be appreciated by inspection of the data in Figure 9.3.

Because the starting point of the reaction isn't known exactly, at first one should approximate the data with any of the described above fitting functions (exponential function 19 for example) and extrapolate the obtained function to conversion 0%. It gives the time t_0 (in example t_0 = -2954 s). While applying the second definition of synthetic half-life one needs to find the time of 50% conversion (7900 s in example). The difference gives the synthetic half-life: τ^{syn} = 7900 – (-954) = 10854 s = **181 min**. On other hand while applying the definition of kinetic half-life τ^{kin}, it should be taken into account, that the full conversion isn't equal to 100% (y_0 = 99.0), probably because of the errors involved in weighting and mixing of the reagents. It means, that one should find the time, when the concentration of alcohol decrease to half of its initial value, i.e. when the conversion is the half of maximal conversion y_0 (49.5%), this time is equal to 7730 s. Then the kinetic half-life is τ^{kin} = 7730 – (-2954) = 10684 s = **178 min**. The difference between these two half-lives is 3 min.

Chapter 9. Kinetics

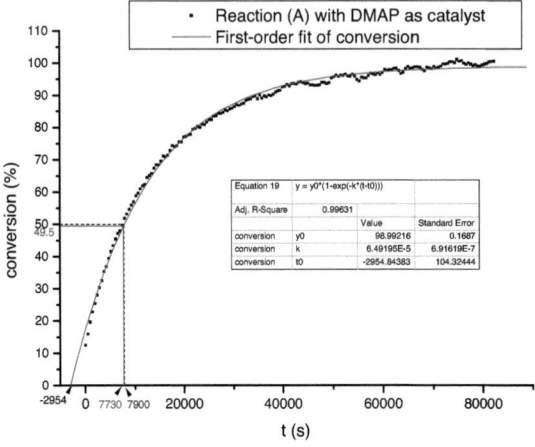

Figure 9.3. Comparison of two definitions of half-life.

In order to get the expressions for half-lives for the described functions, one should solve equations 15, 19, 36 and 37 for $y = 50\%$ (for the synthetic half-life) or $y = y_0/2$ (for the kinetic half-life). Solution of these equations gives equations 38-44 for half-lives.

1. The zero-order reaction (equation 15).

$$\tau^{kin} = \frac{[ROH]_0}{2k_o} \quad (38)$$

2. The first-order reaction (equation 19).

$$\tau^{kin} = \frac{\ln 2}{k_1} \quad (39)$$

$$\tau^{syn} = \frac{\ln\left(\dfrac{y_0}{y_0 - 50}\right)}{k_1} \quad (40)$$

3. The second-order reaction ($n = 2$) (equation 37).

$$\tau^{kin} = \frac{\ln 1.5}{k_2[ROH]_0} \quad (41)$$

294

$$\tau^{syn} = \frac{\ln\left(\dfrac{y_0 - 25}{y_0 - 50}\right)}{k_2 [ROH]_0} \qquad (42)$$

4. The second-order reaction (general case n ≠ 2) (equation 36).

$$\tau^{kin} = \frac{\ln\left(\dfrac{2n-1}{n}\right)}{k_2 [ROH]_0 (n-1)} \qquad (43)$$

$$\tau^{syn} = \frac{\ln\left(\dfrac{y_0 - \dfrac{50}{n}}{y_0 - 50}\right)}{k_2 [ROH]_0 (n-1)} \qquad (44)$$

V. Comparison of the fitting functions

In order to compare the suitability of the rate laws described above lets fit the experimental data for the reaction (A) with DMAP as catalyst using these functions (Figure 9.4). One of the measures of performance is the correlation coefficients R^2. Figure 9.4 shows that the best correlation with $R^2 = 0.9991$ is obtained with function 36. In this example the differences between half-lives, derived from the different fitting functions, appear from the differences in t_0, i.e. from the interpolation of the fitting curve to 0% conversion.

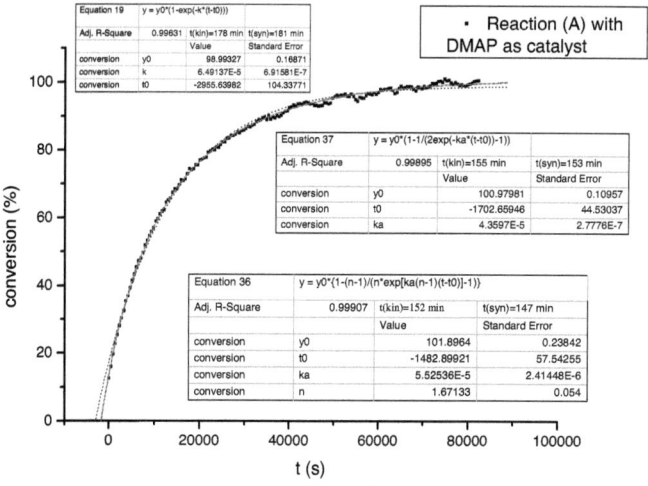

Figure 9.4. Comparison of correlation coefficients for fitting functions 19, 36 and 37.

Chapter 9. Kinetics

The second performance measure involves plotting of the residuals of conversion versus time (Figure 9.5). In the cases of equations 36 and 37 the residuals are statistically distributed. Fitting with equation 19 doesn't give statistical distribution of residuals. This implies that function 19 is not suitable for the fitting of this data on principal grounds.

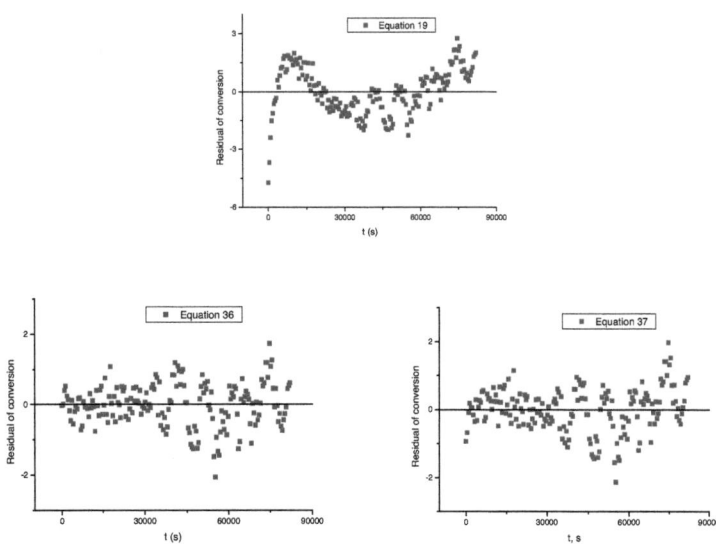

Figure 9.5. Plots of residuals of conversion versus time.

V. Selection of data points

One of the options for improving a correlation is to choose just the data points before conversion reaches 100% and drop points with conversion over 100%. The results for fitting only this part of the data points is shown on Figure 9.6.

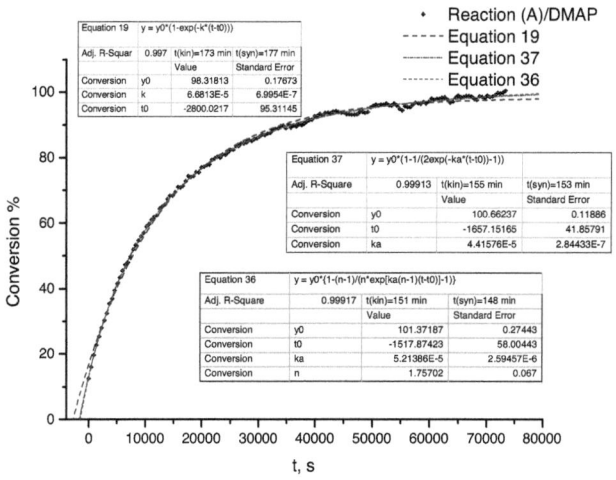

Figure 9.6. Fitting of the data with conversion below 100%.

Comparing Figures 4 and 6 one can see that the half-lives of fitting with equation 19 are more sensitive to the choice of data, than half-lives with equations 36 and 37. Using of equation 37 gives in this case a good correlation coefficient and insensitivity to the choice of data points with three variables, while a qualitatively very similar result is obtained with equation 36 using four variables. Due to the smaller number of variables equation 37 thus represents the preferred fitting function.

References.

[1] (a) Hori, K.; Ikenaga, Y.; Arata, K.; Takahashi, T.; Kasai, K.; Noguchi, Y.; Sumimoto, M.; Yamamoto, H.; *Tetrahedron* **2007**, *63*, 1264-1269; (b) Pliego, J. R.; Riveros, J. M.; *Chem. Eur. J.* **2002**, *8*, 1945-1953; (c) Haffner, F.; Hu, C.-H.; Brinck, T.; Norim, T.; *J. Mol. Struct. (THEOCHEM)* **1999**, *459*, 85-93; (d) Pliego, J. R.; Riveros, J. M.; *J. Am. Chem. Soc.* **2004**, *126*, 2520-2526; (e) Zhan, C.-G.; Landry, D. W.; Ornstein, R. L.; *J. Phys. Chem. A* **2000**, *104*, 7672-7678; (f) Zhan, C.-G.; Landry, D. W.; Ornstein, R. L.; *J. Am. Chem. Soc.* **2000**, *122*, 2621-2627; (g) Pliego, J. R.; Riveros, J. M.; *Chem. Eur. J.* **2001**, *7*, 169-175.

[2] Carey, F. A.; Sundberg, R.J.; *Advanced organic chemistry*. 3nd edn., Plenum Press: New York, **1990**.

[3] (a) Jencks, W. P.; *Chem. Rev.* **1985**, *85*, 511-527; (b) Williams, A.; *Chem. Soc. Rev.* **1994**, *23*,93-100; (c) Castro, C.; Castro, E. A.; *J. Org. Chem.* **1981**, *46*, 2939-2943.

[4] (a) Xie, D.; Xu, D.; Zhang, L.; Guo, H.; *J. Phys. Chem. B* **2005**, *109*, 5259-5266; (b) Hori, K.; Kamimura, A.; Kimoto, J.; Gotoh, S.; Ihara, Y.; *J. Chem. Soc. Perkin Trans. 2* **1994**, 2053-2058; (c) Hu, C.-H.; Brinck, T.; Hult, K.; *Int. J. Quant. Chem.* **1998**, *69*, 89-103; (d) Topf, M.; Richards, W. G.; *J. Am. Chem. Soc.* **2004**, *126*, 14631-14641.

[5] (a) Kevill, D. N.; Foss, F. D.; *J. Am. Chem. Soc.* **1969**, *91*, 5054-5059; (b) Ross, S. D.; *J. Am. Chem. Soc.* **1970**, *92*, 5998-6002; (c) Ba-Saif, S. A.; Maude, A. B.; Williams, A.; *J. Chem. Soc. Perkin Trans. 2* **1994**, 2395-2400; (d) Maude, A. B.; Williams, A.; *J. Chem. Soc. Perkin Trans. 2* **1995**, 691-696; (e) Maude, A. B.; Williams, A.; *J. Chem. Soc. Perkin Trans. 2* **1997**, 179-183.

[6] (a) Kruger, H.; *J. Mol. Struct. (THEOCHEM)* **2002**, *577*, 281-285; (b) Petrova, T.; Okovytyy, S.; Gorb, L.; Leszczynski, J.; *J. Phys. Chem. A* **2008**, *112*, 5224-5235.

[7] (a) Otera, J. *Esterification: Methods, Reactions and Applications*, Wiley-VCH, Weinheim, **2003**; (b) Baer, H. H.; Mateo, F. H.; *Can. J. Chem.* **1990**, *68*, 2055-2059; (c) Uno, H.; Shiraishi, Y.; Matsushima, Y., *Bull. Chem. Soc. Jpn.* **1991**, *64*, 842-850; (d) Fleming, I.; Ghosh, S. K.; *J. Chem. Soc., Chem. Commun.* **1992**, 1775-1777; (e) Ishihara, K.; Kubota, M.; Kurihara, H.; Yamamoto, H.; *J. Am. Chem. Soc.* **1995**, *117*, 4413-4414; (f) Ishihara, K.; Kubota, M.; Kurihara, H.; Yamamoto, H.; *J. Org. Chem.* **1996**, *61*, 4560-4567.

[8] (a) Orita, A.; Tanahashi, C.; Kakuda, A.; Otera, J.; *J. Org. Chem.* **2001**, *66*, 8926-8934; (b) Chen, C.-T.; Kuo, J.-H.; Li, C.-H.; Barhate, N. B.; Hon, S.-W. et al; *Org. Lett.* **2001**, *3(23)*, 3729-3732; (c) Chauhan, K. K.; Frost, C. G.; Love, I.; Waite, D.; *Synlett* **1999**,

1743-1744; (d) Saravanan, P.; Singh, V. K.; *Tetrahedron Lett.* **1999**, *40*, 2611-2614; (e) Iqbal, J.; Srivastava, R. R.; *J. Org. Chem.* **1992**, *57*, 2001-2007; (f) Procopiou, P. A.; Baugh, S. P. D.; Flack, S. S.; Inglis, G. G. A.; *J. Org. Chem.* **1998**, *63*, 2342-2347.

[9] (a) Dumeunier, R.; Marko, I. E.; *Tetrahedron Lett.* **2004**, *45*, 825-829; (b) Bartnicka, H.; Smagowski, H.; *Pol. J. Chem.* **1992**, *66*, 1295-1300.

[10] (a) Tomassy, B.; Zwierzak, A.; *Synth. Commun.* **1998**, *28*, 1201-1214; (b) Sano, T.; Ohashi, K.; Oriyama, T.; *Synthesis* **1999**, 1141-1144; (c) Oriyama, T.; Hori, Y.; Imai, K.; Sasaki, R.; *Tetrahedron Lett.* **1996**, *37*, 8543-8546; (d) Terakado, D.; Koutaka, H.; Oriyama, T.; *Tetrahedron: Asymmetry* **2005**, *16*, 1157-1165.

[11] (a) Litvinenko L. M.; Kirichenko A. I.; *Dokl. Chem.* **1967**, 763-766; *Dokl. Akad. Nauk SSSR Ser. Khim.* **1967**, **176**, 97-100; (b) Steglich, W.; Höfle, G.; *Angew. Chem.* **1969**, *81*, 1001; *Angew. Chem. Int. Ed. Engl.* **1969**, *8*, 981; (c) Höfle, G.; Steglich, W.; Vorbrüggen, H.; *Angew. Chem.* **1978**, *90*, 602-615; *Angew. Chem. Int. Ed. Engl.* **1978**, *17*, 569–583.

[12] (a) Spivey, A. C.; Arseniyadis, S.; *Angew. Chem.* **2004**, *116*, 5552-5557; *Angew. Chem. Int. Ed. Engl.* **2004**, *43*, 5436-5441; (b) Ragnarsson, U.; Grehn L.; *Acc. Chem. Res.* **1998**, *31*, 494-501; (c) Murugan, R.; Scriven, E. F. V.; *Aldrichimica Acta* **2003**, *36*, 21-27; (d) Scriven, E. F. V.; *Chem. Soc. Rev.* **1983**, *12*, 129-161.

[13] (a) Held, I.; Larionov, E.; Bozler, C.; Wagner, F.; Zipse, H.; *Synthesis* **2009**, 2267-2277; (b) Sakakura, A.; Kawajiri, K.; Ohkubo, T.; Kosugi, Y.; Ishihara, K.; *J. Am. Chem. Soc.* **2007**, *129*, 14775-14779.

[14] Xu, S.; Held, I.; Kempf, B.; Mayr, H.; Steglich, W.; Zipse, H.; *Chem. Eur. J.* **2005**, *11*, 4751-4757.

[15] Hassner, A.; Krepski, L. R.; Alexanian, V.; *Tetrahedron* **1978**, *34*, 2069-2076.

[16] Wakselman, M.; Guibe-Jampel, E.; *Tetrahedron Lett.* **1970**, *11*, 1521-1525.

[17] (a) Guibe-Jampel, E.; Le Corre, G.; Wakselman, M.; *Tetrahedron Lett.* **1975**, *16*, 1157-1160; (b) Kattnig, E.; Albert, M.; *Org. Lett.* **2006**, *4(6)*, 945-948.

[18] Lutz, V.; Glatthaar, J.; Würtele, C.; Serafin, M.; Hausmann, H.; Schreiner, P. R.; *Chem. Eur. J.* **2009**, *15*, 8548–8557.

[19] (a) Nederberg, F.; Connor, E. F.; Möller, M.; Glauser, T.; Hedrick, J. L.; *Angew. Chem. Int. Ed.* **2001**, *40*, 2712-2715; (b) Bonduelle , C.; Martín-Vaca, B.; Cossío, F. P.; Bourissou, D.; *Chem. Eur. J.* **2008**, *14*, 5304-5312; (c) Pratt, R. C.; Lohmeijer, B. G. G.; Long, D. A.; Waymouth, R. M.; Hedrick, J. L.; *J. Am. Chem. Soc.* **2006**, *128*, 4556-4557; (d) Chuma, A.; Horn, H. W.; Swope, W. C.; Pratt, R. C.; Zhang, L.; Lohmeijer, B. G. G.;

Wade, C. G.; Waymouth, R. M.; Hedrick, J. L.; Rice, J. E.; *J. Am. Chem. Soc.* **2008**, *130*, 6749-6754.

[20] Wurz, R.; *Chem. Rev.* **2007**, *107*, 5570-5595.

[21] For leading reviews on (thio)urea mediated reactions see: (a) Schreiner, P. R. *Chem. Soc. Rev.* **2003**, 289; (b) Pihko, P. M. *Angew. Chem. Int. Ed.* **2004**, *43*, 2062; (c) Connon, S. J. *Chem. Eur. J.* **2006**, *12*, 5418; (d) Doyle, A. G.; Jacobsen, E. N. *Chem. Rev.* **2007**, *107*, 5713; (e) Zhang, Z. G.; Schreiner, P. R. *Chem. Soc. Rev.* **2009**, *38*, 1187.

[22] (a) Jensen, K. H.; Sigman, M. S. *Angew. Chem.* **2007**, *119*, 4832-4834; *Angew. Chem. Int. Ed. Engl.* **2007**, *46*, 4748-4750; (b) Li, X.; Deng, H.; Zhang, B.; Li, J.; Zhang, L.; Luo, S.; Cheng, J.-P. *Chem. Eur. J.* **2010**, *16*, 450-455; (c) Ghobril, C.; Hammar, P.; Kodepelly, S.; Spiess, B.; Wagner, A.; Himo, F.; Baati, R. *ChemCatChem* **2010**, *2*, 1573-1581.

[23] (a) Mayr, H.; Kempf, B.; Ofial, A. R.; *Acc. Chem. Res.* **2003**, *36*, 66-77; (b) Brotzel, F.; Kempf, B.; Singer, T.; Zipse, H.; Mayr, H.; *Chem. Eur. J.* **2007**, *13*, 336-345.

[24] (a) Shinisha, C. B.; Sunoj, R. B.; *Org. Lett.* **2009**, *11*, 3242-3245; (b) Li, X.; Liu, P.; Houk, K. N.; Birman, V. B.; *J. Am. Chem. Soc.* **2008**, *130*, 13836-13837; (c) A. Hamza, G. Schubert, T. Soos, I. Papai, *J. Am. Chem. Soc.* **2006**, *128*, 13151-13160.

[25] (a) Campodonico P. R.; Aizman, A.; Contreras, R.; *Chem. Phys. Lett.* **2006**, *422*, 204-209; (b) Y. Wei, T. Singer, H. Mayr, G. N. Sastry, H. Zipse, *J. Comput. Chem.* **2008**, *29*, 291-297; (c) Wei, Y.; Sastry, G. N.; Zipse, H.; *J. Am. Chem. Soc.* **2008**, *130*, 3473-3477.

[26] (a) Heinrich, M. R.; Klisa, H. S.; Mayr, H.; Steglich, W.; Zipse, H.; *Angew. Chem.* **2003**, *115*, 4975-4977; *Angew. Chem. Int. Ed.* **2003**, *42*, 4826-4828; (b) Held, I.; Villinger, A.; Zipse, H.; *Synthesis* **2005**, 1425-1431; (c) Held, I.; Xu, S.; Zipse, H.; *Synthesis* **2007**, 1185-1196.

[27] Singh, S.; Das, G.; Singh, O. V.; Han, H.; *Org. Lett.* **2005**, *9(3)*, 401-404.

[28] Singh, S.; Das, G.; Singh, O. V.; Han, H.; *Tetrahedron Lett.* **2007**, *48*, 1983-1986.

[29] (a) Catalyst **5e** was prepared by Florian Achrainer, see: F. Achrainer, Master thesis, **2008**; (b) Catalyst **5c** was prepared by Jowita Humin, see: J. Humin, Master thesis, **2010**.

[30] D'Elia, V.; Liu, Y.; Zipse, H. *Eur. J. Org. Chem.* **2011**, 1527 - 1533.

[31] De Rycke, N.; Berionni, G.; Couty, F.; Mayr, H.; Goumont, R.; David, O. R. P. *Org. Lett.* **2011**, *13*, 530–533.

[32] (a) Becke, A. D. *J. Chem. Phys.* **1993**, *98*, 5648-5652; (b) Lee, C.; Yang, W.; Parr, R. G. *Phys. Rev. B* **1988**, *37*, 785; (c) Stephens, P. J.; Devlin, F. J.; Chabalowski, C. F.; Frisch, M. J. *J. Phys. Chem.* **1994**, *98*, 11623-11627.

References

[33] (a) Schmider, H. L.; Becke, A. D. *J. Chem. Phys.* **1998**, *108*, 9624-9631; (b) Becke, A. D. *J. Chem. Phys.* **1997**, *107*, 8554-8560; (c) Bienati, M.; Adamo, C.; Barone, V. *Chem. Phys. Lett.* **1999**, *311*, 69-76.

[34] Schrödinger, LLC., *MacroModel 9.7*, **2009**.

[35] Gaussian 03, Revision D.01, M. J. Frisch, G. W. Trucks, H.B. Schlegel, G. E. Scuseria, M. A. Robb, J. R. Cheeseman, J. A.Montgomery, Jr., T. Vreven, K. N. Kudin, J. C. Burant, J. M.Millam, S. S. Iyengar, J. Tomasi, V. Barone, B. Mennucci, M.Cossi, G. Scalmani, N. Rega, G. A. Petersson, H. Nakatsuji, M.S3Hada, M. Ehara, K. Toyota, R. Fukuda, J. Hasegawa, M. Ishida,T. Nakajima, Y. Honda, O. Kitao, H. Nakai, M. Klene, X. Li, J.E. Knox, H. P. Hratchian, J. B. Cross, V. Bakken, C. Adamo, J.Jaramillo, R. Gomperts, R. E. Stratmann, O. Yazyev, A. J. Austin,R. Cammi, C. Pomelli, J. W. Ochterski, P. Y. Ayala, K. Morokuma,G. A. Voth, P. Salvador, J. J. Dannenberg, V. G. Zakrzewski, S.Dapprich, A. D. Daniels, M. C. Strain, O. Farkas, D. K. Malick,A. D. Rabuck, K. Raghavachari, J. B. Foresman, J. V. Ortiz, Q.Cui, A. G. Baboul, S. Clifford, J. Cioslowski, B. B. Stefanov,G. Liu, A. Liashenko, P. Piskorz, I. Komaromi, R. L. Martin, D.J. Fox, T. Keith, M. A. Al-Laham, C. Y. Peng, A. Nanayakkara,M. Challacombe, P. M. W. Gill, B. Johnson, W. Chen, M. W. Wong,C. Gonzalez, and J. A. Pople, Gaussian, Inc., Wallingford CT, **2004**.

[36] For a review on light-gated catalyst systems, see: Stoll, R. S.; Hecht, S. *Angew. Chem.* **2010**, *122*, 5176-5200; *Angew. Chem. Int. Ed.* **2010**, *49*, 5054-5075.

[37] Peters, M. V.; Stoll, R. S.; Kühn, A.; Hecht, S. *Angew. Chem.* **2008**, *120*, 6056-6060; *Angew. Chem. Int. Ed.* **2008**, *47*, 5968-5972.

[38] Stoll, R. S.; Hecht, S. *Org. Lett.* **2009**, *11*, 4790-4793.

[39] Sugimoto, H.; Kimura, T.; Inoue, S. *J. Am. Chem. Soc.* **1999**, *121*, 2325-2326.

[40] (a) March, J.; *Advanced organic chemistry*. 3nd edn., John Wiley & Sons, **1985**; (b) Hansch, C.; Leo, A.; Taft, R. W. *Chem. Rev.* **1991**, *91*, 165-195.

[41] Dokić, J.; Gothe, M.; Wirth, J.; Peters, M. V.; Schwarz, J.; Hecht, S.; Saalfrank, P. *J. Phys. Chem. A* **2009**, *113*, 6763-6773.

[42] (a) El-Tamany, S.; Raulfs, F. W.; Hopf, H. *Angew. Chem.* **1983**, *95*, 631; *Angew. Chem. Int. Ed. Engl.* **1983**, *22*, 633; (b) Belokon, Y.; Moskalenko, M.; Ikonikov, N.; Yashikina, L.; Antonov, D.; Vorontsov, E.; Rozenberg, V. *Tetrahedron: Asymmetry* **1997**, *8*, 3245; (c) Danilova, T.; Rozenberg, V.; Starikova, Z. A.; Bräse, S. *Tetrahedron: Asymmetry* **2004**, *15*, 223.

[43] Fu, G. C.; *Acc. Chem Res.* **2004**, *37*, 542-547.

References

[44] Danilova, T. I.; Rozenberg, V. I.; Vorontsov, E. V.; Starikova, Z. A.; Hopf, H. *Tetrahedron: Asymmetry* **2003**, *14*, 1375.

[45] (a) Braddock, D. C.; MacGilp, I. D.; Perry, B. G. *Adv. Synth. Catal.* **2004**, *346*, 1117-1130; (b) Schneider, J. F.; Falk, F. C.; Fröhlich, R.; Paradies, J. *Eur. J. Org. Chem.* **2010**, 2265-2269.

[46] Fisher, C. B.; Xu, S.; Zipse, H. *Chem. Eur. J.* **2006**, *12*, 5779-5784.

[47] (a) Richards, C. J.; Locke, A. J.; *Tetrahedron: Asymmetry* **1998**, *9*, 2377-2407.; (b) Rebière, F.; Riant, O.; Ricard, L.; Kagan, H. B.; *Angew. Chem. Int. Ed. Engl.* **1993**, *32*, 568–570.

[48] (a) Kloetzing, R. J.; Knochel, P.; *Tetrahedron: Asymmetry* **2005**, *17*, 116-123; (b) Thaler, T.; Geittner, F.; Knochel, P.; *Synlett* **2007**, 2655-2658.

[49] (a) Fu, G. C.; Ie, Y.; *Chem. Comm.* **2000**, 119-120; (b) Ruble, J. C.; Fu, G. C.; *J. Am. Chem. Soc.* **1998**, *120*, 11532–11533; (c) Tao, B.; Ruble, J. C.; Hoic, D. A.; Fu, G. C. *J. Am. Chem. Soc.* **1999**, *121*, 5091–5092; (d) Arp, F. O.; Fu, G. C. *J. Am. Chem. Soc.* **2006**, *128*, 14264-14265; (e) Arai, S.; Bellemin-Laponnaz, S.; Fu, G. C. *Angew. Chem., Int. Ed.* **2001**, *40*, 234–236.

[50] Nguyen, H. V.; Motevalli, M.; Richards, C. J.; *Synlett* **2007**, 725-728.

[51] Seitzberg, J. G.; Dissing, C.; Sotofte, I.; Norrby, P.-O.; Johannsen, M.; *J. Org. Chem.* **2005**, *70*, 8332-8337.

[52] Poisson, T.; Penhoat, M.; Papamicaël, C.; Dupas, G.; Dalla, V.; Marsais, F.; Levacher, V.; *Synlett* **2005**, 2285-2288.

[53] Wong, C.-H.; Whitesides, G. M. In *Enzymes in Synthetic Organic Chemistry*; Elsevier Science Ltd.: Oxford, **1994**.

[54] Kagan, H. B.; Fiaud, J. C.; *Top. Stereochem.* **1988**, *18*, 249-330.

[55] (a) Kawabata, T.; Nagato, M.; Takasu, K.; Fuji, K.; *J. Am. Chem. Soc.* **1997**, *119*, 3169-3170; (b) Pelotier, B.; Priem, G.; Macdonald, S. J. F.; Anson, M. S.; Upton, R. J.; Campbell, I. B.; *Tetrahedron Lett.* **2005**, *46*, 9005-9007.

[56] Wei, Y.; Held, I.; Zipse, H. *Org. Biomol. Chem.* **2006**, *4*, 4223-4230.

[57] Dalaigh, C. O.; Hynes, S. J.; O'Brien, J. E.; McCabe, T.; Maher, D. J. Watson, G. W.; Connon, S. J. *Org. Biomol. Chem.* **2006**, *4*, 2785-2793.

[58] (a) Kanta De, C.; Klauber, E. G.; Seidel, D. *J. Am. Chem. Soc.* **2009**, *131*, 17060-17061; (b) Klauber, E. G.; Kanta De, C.; Shah, T. K.; Seidel, D. *J. Am. Chem. Soc.* **2010**, *132*, 13624-13626.

References

[59] (a) Sotohme, Y.; Tanatani, A.; Hashimoto, Y.; Nagasawa, K. *Tetrahedron Lett.* **2004**, *45*, 5589-5592; (b) Lu, L.-Q.; Cao, Y.-J.; Liu, X.-P.; An, J.; Yao, C.-J.; Ming, Z.-H.; Xiao, W.-J. *J. Am. Chem. Soc.* **2008**, *130*, 6946-6948; (c) Zhang, X.-J.; Liu, S.-P.; Lao, J.-H.; Du, G.-J.; Yan, M.; Chan, A. S. C. *Tetrahedron: Asymmetry* **2009**, *20*, 1451-1458.

[60] (a) Shibasaki, M.; Sasai, H.; Arai, T. *Angew. Chem. Int. Ed.* **1997**, *36*, 1236-1256; (b) Shibasaki, M.; Yoshikawa, N. *Chem. Rev.* **2002**, *102*, 2187-2209.

[61] (a) Sawamura, M.; Ito, Y. *Chem. Rev.* **1992**, *92*, 857-871; (b) Steinhagen, H.; Helmchen, G. *Angew. Chem. Int. Ed.* **1996**, *35*, 2339-2342.

[62] Wei, Y. ; Shi, M. *Acc. Chem. Res.* **2010**, *43*, 1005-1018.

[63] A selection of examples: (a) Okino, T.; Hoashi, Y.; Takemoto, Y. *J. Am. Chem. Soc.* **2003**, *125*, 12672-12673; (b) Sohtome, Y.; Hashimoto, Y.; Nagasawa, K. *Adv. Synth. Catal.* **2005**, *347*, 1643-1648; (c) Tsogoeva, S. B.; Yalalov, D. A.; Hateley, M. J.; Weckbecker, C.; Huthmacher, K. *Eur. J. Org. Chem.* **2005**, 4995-5000; (d) Iwabuchi, Y.; Nakatani, M.; Yokoyama, N.; Hatakeyama, S. *J. Am. Chem. Soc.* **1999**, *121*, 10219-10220; (e) List, B. *J. Am. Chem. Soc*, **2002**, *124*, 5656-5657; (f) Kumaragurubaran, N.; Juhl, K.; Zhuang, W.; Bøevig, A.; Jøgensen, K. A. *J. Am. Chem. Soc*, **2002**, *124*, 6254-6255; (g) Bøevig, A.; Juhl, K.; Kumaragurubaran, N.; Zhuang, W.; Jøgensen, K. A. *Angew. Chem. Int. Ed.* **2002**, *41*, 1790-1792.

[64] (a) Matsui, K.; Takizawa, S.; Sasai, H. *J. Am. Chem. Soc.* **2005**, *127*, 3680-3681; (b) Rabalakos, C.; Wulff, W. D. *J. Am. Chem. Soc.* **2008**, *130*, 13524-13525.

[65] (a) Jurecka, P.; Sponer, J.;Cerny, J.; Hobza, P.; *Phys. Chem. Chem. Phys.* **2006**, *8*, 1985; (b) Cybulski, S. M.; Lytle, M. L.; *J. Chem. Phys.* **2007**, *127*, 141102.

[66] Knölker, H-J.; Braxmeier, T.; *Synlett* **1997**, 925.

[67] (a) Hesse, M.; *Spectroscopic Methods in Organic Chemistry*, Thieme, Stuttgart, **2008**; (b) Drakenberg, T.; Dahlqvist, K. I.; Forsen, S. *J. Phys. Chem.* **1972**, *76*, 2178–2183.

[68] Yamada, S.; Misono, T.; Iwai, Y.; Masumizu, A.; Akiyama, Y. *J. Org. Chem.* **2006**, *71*, 6872-6880.

[69] Malardier-Jugroot, C.; Spivey, A. C.; Whitehead, M. A.; *J. Mol. Struct. (THEOCHEM)* **2003**, *623*, 263-276.

[70] (a) Duffey, T. A.; MacKay, J. A.; Vedejs, E. *J. Org. Chem.* **2010**, *75*, 4674–4685; (b) Vedejs, E.; Daugulis, O.; Harper, L. A.; MacKay, J. A.; Powell, D. R. *J. Org. Chem.* **2003**, *68*, 5020-5027.

[71] (a) Müller, C. E.; Wanka, L.; Jewell, K.; Schreiner, P. R.; *Angew. Chem. Int. Ed.* **2008**, *47*, 6180-6183; (b) Miller, S. J; Copeland, G. T.; Papaioannoau, N.; Horstmann, T. E.; Ruel, E. M.; *J. Am. Chem. Soc.* **1998**, *120*, 1629-1630.

[72] (a) Pandit, N. K.; Connors, K. A.; *J. Pharm. Sci.* **1982**, *71*, 485-491; (b) Neveux, M.; Bruneau, C.; Lecolier, S.; Dixneuf, P. H.; *Tetrahedron* **1993**, *49*, 2629-2640; (c) Lapshin, S. A.; Smirnov, Y. I.; Litvinenko, L. M.; Fedorov, V. V.; Kapkan, L. M.; Lange, R.; *Russ. J. Gen. Chem.* **1985**, *55(6)*, 1385-1389.

[73] Spivey, A.; Fekner, T.; Spey, S. E.; *J. Org. Chem.*, **2000**, *65*, 3154-3159.

[74] Mohan, M.; Spivey, A.; **2010**, unpublished results

[75] (a) Satyanarayana, T.; Kagan, H. B.; *Tetrahedron* **2007**, *63*, 6415-6422; (b) Keith, J. M.; Larrow, J. F.; Jacobsen, E. N.; *Adv. Synth. Catal.* **2001**, *343*, 5-26.

[76] Grimme, S. *J. Comput. Chem.* **2006**, *27*, 1787-1799; Grimme, S. *J. Comput. Chem.* **2004**, *25*, 1463-1473.

[77] (a) Spivey, A. C.; Leese, D. P.; Zhu, F.; Davey, S. G.; Jarvest, R. L.; *Tetrahedron* **2004**, *60*, 4513-4525; (b) Spivey, A. C.; Arseniyadis, S.; Fekner, T.; Maddaford, A.; Leese, D. P. *Tetrahedron* **2006**, *62*, 295–301.

[78] Walker, S. D.; Barder, T. E.; Martinelli, J. R.; Buchwald, S. L. *Angew. Chem. Int. Ed. Engl.* **2004**, *43*, 1871-1876.

[79] (a) Dale, J. A.; Mosher, H. S. *J. Am. Chem. Soc.* **1973**, *95*, 512-519; (b) Sullivan, G. R.; Dale, J. A.; Mosher, H. S. *J. Org. Chem.* **1973**, *38*, 2143-2147; (c) Dale, J. A.; Dull, D. L.; Mosher, H. S. *J. Org. Chem.* **1969**, *34*, 2543-2549; (d) Seco, J. M.; Quinoa, E.; Riguera, R. *Chem. Rev.* **2004**, *104*, 17-118.

[80] Wei, Y.; Sastry, G. N.; Zipse, H. *Org. Lett.* **2008**, *10*, 5413-5416.

[81] Neese, F. ORCA 2.6.4, an ab initio density functional and semiempirical program package, **2007**.

[82] Wright, M. R. *An Introduction to Chemical Kinetics*, Wiley-VCH, Weinheim, **2004**.

[83] For discussion of activation parameters for nucleophilic catalysis in acyl-transfer reactions, see: (a) Oakenfull, D. G.; Riley, T.; Gold, V. *Chem. Commun.* **1966**, 385. (b) Milstien, J. B.; Fife, T. H. *J. Am. Chem. Soc.* **1968**, *90*, 2164. (c) Schowen, R. L.; Behn, C. G. *J. Am. Chem. Soc.* **1968**, *90*, 5839. (d) Butler, A. R.; Robertson, I. H. *J. Chem. Soc., Perkins Trans. 2* **1975**, 660.

[84] Koh, H. J.; Lee, J.-W.; Lee, H. W.; Lee, I. *Can. J. Chem.* **1998**, *76*, 710.

Abbreviations

Ac	acetyl	m	multiplet
Ar	aryl	Me	methyl
Bn	benzyl	min	minute
Boc	tert-butoxycarbonyl	mol	mole
br	broad	MS	mass spectrometry
Bu	butyl	NMR	nuclear magnetic resonance
conv.	conversion	NOESY	Nuclear Overhauser effect spectroscopy
CPCM	Conductor-like polarizable continuum model	o	ortho
d	doublet	OPLS	Optimized Potentials for Liquid Simulations
DCM	dichloromethane	PCM	polarizable continuum model
DBU	1,8-Diazabicyclo[5.4.0]undec-7-ene	pent	pentyl
DMAP	4-dimethylaminopyridine	Ph	phenyl
DMSO	dimethyl sulfoxide	PPY	4-pyrrolidinopyridine
equiv.	equivalent	i-Pr	isopropyl
EI	electron-impact	q	quartet
ESI	Electron Spray Ionization	RT	room temperature
FID	Free induction decay	s	singlet
h	hour	t	triplet
HPLC	High-performance liquid chromatography	t-Bu	*tert-butyl*
HRMS	high resolution mass spectroscopy	TEA	triethylamine
IEF PCM	integral equation formalism version of PCM	THF	tetrahydrofuran
IR	infrared	UAHF	United atom Hartree-Fock
J	coupling constant (NMR)	UV	Ultraviolet

Die VDM Verlagsservicegesellschaft sucht für wissenschaftliche Verlage abgeschlossene und herausragende

Dissertationen, Habilitationen, Diplomarbeiten, Master Theses, Magisterarbeiten usw.

für die kostenlose Publikation als Fachbuch.

Sie verfügen über eine Arbeit, die hohen inhaltlichen und formalen Ansprüchen genügt, und haben Interesse an einer honorarvergüteten Publikation?

Dann senden Sie bitte erste Informationen über sich und Ihre Arbeit per Email an *info@vdm-vsg.de*.

Sie erhalten kurzfristig unser Feedback!

VDM Verlagsservicegesellschaft mbH
Dudweiler Landstr. 99   Telefon +49 681 3720 174
D - 66123 Saarbrücken   Fax   +49 681 3720 1749
www.vdm-vsg.de

Die VDM Verlagsservicegesellschaft mbH vertritt

Printed by Books on Demand GmbH, Norderstedt / Germany